Progress in Molecular and Subcellular Biology

Series Editors: W.E.G. Müller (Managing Editor),
Ph. Jeanteur, Y. Kuchino, A. Macieira-Coelho, R. E. Rhoads

41

Volumes Published in the Series

Progress in Molecular and Subcellular Biology

Volume 26
Signaling Pathways for Translation: Insulin and Nutrients
R.E. Rhoads (Ed.)

Volume 27
Signaling Pathways for Translation: Stress, Calcium, and Rapamycin
R.E. Rhoads (Ed.)

Volume 28
Small Stress Proteins
A.-P. Arrigo and W.E.G. Müller (Eds.)

Volume 29
Protein Degradation in Health and Disease
M. Reboud-Ravaux (Ed.)

Volume 30
Biology of Aging
A. Macieira-Coelho

Volume 31
Regulation of Alternative Splicing
Ph. Jeanteur (Ed.)

Volume 32
Guidance Cues in the Developing Brain
I. Kostovic (Ed.)

Volume 33
Silicon Biomineralization
W.E.G. Müller (Ed.)

Volume 34
Invertebrate Cytokines and the Phylogeny of Immunity
A. Beschin and W.E.G. Müller (Eds.)

Volume 35
RNA Trafficking and Nuclear Structure Dynamics
Ph. Jeanteur (Ed.)

Volume 36
Viruses and Apoptosis
C. Alonso (Ed.)

Volume 38
Epigenetics and Chromatin
Ph. Jeanteur (Ed.)

Volume 40
Developmental Biology of Neoplastic Growth
A. Macieira-Coelho (Ed.)

Volume 41
Molecular Basis of Symbiosis
J. Overmann (Ed.)

Subseries: Marine Molecular Biotechnology

Volume 37
Sponges (Porifera)
W.E.G. Müller (Ed.)

Volume 39
Echinodermata
V. Matranga (Ed.)

Jörg Overmann (Ed.)

Molecular Basis of Symbiosis

With 60 Figures, 5 in Color, and 10 Tables

 Springer

Professor Dr. JÖRG OVERMANN
Section of Microbiology
Department of Biology
Maria-Ward-Str. 1a
80638 Munich
Germany

ISSN 0079-6484
ISBN-10 3-540-28210-6 Springer-Verlag Berlin Heidelberg New York
ISBN-13 978-3-540-28210-5

Library of Congress Control Number: 2005934306

This work is subject to copyright. All rights are reserved, whether the whole or part of the material is concerned, specifically the rights of translation, reprinting, reuse of illustrations, recitation, broadcasting, reproduction on microfilm or in any other way, and storage in data banks. Duplication of this publication or parts thereof is permitted only under the provisions of the German Copyright Law of September 9, 1965, in its current version, and permission for use must always be obtained from Springer-Verlag. Violations are liable for prosecution under the German Copyright Law.

Springer-Verlag is a part of Springer Science + Business Media
springeronline.com

© Springer Berlin Heidelberg 2006
Printed in Germany

The use of general descriptive names, registered names, trademarks, etc. in this publication does not imply, even in the absence of a specific statement, that such names are exempt from the relevant protective laws and regulations and therefore free for general use.

Product liability: The publishers cannot guarantee the accuracy of any information about dosage and application contained in this book. In every individual case the user must check such information by consulting the relevant literature.

Production: SPI Publishing Services, Pondicherry
Typesetting: SPI Publishing Services, Pondicherry
Cover design: *design & production* GmbH, Heidelberg, Germany

Printed on add free paper   30/3150Re   5 1 3 2 1 0

Preface

Symbiotic associations involving prokaryotes are known from many different phylogenetic lineages and occur ubiquitously. The significance of these associations is well established for insects, where 15–20% (i.e., up to 190,000) of the species depend on symbioses with bacteria (Buchner 1965). As another example, up to 85% of all prokaryotes in the termite hindgut occur in ecto- or endosymbiotic associations with flagellates (Berchtold et al. 1999). In some aquatic environments, essentially all cells of Green Sulfur Bacteria occur in the symbiotic state (Glaeser and Overmann 2003). Finally, whereas a total of 200 bacterial species are considered human pathogens (Lengeler et al. 1999), the number of autochthonous (commensal and symbiotic) bacterial species associated just with the human intestinal tract is more than three times higher (Suau et al. 1999). Obviously, symbiosis represents a typical way of prokaryotic life.

In comparison to the approximately 6,000 validly described and mostly well-studied species of prokaryotes, much less is known about symbiotic associations. This is largely due to technical limitations. Maintaining and especially growing symbiotic associations in the laboratory is inherently difficult. Since the advent of the era of molecular ecology, however, 16S rRNA-based culture-independent techniques continue to reveal novel phylotypes of symbiotic prokaryotes. A comprehensive search of the Genbank database yielded 937 entries of 16S rRNA gene sequences for symbiotic Bacteria or Archaea (Overmann and Schubert 2002). Yet, this latter number represents only a few percent of the total number of 16S rRNA gene sequences in the database and hence most likely is not representative for the entire diversity of symbiotic prokaryotes.

Where estimated, the age of symbiotic associations ranges from 250 (Baumann et al. 1998) to 20 million years (Dubilier et al. 1995). During these long time periods, co-evolution of the partner organisms occurred (Bandi et al. 1996; Baumann et al. 1998; Sauer et al. 2000), leading to specific mechanisms of organismic interactions, and resulting in novel physiological capabilities of the association as compared to those of the individual partners. Well-known examples include the colonization of nutrient-poor soils by legumes associated with N_2-fixing rhizobia, or the colonization of Arctic and Antarctic extreme habitats by lichens. Mechanisms of symbiotic interaction may have already evolved in association with former, different partner organisms, as has been proposed for opportunistic human pathogens which appear to have developed their virulence factors in interactions with non-mammalian eukaryotes (Hogan and Kolter 2002).

The exchange of specific signals, the reciprocal regulation, and the physiological interactions developed in symbiotic associations have always attracted the interest of researchers. In the past, the complex nature of these interactions and the lack of suitable laboratory models often impeded the understanding of these fundamental processes. However, symbiosis research has recently entered an exciting era because molecular biology now provides us with a wealth of suitable and sophisticated novel techniques. These tools are particularly useful for studying associations in which the partners cannot be separated from each other or cannot be grown in the laboratory.

Recent advances in molecular symbiosis research permit a first comparison across different symbiotic and also pathogenic systems. Indeed, common molecular principles of organismic interaction begin to emerge. With reference to the symbiotic systems covered in this book, the recent discovery of RTX toxin-like components in two very different symbiotic associations, namely phototrophic consortia (Chap. 2) and the association of *Riftia pachyptila* with chemoautotrophic bacteria (Cavanaugh 2004; Chap. 10) was unexpected and suggests that this module is employed in phylogenetically very distant systems. Future research will reveal whether these modules evolved from the same ancestor and thus are homologous, or whether they developed independently by convergent evolution.

It is the goal of this book to contribute towards a broader perspective of the diversity of symbiotic systems. The identification of unifying themes among different systems will stimulate research in this fascinating topic even further. To this end, a set of 14 different model systems have been chosen. They comprise some well-known symbioses for which a considerable amount of information has already been gathered over the past few years. Nevertheless, important novel aspects of symbiotic interactions have recently emerged. Other experimental systems have only recently become amenable to experimental manipulation, but already provided exciting insights into the molecular mechanisms of symbiosis.

It is the belief of the editor as well as the authors, that a better understanding of the molecular mechanisms of symbioses ultimately will also lead to novel strategies for the exploitation of such systems for biotechnological purposes, and potentially also help to improve strategies for the treatment of (human) pathogens.

References

Bandi C, Siron M, Damiani G, Magrassi L, Nalepa CA, Laudani U, Sacchi L (1996) The establishment of intracellular symbiosis in an ancestor of cockroaches and termites. Proc R Soc Lond B 259:293–299

Baumann P, Baumann L, Clark MA, Thao ML (1998) *Buchnera aphidicola*: the endosymbiont of aphids. ASM News 64:203–209

Berchtold M, Chatzinotas A, Schönhuber W, Brune A, Amann R, Hahn D, König H (1999) Differential enumeration and in situ localization of micro-organisms in the hindgut of the lower termite *Mastotermes darwiniensis* by hybridization with rRNA-targeted probes. Arch Microbiol 172:407–416

Buchner P (1965) Endosymbiosis of animals with plant microorganisms. Interscience, New York

Cavanaugh CM (2004) Evolution of chemoautotrophic symbionts: emerging patterns and implications for cospeciation. 10th international symposium on microbial ecology, Cancun, Mexico

Dubilier N, Giere O, Distel DL, Cavanaugh CM (1995) Characterization of chemoautotrophic bacterial symbionts in a gutless marine worm (Oligochaeta, Annelida) by phylogenetic 16S rRNA sequence analysis and in situ hybridization. Appl Environ Microbiol 61:2346–2350

Glaeser J, Overmann J (2003) Characterization and in situ carbon metabolism of phototrophic consortia. Appl Environ Microbiol 69:3739–3750

Hogan D, Kolter R (2002) *Pseudomonas-Candida* interactions: an ecological role for virulence factors. Science 296:2229–2232

Lengeler JW, Drews G, Schlegel HG (1999) Biology of the prokaryotes. Thieme, Stuttgart, 955 pp

Overmann J, Schubert K (2002) Phototrophic consortia: model systems for symbiotic interrelations between prokaryotes. Arch Microbiol 177:201–208

Sauer C, Stackebrandt E, Gadau J, Hölldobler B, Gross R (2000) Systematic relationships and cospeciation of bacterial endosymbionts and their carpenter ant host species: proposal of the new taxon *Candidatus Blochmannia* gen. nov. Int J Syst Evol Microbiol 50:1877–1886

Suau A, Bonnet R, Sutren M, Godon JJ, Gibson GR, Collins MD, Dore J (1999) Direct analysis of genes encoding 16S rRNA from complex communities reveals many novel molecular species within the human gut. Appl Environ Microbiol 65:4799–4807

Contents

Syntrophic Associations in Methanogenic Degradation 1
B. Schink
1 Introduction 1
2 Types of Cooperation among Anaerobic Microorganisms 2
3 Energetical Aspects 5
4 Concept of Syntrophic Energy Metabolism 6
5 Energy Metabolism in Syntrophically Fermenting Bacteria 7
    5.1 Butyrate Oxidation 7
    5.2 Syntrophic Propionate Oxidation 8
    5.3 Syntrophic Ethanol Oxidation 9
    5.4 Fermentation of Acetone 9
    5.5 Syntrophic Oxidation of Hexoses 10
    5.6 Anaerobic Oxidation of Methane 11
6 Types of Interspecies Metabolite Transfer 13
7 Outlook 14
References 15

Symbiosis between Non-Related Bacteria in Phototrophic Consortia 21
J. Overmann
1 Introduction 21
2 Types of Phototrophic Consortia 23
3 Identification of the Bacteria which Constitute Phototrophic Consortia 24
    3.1 The Epibiont 24
    3.2 The Central Bacterium 27
4 Modes and Specificity of Interaction between the Bacterial Partners 29
5 Selective Advantage of the Interaction 31
6 "Symbiosis Genes" of the Epibiont 33
7 Outlook 33
References 34

Prokaryotic Symbionts of Termite Gut Flagellates: Phylogenetic and Metabolic Implications of a Tripartite Symbiosis 39
A. Brune, U. Stingl
1 Introduction 39
2 The Termite Gut Microecosystem 40
    2.1 Lignocellulose Degradation 40
    2.2 The Gut Microenvironment 40
    2.3 Prokaryotic Diversity 41
3 Hindgut Protozoa 42
    3.1 Phylogeny 43
    3.2 Physiology and Function 43

4 Symbiotic Associations with Prokaryotes 45
    4.1 Prokaryotic Epibionts 45
        4.1.1 Methanogens 48
        4.1.2 Spirochaetes 48
        4.1.3 Bacteroidales 49
    4.2 Prokaryotic Endosymbionts 50
        4.2.1 Methanogens 50
        4.2.2 The 'Endomicrobia' 50
        4.2.3 Endonuclear Bacteria 52
5 Functional and Metabolic Implications 52
    5.1 Hydrogen Metabolism 52
    5.2 Other Possible Functions 53
6 Conclusions 54
References 55

**Towards an Understanding of the Killer Trait: *Caedibacter*
Endocytobionts in *Paramecium* 61**
J. Kusch, H.-D. Görtz
1 Introduction 61
2 Evolutionary Ecology of the *Caedibacter* Symbiosis and R Body
   Production 63
3 Host Specificity of *Caedibacter* 67
4 Molecular Evolution of R body Production 69
5 Conclusions 72
References 72

**Bacterial Ectosymbionts which Confer Motility:
Mixotricha paradoxa from the Intestine of the Australian
Termite *Mastotermes darwiniensis* 77**
Helmut König, Li Li, Marika Wenzel, Jürgen Fröhlich
1   Introduction 77
2   The Intestinal Microbiota of *Mastotermes darwiniensis* 78
3   Symbiotic Interactions between Spirochetes and Flagellates 81
4   Morphology of *Mixotricha paradoxa* 82
5   Phylogenetic Position of *Mixotricha paradoxa* 84
6   Glycolytic Activities of *Mastotermes darwiniensis* and its Flagellates 84
7   Identification of the Ectosymbiotic Spirochetes of *Mixotricha paradoxa* 87
8   Identification of the Ectosymbiotic Rod-Shaped Bacterium of
     Mixotricha paradoxa 88
9   Assignment of 16S rDNA Sequences to the Corresponding
     Ectosymbiotic Bacterial Morphotypes 88
10   Conclusions 91
References 91

Extrusive Bacterial Ectosymbiosis of Ciliates 97
G. Rosati
1 Introduction 97
2 The Ciliate Host 98
3 The Epixenososomal Band 99
4 Epixenosomes 100
   4.1 Nature and Phylogenetic affiliation 100
   4.2 Morphological Characteristics 102
   4.3 The Dome-Shaped Zone and Inclusion Body 104
   4.4 The Extrusive Apparatus 104
   4.5 The Basket Tubules 106
5 Can *Euplotidium* and Epixenosomes Survive on their Own? 107
6 The Ejection Mechanism and its Significance 108
7 Why Does *Euplotidium* Maintain its Epixenosomes? 109
8 Concluding Remarks and Future Perspectives 110
References 112

Hydrogenosomes and Symbiosis 117
J.H.P. Hackstein, N. Yarlett
1 Introduction 117
2 Hydrogenosomes and Mitochondrial Remnant Organelles
   Evolved Repeatedly 118
3 Hydrogenosomes: Organelles that Can Use Protons
   as Electron Acceptors 122
   3.1 Hydrogenosomes of *Trichomonas vaginalis* 122
   3.2 Hydrogenosomes of Anaerobic Ciliates: at Least One Appears
       to be a Mitochondrion that Produces Hydrogen 126
   3.3 Hydrogenosomes of Anaerobic Chytrids: an Alternative Way to
       Adapt to Anaerobic Environments 128
4 A Common Ancestor that Produced Hydrogen 131
5 Symbiotic Associations that Depend on Intracellularly
   Generated Hydrogen 132
6 Conclusions 135
References 135

Molecular Interactions between *Rhizobium* and Legumes 143
P. Skorpil, W.J. Broughton
1 Introduction 143
2 Nod-Factors 144
3 Nod-Factor Perception 146
4 Surface Polysaccharides 149
5 Secreted Proteins 152
6 Conclusions 153
References 154

Molecular Mechanisms in the Nitrogen-Fixing *Nostoc*-Bryophyte Symbiosis 165
J.C. Meeks
1 Introduction 165
2 The *Nostoc*-Bryophyte Symbiotic Experimental System........................167
3 Establishment of the Association – Differentiation and Behavior of Hormogonia 172
   3.1 Substage 1. Induction of Hormogonium Differentiation.................172
   3.2 Substage 2. Control of Hormogonium Behavior............................173
   3.3 Substage 3. Colonization and the Repression of Hormogonium Differentiation 176
4 Development of a Functional Nitrogen-Fixing Association Differentiation and Behavior of Heterocysts..179
   4.1 Substage 4. Growth and Metabolic Control 179
   4.2 Substage 5. Heterocyst Differentiation and Behavior 183
      4.2.1 Differentiation in the Free-Living State 184
      4.2.2 Differentiation in Symbiosis..187
5 Future Directions – Genome and Genetic Analyses 189
References 190

Symbiosis of Thioautotrophic Bacteria with *Riftia pachyptila* 197
F.J. Stewart, C.M. Cavanaugh
1 Introduction..197
   1.1 Discovery of the *Riftia pachyptila* Symbiosis................................198
   1.2 Vent Habitat 203
2 Anatomy and Ultrastructure...204
3 Nutritional Basis of the Symbiosis...207
   3.1 Thioautotrophy..207
   3.2 Sulfide Acquisition 208
   3.3 Inorganic Carbon Acquisition 209
   3.4 Nitrogen Acquisition...211
   3.5 Organic Compound Transfer and Symbiont Growth 212
4 Symbiont Transmission and Evolution 214
5 Future Directions..216
References 218

Symbioses of Methanotrophs and Deep-Sea Mussels (Mytilidae: Bathymodiolinae)..227
E.G. DeChaine, C.M. Cavanaugh
1 Introduction..227
2 Methanotrophic Symbioses..228
   2.1 Other Invertebrate Hosts 231
   2.2 Methane-Utilizing Bacteria...233
   2.3 Known Environments Inhabited by Methanotrophic Symbioses.....233

3 Bathymodioline Symbioses 235
    3.1 Bacterial Symbionts 236
    3.2 Distribution of Symbionts within Mussel Gill Tissue........................238
4 Nutrient Assimilation..239
    4.1 Carbon Assimilation 239
    4.2 Nitrogen and Other Essential Nutrients 240
5 Evolution and Biogeography of Bathymodioline Symbioses 241
6 Summary and Conclusions 242
References 243

Symbioses between Bacteria and Gutless Marine Oligochaetes 251
N. Dubilier, A. Blazejak, C. Rühland
1 Introduction 251
2 Biogeography of the Hosts 252
    2.1 Geographic Distribution 252
    2.2 Phylogeny of the Hosts 253
3 Environment..254
    3.1 Reduced Sulfur Compounds 254
    3.2 Oxygen and Other Electron Acceptors 255
    3.3 Other Environmental Factors 256
4 Structural Aspects 257
    4.1 Morphology of the Symbiosis 257
    4.2 Multiple Bacterial Morphotypes 258
    4.3 Transmission of the Symbionts 259
5 Molecular Identification and Phylogeny of the Symbionts 261
    5.1 Detection of Multiple Symbiont Phylotypes 261
    5.2 Phylogeny 262
        5.2.1 Gamma Proteobacterial Symbionts 263
        5.2.2 Delta Proteobacterial Symbionts 264
        5.2.3 Alpha Proteobacterial symbionts...265
        5.2.4 Spirochaete Symbionts 267
6 Functional Aspects..267
7 Outlook 270
References 271

Roles of Bacterial Regulators in the Symbiosis between
***Vibrio fischeri* and *Euprymna scolopes*..277**
K. Geszvain, K. Visick
1 Introduction 277
2 Early Events in the *Euprymna scolopes – Vibrio fischeri* Symbiosis 277
    2.1 *Vibrio fischeri* Strains are Specifically Recruited from the
         Seawater 277
    2.2 *Vibrio fischeri* Cells Navigate Physical and Chemical
         Barriers to Colonize *Euprymna scolopes* 279

 2.3 Both Organisms Undergo Developmental Changes
   in Response to the Symbiosis 279
3 Regulatory Systems Employed by *Vibrio fischeri* to Promote
 the Symbiosis 281
 3.1 Two-Component Signal Transduction Systems 281
 3.2 Quorum-Sensing Regulatory Systems 285
4 Future Directions 287
References 288

Molecular Requirements for the Colonization of
***Hirudo medicinalis* by *Aeromonas veronii* 291**
J. Graf
1 Introduction 291
2 The Digestive Tract Symbiosis of *Hirudo medicinalis* 292
3 Characterization of the Microbiota in the Crop of the Leech 294
4 Specificity of the Symbiosis 295
5 Importance of the Vertebrate Complement System 296
6 Conclusions and Outlook 300
References 301

Index 305

Syntrophic Associations in Methanogenic Degradation

Bernhard Schink

1
Introduction

For many decades, microbiology has tried to understand the activities of microorganisms in nature on the basis of pure culture studies which allow reliable identification of the actors in play and reproducible assessments of their activities under defined conditions. This approach has undoubtedly been successful, but it has overlooked that microbes in nature interact with each other and may depend on these interactions to a various extent. The mutual relationship of partner organisms to each other may vary from only marginal interaction to absolute mutual dependence on each other. Some microorganisms excrete metabolites, e.g., precursors of vitamins or certain amino acids, which are used by a partner organism that lacks specific synthesis pathways and profits from this support, even if it could synthesize the respective compound on its own and this way only saves biosynthetic energy. Types of more intense cooperation and mutual interdependence are found preferentially among anaerobic bacteria, although we have to admit that our view is probably constrained by the cultures we know: Since especially aerobic bacteria are usually isolated with simple media that select for easy-to-cultivate organisms degrading a simple cocktail of substrates on their own, we may overlook other bacteria that are outcompeted under such conditions, and may display more-refined types of interaction with others. Since we know of only a small fraction of all the microorganisms present in the environment, we cannot exclude that other bacteria in the natural environment might depend to a large extent upon cooperations with partners, and perhaps this is just one of the reasons why we so far have failed to cultivate them.

B. Schink (e-mail: Bernhard.Schink@uni-konstanz.de)
Lehrstuhl für Mikrobielle Ökologie, Fakultät für Biologie, Universität Konstanz, Universitätsstr. 10, 78457 Konstanz, Germany

2
Types of Cooperation Among Anaerobic Microorganisms

Whereas aerobic bacteria are usually considered to be able to degrade complex organic matter completely to CO_2 and H_2O, this is true in the anaerobic world only in some exceptional cases. Complex biomass is typically degraded in several steps, including classical (primary) fermentations, with subsequent further oxidation by sulfate reduction or iron reduction, or by coupling primary fermentations with secondary fermentations to methanogenesis at the very end (Bryant 1979; McInerney 1988; Stams 1994; Schink 1997). This kind of job-sharing among anaerobic microorganisms makes the whole process more complicated at first sight, but ascribes to every single organism only a limited task it has to fulfill and, with this, far less effort is needed for regulation of its metabolism.

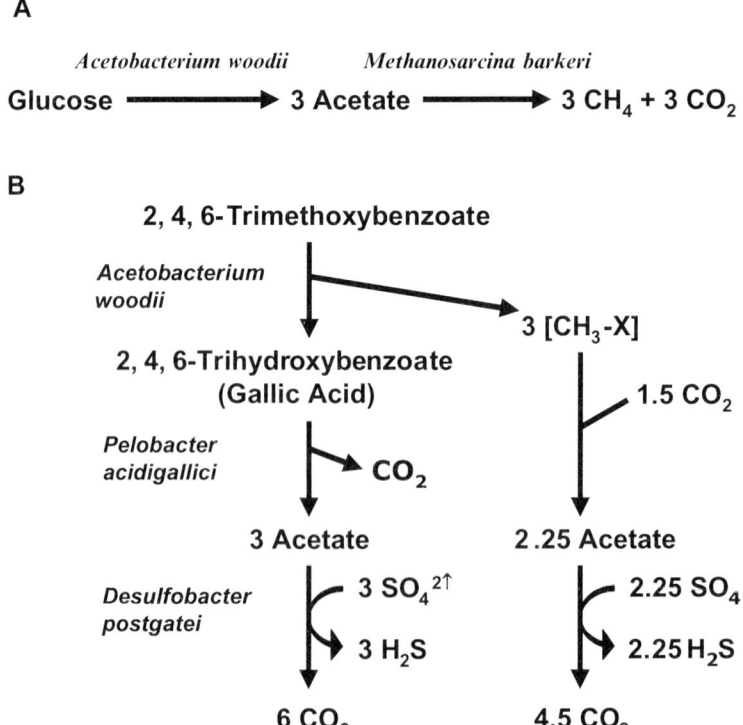

Fig. 1. Metabiotic cooperations in defined cocultures degrading glucose (**A**) and trimethoxybenzoate to methane and CO_2 (**B**)

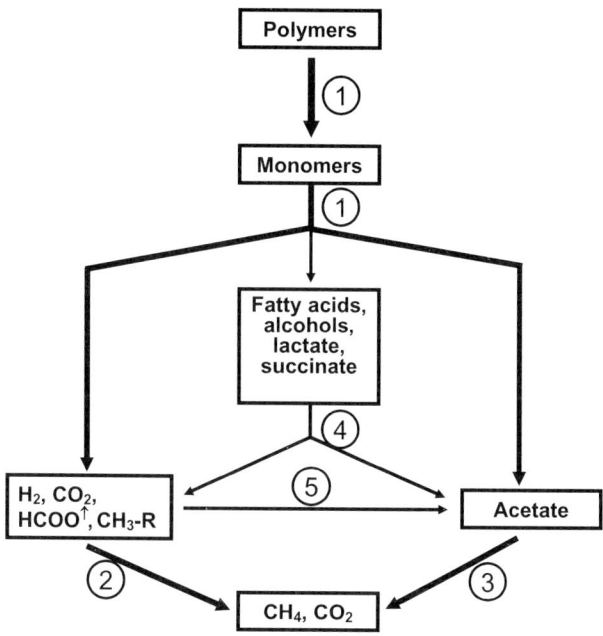

Fig. 2. Carbon and electron flow in the methanogenic degradation of complex organic matter. Groups of prokaryotes involved: *1* primary fermentative bacteria; *2* hydrogen-oxidizing methanogens; *3* acetate-cleaving methanogens; *4* secondary fermenting bacteria (syntrophs); and *5* homoacetogenic bacteria

The interdependence among these partners may vary from an "assembly line"-type of cooperation called metabiosis in which only the later partner in the line profits from the former one but the advantage to the former members in the line by the later partners is negligible. Examples of this kind are degradation of glucose via acetate to methane by cooperation of *Acetobacterium woodii* and *Methanosarcina barkeri* (Fig. 1a; Winter and Wolfe 1979), or complete oxidation of trimethoxybenzoate via gallic acid and acetate by a triculture consisting of *A. woodii*, *Pelobacter acidigallici*, and *Desulfobacter postgatei* (Fig. 1b; Kreikenbohm and Pfennig 1985). Degradation of sugars and polysaccharides by clostridia is influenced positively by cooperation with hydrogen-consuming methanogens that shift the fermentation pattern to more acetate formation and, with this, to higher ATP yields (Schink 1997). Degradation of such compounds in sediments or in well-balanced sludge digestors may proceed nearly exclusively through acetate plus hydrogen, i.e., through the bold arrows in Fig. 2, with very little production of reduced side products such as butyrate and other fatty acids. Excessive production of these reduced side products is typically found in pure cultures or in unbalanced reactors receiving easily fermentable substrates at high rates that cannot be counterbalanced by sufficient growth of methanogenic partners. Cooperative interactions between fermenting bacteria and methanogenic partners have been found to be involved also in the anaerobic degradation of amino acids (Wildenauer and Winter 1986; Winter et al. 1987). The extent of cooperation between the partners degrading amino acids varies dramatically depending on the degradation pathways, from total independence of each other to obligately syntrophic relationships. Quite often, different degradation pathways are used for one specific amino acid in the presence or absence of hydrogen-scavenging partner organisms (Schink and Stams 2001). Finally, there are the strictly syntrophic relationships in which both partners depend on each other for energetic reasons and together perform a fermentation process that neither could run on its own, as is typical of syntrophic associations.

Syntrophy is a special case of symbiotic cooperation between two metabolically different types of bacteria that depend on each other for degradation of a certain substrate, typically through transfer of one or more metabolic intermediate(s) between the partners. The pool size of the intermediate shuttled between the partners has to be kept low to allow efficient cooperation. The term "syntrophy" should be restricted to those cooperations in which both partners depend on each other to perform the metabolic activity observed, and in which the mutual dependence cannot be overcome by simply adding a cosubstrate or any type of nutrient. A classical example is the "*Methanobacillus omelianskii*" culture, which was later shown to be a coculture of two partner organisms, the S strain and the strain M.o.H. (Bryant et al. 1967). Both strains cooperate in the conversion of ethanol to acetate and methane by interspecies hydrogen transfer, as follows:

S Strain:

$$2 \ CH_3CH_2OH + 2 \ H_2O \rightarrow 2 \ CH_3COO^- + 2 \ H^+ + 4 \ H_2$$
$$\Delta G_0' = +19 \text{ kJ per 2 mol of ethanol}$$

Strain M.o.H.:

$$4 \ H_2 + CO_2 \rightarrow CH_4 + 2 \ H_2O$$
$$\Delta G_0' = -131 \text{ kJ per mol of methane}$$

Coculture:

$$2 \ CH_3CH_2OH + CO_2 \rightarrow 2 \ CH_3COO^- - +2 \ H^+ + CH_4$$
$$DG_0' = -112 \text{ kJ per mol of methane}$$

Thus, the fermenting bacterium cannot grow with ethanol in the absence of the hydrogen-scavenging partner because it carries out a reaction that is endergonic under standard conditions. The first reaction can provide energy for the first strain only if the hydrogen partial pressure is kept low enough ($<10^{-3}$ bar) by the methanogen. Therefore, neither partner can grow with ethanol alone, and ethanol degradation depends on the cooperating activities of both.

In this article, we avoid the term "consortium", which is often used to describe any kind of enrichment cultures cooperating in whatever way. This term was originally coined for the structured phototrophic aggregates like *"Pelochromatium"* and *"Chlorochromatium"* and should be restricted to such spatially well-organized systems (see Chap. 2).

3
Energetical Aspects

Anaerobes grow with small amounts of energy, and syntrophically cooperating anaerobes are extremely skilled in the exploitation of minimal energy spans. Synthesis of ATP under the conditions prevailing in an actively growing cell requires +49 kJ per mol (Thauer et al. 1977). Since part of the total energy is always lost in irreversible reaction steps as heat (on average about 20 kJ per mol ATP) a total of about 70 kJ per mol ATP synthesized has to be calculated for ATP synthesis in a living cell (Schink 1990). One may argue that (especially under conditions of energy limitation) an organism may waste less energy in heat production, or that it may operate at an energy charge considerably lower than that quoted for well-growing cells. Nonetheless, one cannot expect the energy requirement for irreversible ATP synthesis to go substantially below about +60 kJ per mol.

According to the Mitchell theory of respirative ATP synthesis, ATP formation is coupled to a vectorial transport of charged groups, typically protons, across a semipermeable membrane. If the ratio of proton translocation over ATP synthesized is 3, the smallest quantum of metabolically convertible energy, equivalent to the transport of a monovalent cation across the charged cytoplasmic membrane, is equivalent to one-third of an ATP unit. This means that a bacterium needs a minimum of about –20 kJ per mol reaction to exploit a reaction's free energy change (Schink and Thauer 1988; Schink and Stams 2001).

On the basis of studies on the structure and function of F_1–F_0 ATPases in recent years, the stoichiometry of ATP synthesis versus proton translocation appears not to be as strictly fixed as assumed so far. Rather, the system may operate like a sliding clutch, meaning that at very low energy input, the energy transfer into ATP synthesis may be substoichiometric. Moreover, the stoichiometry is not necessarily three protons per one ATP, but is determined by the number of subunits arranged in the F_0 versus the F_1 complex. This concept would allow also stoichiometries of 4 to 1, perhaps even 5 to 1

Engelbrecht and Junge 1997; Cherepanov et al. 1999; Stock et al. 1999; Dimroth 2000; Seelert et al. 2000). As a consequence, the minimum energy increment that can still be used for ATP synthesis may be as low as –15 or –12 kJ per mol reaction. In some cases, to make their living, bacteria cooperating in syntrophic fermentations are limited to this range of energy; Hoehler et al. (2001) calculated from metabolite concentrations in natural habitats for the partner bacteria cooperating in syntrophic conversions minimum amounts of exploitable energy in the range of –10 to –19 kJ per mol reaction.

The postulate that there is a minimum amount of approximately –12 to –15 kJ per reaction needed to drive ATP synthesis has been questioned recently in a paper where the remnant energy in starving syntrophically fermenting bacteria had been determined to be as low as –4 kJ per mol (Jackson and McInerney 2002). However, the authors showed only that the "battery" can burn down to such low values at low energy supply; they did not prove that the system can produce ATP under these conditions and thus they did not disprove the concept of a minimum amount of energy for ATP synthesis in the range discussed above.

4
Concept of Syntrophic Energy Metabolism

The effect of hydrogen- and acetate-scavenging partner organisms on syntrophic fermentations becomes obvious if one compares such fermentations under standard conditions (which are roughly comparable to those prevailing in pure cultures) and under conditions that are similar to those prevailing in natural or semi-natural environments. As shown in Table 1, the endergonic fermentations of ethanol, alanine and butyrate turn into exergonic reactions if conditions are assumed that are comparable to those we find in a lake sediment or a sewage sludge digestor; for these calculations, I used hydrogen and acetate concentrations that are at the energetical limit to provide an energy minimum to the methanogens consuming these intermediates. It is obvious that alanine fermentation even allows synthesizing a full ATP unit under these conditions and is, with this, a little atypical.

In most other instances of syntrophic cooperations, the partner organisms have to live with only fractions of an ATP equivalent per reaction run. This can be accomplished by combinations of substrate level phosphorylation steps with a reinvestment of ATP fractions, typically in one or more reversed electron transport processes. The situation is most delicate with syntrophic associations degrading fatty acids such as butyrate, long-chain fatty acids, propionate, or acetate which leave the absolute minimum of 1/3 – 1/4 ATP equivalent (corresponding to 15 – 20 kJ per reaction run) to every partner.

Syntrophic acetate conversion to methane and CO_2 can yield this minimum amount of energy only at enhanced temperature: the reaction operates at its lower temperature limit at 37°C (Schink and Stams 2001) but runs far better at 55–60°C (Zinder and Koch 1984; Hattori et al. 2000). The energetical situation of syntrophic ethanol conversion to methane plus CO_2 is considerably easier, but so far we do not have a convincing concept how energy sharing between the partners is accomplished at the biochemical level.

5
Energy Metabolism in Syntrophically Fermenting Bacteria

5.1
Butyrate Oxidation

In syntrophically butyrate-oxidizing bacteria (McInerney et al. 1979), 1 ATP is synthesized by substrate-level phosphorylation through thiolytic acetoacetyl-CoA cleavage (Wofford et al. 1986), but part of this energy has to be reinvested in reverse electron transport to allow proton reduction with electrons from the butyryl CoA dehydrogenase reaction at a hydrogen partial pressure of 10^{-4} to 10^{-5} bar (Thauer and Morris 1984). Experimental evidence of a reverse electron transport system between the crotonyl-CoA/butyryl-CoA couple ($E^{\circ\prime} = -125$ mV) and the H^+/H_2 couple has been provided with Syntrophomonas wolfei (Wallrabenstein and Schink 1994). If two protons are transferred in this reverse electron transport system, one-third of the ATP synthesized by substrate-level phosphorylation (equivalent to -20 kJ per mol) would remain for growth and maintenance of the fatty acid-oxidizing bacterium, in accordance with the above assumptions.

The energetic situation of a binary mixed culture degrading butyrate to acetate and methane is considerably more difficult.

$$2\ CH_3CH_2CH_2COO^- + CO_2 + 2\ H_2O \rightarrow 4\ CH_3COO^- + 2\ H^+ + CH_4$$

$$\Delta G_0' = -35\ \text{kJ per 2 mol of butyrate}$$

The energy gain increases to -46 kJ at butyrate and acetate concentrations in the range of 10 mM as typically used in laboratory cultures. Under these conditions, the energetic situation for the partners gets tough (-15 kJ per mol partial reaction), especially at the end of the substrate conversion process. Very slow, often non-exponential growth and substrate turnover as usually observed with such binary mixed cultures (Dwyer et al. 1988) indicates that the energy supply is insufficient. We have often observed, as did other authors, that accumulating acetate (>10 mM) inhibits butyrate degradation in such cultures substantially, either thermodynamically via product accumulation or by uncoupling caused by accumulating acid residues.

The energetic difference between the ternary mixed culture and an artificial binary mixed culture exemplifies that the acetate-cleaving methanogens fill an important function in removal of acetate, and with this, "pull" the butyrate oxidation reaction. It also explains why the addition of an acetate-cleaving methanogen to a defined binary mixed culture enhances growth and substrate turnover considerably (Ahring and Westermann 1988; Beaty and McInerney 1989).

5.2
Syntrophic Propionate Oxidation

For syntrophic propionate oxidation according to the equation

$$4 \, CH_3CH_2COO^- + 4 \, H^+ + 2 \, H_2O \rightarrow 7 \, CH_4 + 5 \, CO_2$$

$$\Delta G_0' = -249 \text{ kJ per 4 mol of propionate}$$

a metabolic flow scheme can be drawn, leaving a free energy change in the range of –22 to –23 kJ per mol reaction (11 partial reactions) to all partners involved under standard conditions (Stams et al. 1989; Schink 1991), and this values decreases to –19 kJ in laboratory cultures and to –12 kJ at propionate concentrations prevailing in sediments or sludge digestors (Table 1). Studies in defined mixed cultures and in undefined communities in rice field soil have basically confirmed these calculations (Scholten and Conrad 2000; Fey and Conrad 2000). The pathway of propionate oxidation in such bacteria is basically a reversal of fermentative propionate formation, including methylmalonyl CoA, succinate, malate, pyruvate, and acetyl CoA as intermediates (Koch et al. 1983; Schink 1985, 1991; Houwen et al. 1987, 1990). Propionate is activated by CoA transfer from acetyl CoA (Houwen et al. 1990; Plugge et al. 1993) or succinyl CoA. Of the redox reactions involved, succinate oxidation and malate oxidation are the most difficult ones to couple to proton reduction. The enzymes and electron transfer components involved in propionate oxidation were studied with Syntrophobacter wolinii (Boone and Bryant 1980; Houwen et al. 1990; Plugge et al. 1993) and Syntrophobacter pfennigii (Wallrabenstein et al. 1995). Experiments on hydrogen formation in the presence and absence of protonophores or the ATPase inhibitor DCCD indicated that an intact proton motive force maintained by ATP hydrolysis is required for hydrogen release, probably in the first oxidation step (Dörner 1992).

Studies with *Syntrophobacter fumaroxidans* have shown that not only hydrogen but also formate contributes to interspecies electron transport in this system: Syntrophic propionate oxidation was possible only in cooperation with formate- and hydrogen-oxidizing methanogens, not with *Methanobrevibacter* strains that are unable to oxidize formate (Dong et al. 1994). Enzyme measurements in cells grown syntrophically with propionate contained tenfold higher formate dehydrogenase activity in comparison with cells grown in pure culture with fumarate; the hydrogenase activity was unchanged (de Bok et al. 2002a,b, 2004).

5.3
Syntrophic Ethanol Oxidation

Although the case of "*Methanobacillus omelianskii*" is the classical example of interspecies hydrogen transfer, numerous further syntrophically ethanol-oxidizing bacteria have been isolated, such as *Thermoanaerobium brockii* (Ben-Bassat et al. 1981) and various *Pelobacter* strains (Eichler and Schink 1986). Also certain ethanol-oxidizing sulfate reducers such as *Desulfovibrio vulgaris* oxidize ethanol in the absence of sulfate by hydrogen transfer to a hydrogen-oxidizing methanogenic partner.

Unfortunately, the biochemistry of this syntrophic cooperation is still unclear. The total reaction

$$2\ CH_3CH_2OH + CO_2 \rightarrow 2\ CH_3COO^- + 2\ H^+ + CH_4$$

yields –112 kJ per 2 mol ethanol under standard conditions. On the side of the ethanol oxidizer, e.g., the "S-strain" of "*Methanobacillus omelianskii*" or *Pelobacter acetylenicus*, ethanol dehydrogenase, acetaldehyde acceptor oxidoreductase (acetyl CoA-forming), phosphotransacetylase, and acetate kinase have been detected, forming one ATP per ethanol through substrate-level phosphorylation. Since the methanogenic hydrogen oxidizer requires at least one third of an ATP unit for growth (–20 kJ per reaction run, see above), only –45 kJ is available to the ethanol oxidizer per mol ethanol oxidized which is too little energy to form one full ATP. Therefore, part of the energy bound in ATP has to be reinvested to push the overall reaction and balance the energy budget, but this reverse electron transport system has not yet been identified. It could be the ethanol dehydrogenase enzyme itself which couples ethanol oxidation to acetaldehyde ($E^{o\prime}$= –195 mV) with NAD+ reduction ($E^{o\prime}$= –320 mV), electron transfer from NAD+ to a ferredoxin-like electron carrier, or the hydrogen-releasing hydrogenase itself of which some have recently been reported to couple to proton translocations (Sapra et al. 2003; Hedderich 2004).

5.4
Fermentation of Acetone

A special situation is the fermentative conversion of acetone to methane and CO_2, which is catalyzed by syntrophically cooperating bacteria as well, with acetate as the only intermediate transferred between both partners:

$$CH_3COCH_3 + CO_2 + H_2O \rightarrow 2\ CH_3COO^- + 2\ H^+$$

$$\Delta G_0' = -25.8\ kJ \cdot mol^{-1}$$

$$2\ CH_3COO^- + 2\ H^+ \rightarrow 2\ CH_4 + 2\ CO_2$$

$$\Delta G_0' = -71.8\ kJ \cdot mol^{-1}$$

$$CH_3COCH_3 + H_2O \rightarrow 2\ CH_4 + CO_2$$

$$\Delta G_0' = -97.6\ kJ \cdot mol^{-1}$$

Although in this case all partial reactions are exergonic under standard conditions, the primary fermenting bacterium depends on the methanogenic partner, and acetone degradation in the mixed culture is substantially impaired in the presence of acetylene as an inhibitor of methanogens (Platen and Schink 1987). Experiments with the primary acetone-fermenting bacterium in dialysis cultures revealed that acetate accumulation at concentrations higher than 10 mM inhibited growth and acetone degradation (Platen et al. 1994). Under these conditions, the free energy available to the acetone fermenter is still about –40 kJ per mol. Since acetone metabolism by these bacteria starts with an endergonic carboxylation reaction, this might be the amount of energy that they need to invest into this primary substrate activation reaction, perhaps through a membrane-associated enzyme system (Dimroth 1987). Another reason could be that these bacteria cannot operate with energy fractions smaller than the equivalent of one ATP unit. Unfortunately, the acetone-fermenting bacterium was never obtained in pure culture or in defined coculture, and hence detailed studies on its biochemistry and energetics could never be performed.

5.5
Syntrophic Oxidation of Hexoses

Fermentation of hexose to 2 acetate, 2 CO_2 and 4 H_2 as sole products is exergonic (–216 kJ per mol) but does not yield sufficient energy to synthesize 4 ATP by substrate level phosphorylation that are directly linked with this fermentation via glycolysis. Hydrogen removal to a low concentration makes this reaction exergonic enough to allow fermentation according to this pattern. We could recently show that there is a considerable number of primary fermenting bacteria in a lake sediment that can ferment sugars only to acetate, CO_2 and H_2, but they depend on a hydrogen-scavenging partner (Schink and Stingl, unpubl.) and have been overlooked because our usual isolation strategies select those organisms that can switch to a different fermentation pattern in pure culture. The biochemistry of this metabolism has still to be worked out, but the evidence at hand suggests that these organisms, similar to the acetone fermenters mentioned above, may be unable to subfractionate the ATP unit by reversed electron transport systems.

5.6
Anaerobic Oxidation of Methane

Anaerobic, sulfate-dependent methane oxidation is an important process in anoxic marine sediments which involves again two different organisms cooperating in a syntrophic way, at least in most systems studied so far.

Sulfate-dependent methane oxidation is exergonic under standard conditions

$$CH_4 + SO_4^{2-} + 2\,H^+ \rightarrow CO_2 + H_2S + 2\,H_2O$$
$$\Delta G_0' \text{ of } -18 \text{ kJ per mol}$$

Concentrations of the reaction partners in situ in the active sediment layers are in the range of 10^{-2} bar methane, and 1–3 mM of both sulfate and free hydrogen sulfide. Thus, the overall energetics become only slightly more favorable if in situ conditions are taken into consideration. This amount of energy can feed only one bacterium, provided that it is able to exploit this biological minimum energy quantum. Based on the observation that methanogens can catalyze an oxygen-independent methane oxidation (Zehnder and Brock 1979) and the description of a reversal of homoacetogenic fermentation (Zinder and Koch 1984; see Table 1), it was speculated that "reversed methanogenesis" may be the key to this process (Hoehler et al. 1994; Schink 1997). If the overall reaction is actually a syntrophic cooperation involving a methanogen running methane formation backwards and a sulfate-reducing bacterium, it is obvious that only one of the partners can gain metabolic energy from the reaction, and the other one has to run this process only as a cometabolic activity. This would explain why scientists have always failed to enrich for methane-oxidizing sulfate reducers in the past, simply because one cannot enrich for a bacterium on the basis of a cometabolic activity.

Evidence of such syntrophic cooperation in sulfate-dependent methane oxidation was recently obtained through analysis of lipids of marine archaea and sulfate-reducing bacteria in anoxic sediment layers feeding on (^{13}C-depleted) methane (Pancost et al. 2000), and similar findings, combined with molecular population analysis, were reported for archaeal/bacterial communities in marine sediments and close to submarine methane seeps and gas hydrates (Hinrichs et al. 1999; Orphan et al. 2001; Thomsen et al. 2001).

Table 1. Energetics of some key reactions in methanogenic degradation

Reaction	G_0' kJ per mol rct.	G' in situ[a] kJ per mol rct.
$CH_3CH_2OH + H_2O \rightarrow CH_3COO^- + H^+ + 2 H_2$	+9.6	−42
Alanine $+ 2 H_2O \rightarrow CH_3COO^- + CO_2 + NH_4^+ + 2 H_2$	+2.7	−77
$CH_3CH_2CH_2COO^- + 2 H_2O \rightarrow 2 CH_3COO^- + 2 H^+ + 2 H_2$	+48.3	−26
$CH_3CH_2COO^- + 2 H_2O \rightarrow CH_3COO^- + CO_2 + 3 H_2$	+76	−12
$CH_3COO^- + H^+ + 2 H_2O \rightarrow 2 CO_2 + 4 H_2$	+94.9	−8

All calculations are based on published tables (Thauer et al. 1977).
[a]For in situ conditions, the following concentrations were assumed: substrate and acetate 10^{-4} M, H_2 $10^{-4.5}$ atm., CO_2 10^{-2} atm

In sediments overlying methane hydrates off the coast of Oregon, United States, active anaerobic methane oxidation was found to be associated with discrete, spherical microbial aggregates, which consisted, according to fluorescent in situ hybridization analysis (FISH), of *Methanosarcina*-like archaea in the center, surrounded by *Desulfosarcina*-related sulfate-reducing bacteria (Boetius et al. 2000). The energetics of sulfate-dependent methane oxidation at these gas hydrate sites (with methane pressures of about 80 bar) are considerably more favorable than in deep-lying marine sediments, and the overall free energy change of the reaction in situ (–40 kJ per mol) may really allow energy conservation and growth for both partners in this cooperation. Thus, these aggregates represent a model system to understand sulfate-dependent methane oxidation as a syntrophic cooperation phenomenon, but it still needs to be proven whether this model can also be applied to methane oxidation in deep-lying, methane-poor marine sediments. It is also still open which metabolites are transferred between the partners; preliminary evidence indicates that it it neither hydrogen nor formate, methanol or acetate (Nauhaus et al. 2002).

6
Types of Interspecies Metabolite Transfer

In most syntrophic methanogenic associations, hydrogen is the electron carrier between oxidative and reductive metabolic processes. Its small size and ease of diffusibility make it an excellent candidate for such interspecies electron transfer reactions. Nonetheless, in several cases, formate has also been shown to act as an electron carrier through a formate/CO_2 cycle. Theoretical considerations indicate that the formate system has certain advantages in an aqueous phase, whereas hydrogen might be better suited as carrier in densely packed microbial aggregates (Boone et al. 1989). Both carrier systems might also operate simultaneously in one degradative process, as observed in syntrophic propionate oxidation (see above), or the bacteria may switch between both electron transfer channels depending on the prevailing environmental conditions. Syntrophic oxidation of long chain fatty acids profits as well by efficient removal of the coproduct acetate through the activity of acetotrophic methanogens, and the same appears to be true for all fatty acid degrading systems examined so far. Methanogenic acetone degradation gives an example of interspecies transfer of acetate only (see above).

An artificially combined syntrophically acetate-degrading culture described recently consisting of the iron(III) reducing *Geobacter sulfurreducens* and the fumarate- or nitrate-reducing bacterium *Wolinella succinogenes* (Cord-Ruwisch et al. 1998) oxidizes acetate with nitrate as electron acceptor at high rate, obviously independent of interspecies hydrogen transfer. The interspecies electron transfer in this coculture is accomplished by cysteine, which establishes a cysteine/cystine cycle for electron transfer between both partners or, alternatively, through a H_2S/S^0 cycle (Kaden et al. 2002). This electron transfer through sulfur compounds is similar to interspecies electron transfer between a green phototroph and a chemotrophic sulfur-reducing bacterium in the association "*Chloropseudomonas ethylica*" (Biebl and Pfennig 1978), which also cooperates through an H_2S/S_0 cycle.

Interspecies electron transfer has gained new interest through the discovery that microbial iron(III) reduction in natural environments can be mediated by humic compounds (Lovley et al. 1996); in the laboratory, usually anthraquinone-2,6-disulfonate is used as a model substrate. Several fermenting bacteria, e.g., *Propionibacterium* sp. can reduce such external electron carriers (Benz et al. 1998; Emde and Schink 1990) and can deliver electrons this way indirectly to Fe(III) minerals as well although they have never been regarded as iron-reducing bacteria. Electrons from quinoid carriers can as well be taken up by, e.g., nitrate-reducing bacteria or others (Lovley et al. 1999), and humic compounds can thus mediate electron transfer systems between rather different types of bacteria that would usually not be thought of as cooperation partners.

7
Outlook

The few examples mentioned here should illustrate that there are many different types of cooperation between prokaryotes, especially among anaerobic bacteria. Especially in syntrophic associations of fermenting bacteria and methanogenic partners, which catalyze the last steps in methanogenic degradation of organic matter, the partner organisms are forced to very close cooperation in order to exploit the very small amounts of energy available to these steps, which range at the lowermost limit of energy that can be converted into ATP at all. New types of syntrophic associations have also shown that beyond hydrogen, other metabolites may also be exchanged between the partners involved, and the concept of extracellular electron transfer between different bacterial species resembles other types of extracellular electron transfer such as reduction of insoluble ferric iron oxides, be they catalyzed via intermediate dissolved carrier systems or not. The recently discovered cases of syntrophic anaerobic oxidation of methane and of sugars are only two more examples of this kind. Thus, interspecies interactions as we studied them with our syntrophic associations in the past may be good model systems to open our eyes also for unexpected types of interactions that we may not have considered before.

Acknowledgements. This overview is based on some earlier reviews by the same author on this subject (Schink 1990, 1997, 2002; Schink and Stams 2001). I want to thank many colleagues for fruitful discussions in the field of syntrophic interactions, and especially my coworkers who spent so much effort on these fastidious bacteria that are not always fun to deal with.

References

Ahring BK, Westermann P (1988) Product inhibition of butyrate metabolism by acetate and hydrogen in a thermophilic coculture. Appl Environ Microbiol 54:2393–2397

Beaty PS, McInerney MJ (1989) Effect of organic acid anions on the growth and metabolism of *Syntrophomonas wolfei* in pure culture and in defined consortia. Appl Environ Microbiol 55:977–983

Ben-Bassat A, Lamed R, Zeikus JG (1981) Ethanol production by thermophilic bacteria: metabolic control of end product formation in *Thermoanaerobium brockii*. J Bacteriol 146:192–199

Benz M, Schink B, Brune A (1998) Humic acid reduction by *Propionibacterium freudenreichii* and other fermenting bacteria. Appl Environ Microbiol 64:4507–4512

Biebl H, Pfennig N (1978) Growth yields of green sulfur bacteria in mixed cultures with sulfur and sulfate reducing bacteria. Arch Microbiol 117:9–16

Boetius A, Ravenschlag K, Schubert CJ, Rickert D, Widdel F, Giesecke A, Amann R, Jorgensen BB, Witte U, Pfannkuche O (2000) A marine microbial consortium apparently mediating anaerobic oxidation of methane. Nature 407:623–626

Boone DR, Bryant MP (1980) Propionate-degrading bacterium, *Syntrophobacter wolinii* sp. nov. gen. nov, from methanogenic ecosystems. Appl Environ Microbiol 40:626–632

Boone DR, Johnson RL, Liu Y (1989) Diffusion of the interspecies electron carriers H_2 and formate in methanogenic ecosystems, and implications in the measurement of K_M for H_2 or formate uptake. Appl Environ Microbiol 55:1735–1741

Bryant MP (1979) Microbial methane production – theoretical aspects. J Anim Sci 48:193–201

Bryant MP, Wolin EA, Wolin MJ, Wolfe RS (1967) *Methanobacillus omelianskii*, a symbiotic association of two species of bacteria. Arch Microbiol 59:20–31

Cherepanov DA, Mulkidjanian AY, Junge W (1999) Transient accumulation of elastic energy in proton translocating ATP synthase. FEBS Lett 449:1–6

Cord-Ruwisch R, Lovley DR, Schink B (1998) Growth of *Geobacter sulfurreducens* with acetate in syntrophic cooperation with hydrogen-oxidizing anaerobic partners. Appl Environ Microbiol 64:2232–2236

De Bok FA, Luijten ML, Stams AJ (2002a) Biochemical evidence for formate transfer in syntrophic propionate-oxidizing cocultures of *Syntrophobacter fumaroxidans* and *Methanospirillum hungatei*. Appl Environ Microbiol 68:4247–4252

De Bok FA, Roze EH, Stams AJ (2002b) Hydrogenases and formate dehydrogenases of *Syntrophobacter fumaroxidans*. Antonie van Leeuwenhoek 81:283–291

De Bok FA, Plugge CM, Stams AJ (2004) Interspecies electron transfer in methanogenic propionate degrading consortia. Water Res 38:1368–1375

Dimroth P (1987) Sodium transport decarboxylases and other aspects of sodium ion cycling in bacteria. Microbiol Rev 51:320–340

Dimroth P (2000) Operation of the F_0 motor of the ATP synthase. Biochem Biophys Acta 1458:374–386

Dong X, Plugge CM, Stams AJM (1994) Anaerobic degradation of propionate by a mesophilic acetogenic bacterium in co- and triculture with different methanogens. Appl Environ Microbiol 60:2834–2838

Dörner C (1992) Biochemie und Energetik der Wasserstofffreisetzung in der syntrophen Vergärung von Fettsäuren und Benzoat. Thesis, Universität Tübingen

Dwyer DF, Weeg-Aerssens E, Shelton DR, Tiedje JM (1988) Bioenergetic conditions of butyrate metabolism by a syntrophic, anaerobic bacterium in coculture with hydrogen-oxidizing methanogenic and sulfidogenic bacteria. Appl Environ Microbiol 54:1354–1359

Eichler B, Schink B (1986) Fermentation of primary alcohols and diols, and pure culture of syntrophically alcohol-oxidizing anaerobes. Arch Microbiol 143:60–66

Emde R, Schink B (1990) Oxidation of glycerol, lactate, and propionate by *Propionibacterium freudenreichii* in a poised-potential amperometric culture system. Arch Microbiol 153:506–512

Engelbrecht S, Junge W (1997) ATP synthase: a tentative structural model. FEBS Lett 414:485–491

Hattori S, Kamagata Y, Hanada S, Shoun H (2000) *Thermacetogenium phaeum* gen. nov, sp. nov, a strictly anaerobic, thermophilic, syntrophic acetate-oxidizing bacterium. Int J Syst Evol Microbiol 50:1601–1609

Hedderich R (2004) Energy-converting [NiFe] hydrogenases from archaea and extremophiles: ancestors of complex I. Bioenerg Biomembr 36:65–75

Hinrichs KU, Hayes JM, Sylva SP, Brewer PG, DeLong EF (1999) Methane-consuming archaebacteria in marine sediments. Nature 398:802–805

Hoehler TM, Alperin MJ, Albert DB, Martens CS (1994) Field and laboratory studies of methane oxidation in an anoxic marine sediment: evidence for a methanogen-sulfate reducer consortium. Global Biochem Cycl 8:451–463

Hoehler TM, Alperin MJ, Albert DB, Martens CS (2001) Apparent minimum free energy requirements for methanogenic Archaea and sulfate-reducing bacteria in an anoxic marine sediment. FEMS Microbiol Ecol 38:33–41

Houwen FP, Dijkema C, Schoenmakers CHH, Stams AJM, Zehnder AJB (1987) 13C-NMR study of propionate degradation by a methanogenic coculture. FEMS Microbiol Lett 41:269–274

Houwen FP, Plokker J, Stams AJM, Zehnder AJB (1990) Enzymatic evidence for involvement of the methylmalonyl-CoA pathway in propionate oxidation by *Syntrophobacter wolinii*. Arch Microbiol 155:52–55

Jackson BE, McInerney MJ (2002) Anaerobic microbial metabolism can proceed close to thermodynamic limits. Nature 415:454–456

Kaden J, Galushko AS, Schink B (2002) Cysteine-mediated electron transfer in syntrophic acetate oxidation by cocultures of *Geobacter sulfurreducens* and *Wolinella succinogenes*. Arch Microbiol 178:53–58

Koch M, Dolfing J, Wuhrmann K, Zehnder AJB (1983) Pathway of propionate degradation by enriched methanogenic cultures. Appl Environ Microbiol 45:1411–1414

Kreikenbohm R, Pfennig N (1985) Anaerobic degradation of 3.4.5-trimethoxybenzoate by a defined mixed culture of *Acetobacterium woodii*, *Pelobacter acidigallici* and *Desulfobacter postgatei*. FEMS Microbiol Ecol 31:29–38

Lovley DR, Coates JD, Blunt-Harris EL, Phillips EJP, Woodward JC (1996) Humic substances as electron acceptors for microbial respiration. Nature 382:445–448

Lovley DR, Fraga JL, Coates JD, Blunt-Harris EL (1999) Humics as an electron donor for anaerobic respiration. Environ Microbiol 1:89–98

McInerney MJ (1988) Anaerobic hydrolysis and fermentation of fats and proteins. In: Zehnder AJB (ed) Biology of anaerobic microorganisms. Wiley, New York, pp 373–415

McInerney MJ, Bryant MP, Pfennig N (1979) Anaerobic bacterium that degrades fatty acids in syntrophic association with methanogens. Arch Microbiol 122:129–135

Nauhaus K, Boetius A, Krüger M, Widdel F (2002) In vitro demonstration of anaerobic oxidation of methane coupled to sulphate reduction in sediment from a marine gas hydrate area. Environ Microbiol 4:296–305

Orphan VJ, Hinrichs K-U, Ussler W, Paull CK, Taylor LT, Sylva SP, Hayes JM, DeLong EF (2001) Comparative analysis of methane-oxidizing archaea and sulfate-reducing bacteria in anoxic marine sediments. Appl Environ Microbiol 67:1922–1934

Pancost RD, Damsté JSS, de Lint S, van der Maarel MJEC, Gottschal JC, and the Medinaut Shipboard Scientific Party (2000) Biomarker evidence for widespread anaerobic methane oxidation in Mediterranean sediments by a consortium of methanogenic archaea and bacteria. Appl Environ Microbiol 66:1126–1132

Platen H, Schink B (1987) Methanogenic degradation of acetone by an enrichment culture. Arch Microbiol 149:136–141

Platen H, Janssen PH, Schink B (1994) Fermentative degradation of acetone by an enrichment culture in membrane-separated culture devices and in cell suspensions. FEMS Microbiol Lett 122:27–32

Plugge CM, Dijkema C, Stams AJM (1993) Acetyl-CoA cleavage pathway in a syntrophic propionate oxidizing bacterium growing on fumarate in the absence of methanogens. FEMS Microbiol Lett 110:71–76

Sapra R, Bagramyan K, Adams MW (2003) A simple energy-conserving system: proton reduction coupled to proton translocation. Proc Natl Acad Sci USA 100:7545–7550

Schink B (1985) Mechanism and kinetics of succinate and propionate degradation in anoxic freshwater sediments and sewage sludge. J Gen Microbiol 131:643–650

Schink B (1990) Conservation of small amounts of energy in fermenting bacteria. In: Finn RK, Präve P (eds) Biotechnology, Focus 2. Hanser Publ, Munich, pp 63–89

Schink B (1991) Syntrophism among prokaryotes. In: Balows A, Trüper HG, Dworkin M, Schleifer KH (eds) The Prokaryotes, 2nd edn, chap 11. Springer, Berlin Heidelberg New York, pp 276–299

Schink B (1997) Energetics of syntrophic cooperations in methanogenic degradation. Microbiol Mol Biol Rev 61:262–280

Schink B (2002) Synergistic interactions in the microbial world. Antonie van Leeuwenhoek 81:257–261

Schink B, Stams AJM (2001) Syntrophism among prokaryotes. In: Dworkin M, Falkow S, Rosenberg E, Schleifer K-H, Stackebrandt E (eds) The Prokaryotes: an evolving electronic resource for the microbiological community, 3rd edn, (latest update release 3.8, December 2001). Springer, Berlin Heidelberg New York

Schink B, Thauer RK (1988) Energetics of syntrophic methane formation and the influence of aggregation. In: Lettinga G, Zehnder AJB, Grotenhuis JTC, Hulshoff Pol LW (eds) Granular anaerobic sludge; microbiology and technology. Pudoc, Wageningen, pp 5–17

Seelert H, Poetsch A, Dencher NA, Engel A, Stahlberg H, Müller DJ (2000) Proton-powered turbine of a plant motor. Nature 405:418–419

Scholten JCM, Conrad R (2000) Energetics of syntrophic propionate oxidation in defined batch and chemostat cocultures. Appl Environ Microbiol 66:2934–2942

Stams AJM (1994) Metabolic interactions between anaerobic bacteria in methanogenic environments. Antonie van Leeuwenhoek 66:271–294

Stams AJM, Grotenhuis JTC, Zehnder AJB (1989) Structure-function relationship in granular sludge. In Hattori T, Ishida Y, Maruyama Y, Morita RY, Uchida A (eds) Recent advances in microbial ecology. Japan Sci Soc Press Tokyo Japan, pp 440–445

Stams AJM, van Dijk JB, Dijkema C, Plugge CM (1993) Growth of syntrophic propionate-oxidizing bacteria with fumarate in the absence of methanogenic bacteria. AEM 59:1114–1119

Stock D, Leslie AGW, Walker JE (1999) Molecular architecture of the rotary motor in ATP synthase. Science 286:1700–1705

Thauer RK, Morris JG (1984) Metabolism of chemotrophic anaerobes: old views and new aspects. In: Kelly DP, Carr NG (eds) The microbe 1984, part II. Prokaryotes and eukaryotes. Cambridge Univ Press, Cambridge, pp 123–168

Thauer RK, Jungermann K, Decker K (1977) Energy conservation in chemotrophic anaerobic bacteria. Bacteriol Rev 41:100–180

Thomsen TR, Finster K, Ramsing NB (2001) Biogeochemical and molecular signatures of anaerobic methane oxidation in a marine sediment. Appl Environ Microbiol 67:1646–1656

Wallrabenstein C, Schink B (1994) Evidence of reversed electron transport involved in syntrophic butyrate and benzoate oxidation by *Syntrophomonas wolfei* and *Syntrophus buswellii*. Arch Microbiol 162:136–142

Wallrabenstein C, Hauschild E, Schink B (1995) *Syntrophobacter pfennigii* sp. nov, a new syntrophically propionate-oxidizing anaerobe growing in pure culture with propionate and sulfate. Arch Microbiol 164:346–352

Wildenauer FX, Winter J (1986) Fermentation of isoleucine and arginine by pure and syntrophic cultures of *Clostridium sporogenes*. FEMS Microbiol Ecol 38:373–379

Winter J, Wolfe RS (1979) Complete degradation of carbohydrate to carbon dioxide and methane by syntrophic cultures of *Acetobacterium woodii* and *Methanosarcina barkeri*. Arch Microbiol 121:97–102

Winter J, Schindler F, Wildenauer FX (1987) Fermentation of alanine and glycine by pure and syntrophic cultures of *Clostridium sporogenes*. FEMS Microbiol Ecol 45:153–161

Wofford NQ, Beaty PS, McInerney MJ (1986) Preparation of cell-free extracts and the enzymes involved in fatty acid metabolism in *Syntrophomonas wolfei*. J Bacteriol 167:179–185

Zehnder AJB, Brock TD (1979) Methane formation and methane oxidation by methanogenic bacteria. J Bacteriol 137:420–432

Zeikus JG, Winfrey M (1976) Temperature limitation of methanogenesis in aquatic sediments. Appl Environ Microbiol 31:99–107

Zinder SH, Koch M (1984) Non-aceticlastic methanogenesis from acetate: acetate oxidation by a thermophilic syntrophic coculture. Arch Microbiol 138:263–272

Symbiosis between Non-Related Bacteria in Phototrophic Consortia

Jörg Overmann

1
Introduction

Consortia are defined as close associations of microbial cells in which two or more different microorganisms maintain a permanent cell-to-cell contact and form an organized structure (Schink 1991). Currently, 19 different morphological types of bacterial consortia are recognized based on the taxonomy and the arrangement of the participating bacteria (Overmann 2001a,b; Huber et al. 2002; Glaeser and Overmann 2004). The habitats of consortia range from the human oral cavity, which is colonized by the so-called corn-cob bacterial formations, to deep sea sediments which harbor anaerobic methane-oxidizing consortia. Even more frequently observed are irregular aggregates, biofilms, and patches of free-living bacterial cells containing non-related prokaryotes (e.g., Jacobi et al. 1997; Rudolph et al. 2001).

In contrast to monospecific associations like those of autoinducer-producing bacteria (Bassler 2002) or myxobacteria (Reichenbach and Dworkin 1992), only little is known about the significance, specificity, and the evolutionary origin of bacterial interactions in heterogeneous assemblages. To date, syntrophic associations represent the only type of heterogenous assemblage investigated in sufficient detail to permit a functional understanding of the interaction. Metabolites like H_2, formate, acetate, or sulfur are transferred from one partner to the other (Chap. 1), with the efficiency of metabolite transfer depending on the diffusion distance (Schink 1991). Considerable evidence has accumulated that a close juxtaposition of the metabolite-producing and -consuming bacteria occurs, leading to flocs densely populated by the intermixed bacteria (Conrad et al. 1985; Schink 1991). Another type of interaction is an interspecies signal exchange based on AI-2 type autoinducers, which has been proposed to occur in mixed bacterial communities as a type of universal interspecies chemical language (Bassler et al. 1997). The physiological and genetic interactions in other heterogeneous associations, particularly in the most highly structured bacterial consortia, are still largely unknown. This is mostly due to the lack of a suitable laboratory model system.

J. Overmann (e-mail: j.overmann@LRZ.uni-muenchen.de)
Bereich Mikrobiologie, Department Biologie I, Maria-Ward-Str. 1a,
80638 München, Germany

Among all consortia, phototrophic consortia probably represent the highest degree of mutual interdependence between non-related bacteria. Phototrophic consortia consist of green- or brown-colored cells, which are associated with colorless bacteria. The cells are tightly packed and arranged in a highly regular fashion (Fig. 1A, Fig. 2). In the free water column of stratified freshwater lakes, the biomass of phototrophic consortia can amount to as much as two-thirds of the total bacterial biomass (Gasol et al. 1995). Although phototrophic consortia had already been discovered at the beginning of the last century (Lauterborn 1906), they could not be cultivated until more than 90 years later (Fröstl and Overmann 1998). In the recently established enrichment cultures, phototrophic consortia have become amenable to detailed investigations of the phylogenetic composition, physiology and morphology of the cells and, more recently, to molecular studies of the functional basis of the symbiosis.

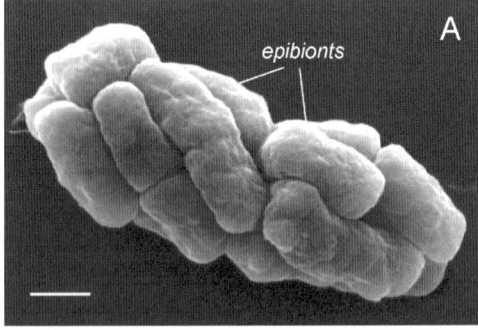

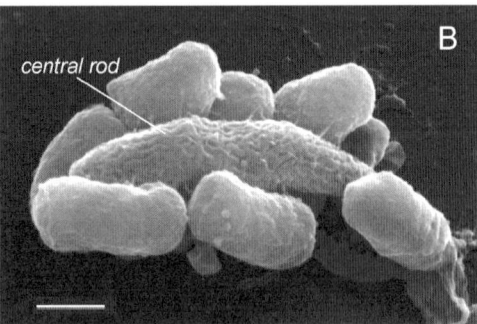

Fig. 1. Transmission electron photomicrographs of "*Chlorochromatium aggregatum*". **A** Intact consortium after fixation in 2% glutardialdehyde. The outer layer of epibionts covers the central rod entirely (courtesy of Martina Schlickenrieder and Gerhard Wanner). **B** "*C. aggregatum*" without glutardialdehyde fixation and after exposure to air and a rinse with distilled water (courtesy of Kajetan Vogl and Gerhard Wanner). Partial disaggregation of the phototrophic consortia result in exposure of the single central rod-shaped bacterium. Note the rough cell surface structure of the latter as compared to the epibionts. *Bars*, 0.5 μm

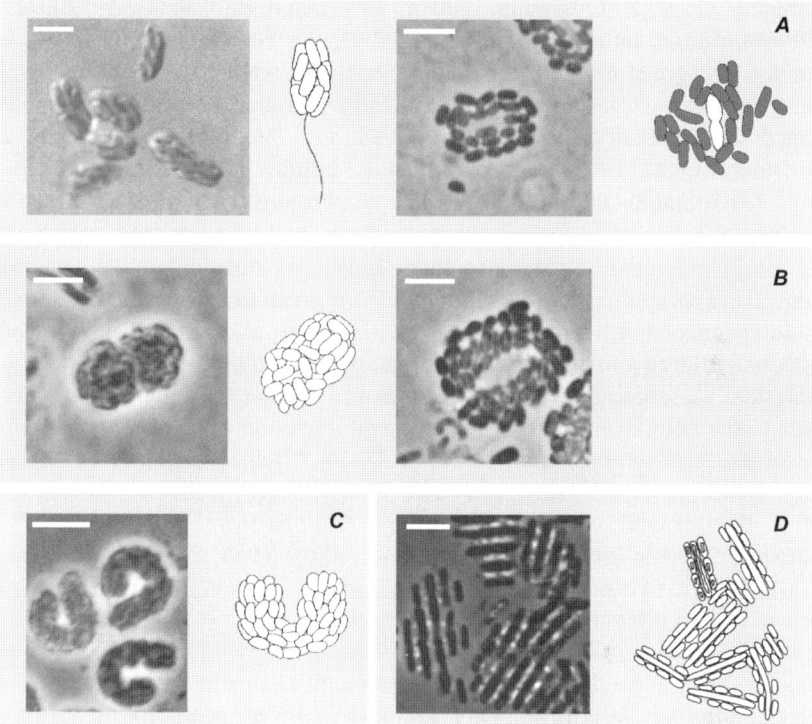

Fig. 2. Light microscopic and schematic views of five different types of phototrophic consortia. **A** Differential interference contrast image and schematic view of "*Chlorochromatium aggregatum*", phase contrast photomicrograph of the disaggregated state and schematic view of "*Pelochromatium roseum*" (*from left to right*). **B** Phase contrast photomicrograph in the intact state, schematic view, and the disaggregated state of "*Pelochromatium latum*". **C** Phase contrast photomicrograph and schematic view of "*Chlorochromatium glebulum*". **D** Phase contrast photomicrograph and schematic view of "*Chloroplana vacuolata*". *Bars*, 5 μm

2
Types of Phototrophic Consortia

To date, a total of 10 morphologically different types of consortia have been detected. Most of them have repeatedly been observed in freshwater habitats. The majority of the phototrophic consortia are round to spindle-shaped, motile, and consist of 13–69 colored cells that form a layer surrounding one colorless central bacterium in the center (Fig. 1B, Fig. 2A,B).

Previously, the different types of phototrophic consortia were distinguished based on the color of their epibionts and the morphology of the entire cell aggregate (Fig. 2) (Overmann 2001a; Overmann and Schubert 2002). In "*Chlorochromatium aggregatum*" (Fig. 2A), the colorless bacterium is surrounded by green-colored rod-shaped bacteria while brown epibionts are found in "*Pelochromatium roseum*". These smaller consortia are spindle-shaped and typically harbor between 12 and 20 epibiont cells (Fig. 2A) (Overmann et al. 1998). The significantly larger "*Chlorochromatium magnum*" (Fröstl and Overmann 2000), "*Pelochromatium latum*" (Glaeser and Overmann 2004) and "*Pelochromatium roseo-viride*" (Gorlenko and Kusnezov 1972) are rather globular in shape (Fig. 2B) and contain ≥40 epibionts (Fig. 2B). The epibionts of "*C. magnum*" are green-colored, whereas those in "*P. latum*" are brown. Interestingly, "*P. roseo-viride*" contains an inner layer of brown-colored and an outer layer of green-colored epibionts.

Besides the color and number of epibionts, further variations exist with respect to the cellular morphology of the epibionts and the overall shape of the consortia. "*Chlorochromatium lunatum*" and "*Pelochromatium selenoides*" carry crescent-shaped green or brown epibionts, respectively (Abella et al. 1998). Phototrophic consortia of the type "*Chlorochromatium glebulum*" are bent (Fig. 2C) and contain green epibionts (Skuja 1957; Fröstl and Overmann 2000). Finally, two non-flagellated morphotypes of phototrophic consortia are known that exhibit a different arrangement of the two associated bacteria. "*Chloroplana vacuolata*" has a sheath-like structure in which long slender colorless and gas-vacuolated rods alternate with chains of rod-shaped and gas-vacuolated green cells (Fig. 2D) (Dubinina and Kusnetsov 1976). Large aggregates of this consortium may contain up to 400 cells of green bacteria. "*Cylindrogloea bacterifera*" consists of green-colored bacteria surrounding a central chain of colorless bacteria with thick capsules (Perfiliev 1914; Skuja 1957).

Since they consist of two different bacteria, the binary names of phototrophic consortia are without standing in nomenclature (Trüper and Pfennig 1971) and, consequently, are given here in quotation marks.

3
Identification of the Bacteria which Constitute Phototrophic Consortia

3.1
The Epibiont

Early electron microscopic studies of phototrophic consortia from the natural samples revealed the presence of chlorosomes in the epibiont cells (Trüper and Pfennig 1971; Caldwell and Tiedje 1974), suggesting that they belonged to the group of green sulfur bacteria. This conclusion was supported by the observation that maximum concentrations of green sulfur bacterial pigments (bacteriochlorophylls *c*, *d* and *e*) in freshwater lakes coincided with population maxima of phototrophic consortia. The phylogenetic affiliation of the epibionts with the Chlorobiaceae could be verified by fluorescence in situ hybridization after a highly specific oligonucleotide probe for green sulfur bacteria had become available and an improved hybridization protocol had been developed (Tuschak et al. 1999).

Recently, repeated cultivation attempts led to the isolation in pure culture of the epibiont from a "*Chlorochromatium aggregatum*" (Vogl et al. 2005). Exhaustive physiological testing of this bacterium did not reveal any unusual physiological capabilities as compared to the known strains of Chlorobiaceae, suggesting that epibionts of phototrophic consortia may grow photolithoautotrophically like their free-living counterparts. Indeed, in situ measurements of light-dependent $H^{14}CO_3^-$ fixation in a natural population of phototrophic consortia and determination of the stable carbon isotope ratios ($\delta^{13}C$) of their bacteriochlorophyll molecules revealed that photoautotrophic growth of the epibionts occurs under natural conditions (Glaeser and Overmann 2003a). It is to be concluded that the epibionts are obligate photolithoautotrophs as all other known Chlorobiaceae (Overmann 2001a). The only unusual feature of green sulfur bacteria associated with phototrophic consortia detected so far is the low cellular concentration of carotenoids in the epibiont of "*Chlorochromatium aggregatum*" (Vogl et al. 2005) and in the brown epibiont from "*Pelochromatium roseum*" (Glaeser et al. 2002; Glaeser and Overmann 2003a). More subtle differences were observed with respect to the light-dependence of growth. In cultures of phototrophic consortium "*Chlorochromatium aggregatum*", light limitation of growth was observed only at light intensities as low as ≤ 3 μmol quanta m^{-2} s^{-1}, while maximum growth rates (doubling times of 1 day) were observed between 5 and 20 μmol quanta m^{-2} s^{-1}. In contrast, the free-living green sulfur bacteria tested reach light saturation of growth at higher light intensities (~10 μmol quanta m^{-2} s^{-1}) (Overmann et al. 1991, 1992, 1998).

Subsequent studies explored the biodiversity and biogeography of phototrophic consortia by determining 16S rRNA gene sequences of epibionts from aquatic environments worldwide (Glaeser and Overmann 2003a, 2004). Epibiont 16S rRNA gene sequences originating from 41 different consortia were obtained after sorting individual consortia by micromanipulation of samples collected in 14 different freshwater lakes. The 16S rRNA genes were amplified by a highly sensitive group-specific PCR, and the amplification products were separated by denaturing gradient gel electrophoresis (DGGE) and sequenced. Most importantly, all epibiont cells in a particular type of phototrophic consortium invariably belonged to one single phylotype. Phylogenetic analyses further demonstrated that the epibionts of each particular type of consortium represent a distinct and novel branch within the radiation of green sulfur bacteria (Fröstl and Overmann 2000; Glaeser and Overmann 2003a, 2004).

Interestingly, morphologically indistinguishable phototrophic consortia, when collected from different lakes, were found to harbor genetically different epibionts. Thus, phylogenetic analyses demonstrated that the "*Chlorochromatium aggregatum*" sampled from European and North American lakes contained seven different epibiont phylotypes depending on the lake, although these consortia were identical with respect to their shape, and the arrangement and color of the epibionts (Glaeser and Overmann 2004). It was concluded that morphologically indistinguishable consortia which occur in geographically distant locations frequently harbor distinct epibionts. In addition, even epibionts with identical partial 16S rRNA gene sequences exhibit considerable differences in morphology and pigmentation and hence genetically clearly differ from each other. Therefore, phototrophic consortia are significantly more diverse than the seven different morphotypes recognized so far. The current estimate amounts to 19 different types of epibionts. Novel types of phototrophic consortia continue to be described (Overmann et al. 1998; Glaeser and Overmann 2004) and future new discoveries are most likely to be made.

It has been postulated that bacteria are ubiquitous (Beijerinck 1913; Baas-Becking 1934) and it was suggested that the high population densities of microorganisms drives a rapid, large-scale dispersal across the physical and geographical barriers (Finlay and Clarke 1999; Finlay 2002). Under these conditions, competitive exclusion of species with an identical ecological niche would be expected to result in a low overall diversity. On the contrary, if endemism occurs among microorganisms, it would result in a significantly higher global diversity, since the latter is maintained by geographic isolation (Staley 1999). Phototrophic consortia are assumed to occupy a narrow and well-defined ecological niche (see below) and therefore represent well-suited model systems to study bacterial biogeography. It was recently demonstrated that epibionts of phototrophic consortia show a nonrandom global distribution (Fig. 3). In fact, the composition of epibionts in consortia from lakes within one geographic region (Europe or North America) was very similar, whereas only two of the 19 epibiont types known were recovered from lakes on both, the European and the North American continents (Glaeser and Overmann 2004) (Fig. 3). While many other bacteria investigated to date indeed appear to be ubiquitous, the dispersal of phototrophic consortia may be much slower than for other bacteria, at least over larger geographical distances (Glaeser and Overmann 2004). Dispersal is certainly limited by the high sensitivity of intact consortia towards molecular oxygen, which leads to a rapid disaggregation of the cell association.

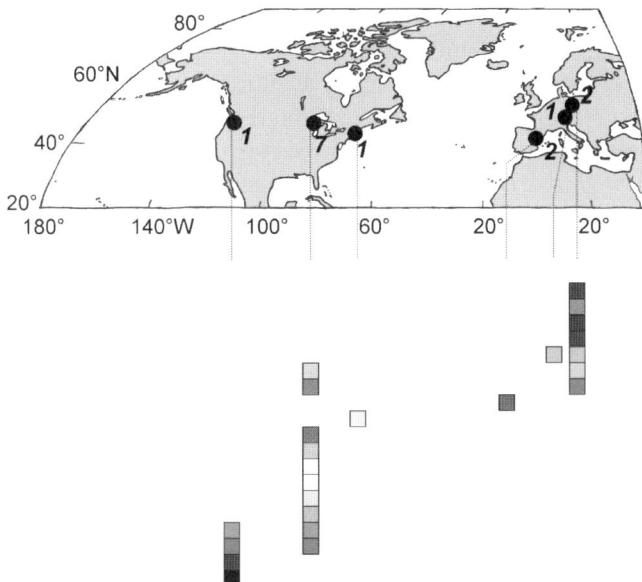

Fig. 3. Biogeography of green sulfur bacterial epibionts of phototrophic consortia based on analyses of partial 16S rRNA gene sequences. Values at each study region give the numbers of lakes investigated. Each *square* represents a particular type of epibiont. Numbers of different epibionts detected in each study region are given in vertical columns. *Squares* at the same horizontal positions designate the same type of epibiont

3.2
The Central Bacterium

In the light microscope, the central rod-shaped bacterium can be distinguished from the green or brown epibionts by its greater size, tapered ends, and low phase contrast (compare Fig. 2A, B).

It had been proposed that the central bacterium in phototrophic consortia might belong to the sulfate- or sulfur-reducing bacteria (Pfennig 1980). This was based on the experience with cocultures of green sulfur bacteria with sulfur- and sulfate-reducing bacteria, in which the phototrophic component could maintain rapid growth rates even at growth-limiting concentrations of sulfide. In these cocultures, a syntrophic sulfur cycle is established providing sulfide as the electron-donating substrate for anoxygenic photosynthesis. Similarly, the association with the central bacterium may render the anoxygenic phototrophic epibionts largely independent of exogenous sulfide, which is usually present in limiting concentrations in the natural habitat. In fact, mixed aggregates of sulfate-reducing bacteria and anoxygenic phototrophic bacteria were recently observed in the chemocline of a meromictic alpine lake, but these aggregates do not display the highly ordered structure which is typical for the phototrophic consortia (Tonolla et al. 2000).

The typical mesophilic Gram-negative sulfur- or sulfate-reducing bacteria belong to the δ-subgroup of the Proteobacteria and to the Gram-positive bacteria with low G+C-content. It therefore came as a surprise when fluorescent in situ hybridization identified the central bacteria present in two different types of consortia as members of the β-Proteobacteria (Fröstl and Overmann 2000). Recently, phylogenetic analyses of the almost complete 16S rRNA sequence revealed that the central bacterium of "*C. aggregatum*" represents a so far isolated phylogenetic lineage (<95% sequence similarity to any other known sequence in the database) and falls into a sister group of the genus *Variovorax* within the family *Comamonadaceae*. The majority of relatives is not-yet-cultured and were found in low-temperature aquatic environments, or aquatic environments containing xenobiotica or hydrocarbons. In CsCl-bisbenzimidazole equilibrium density gradients, the mol% G+C content of the central bacterium was determined to be 55.6% (Kanzler et al. 2005). Since little can be inferred from the phylogenetic affiliation of the central bacterium, information on the physiology of this partner comes from independent experiments.

Intact consortia display a pronounced chemotaxis against 2-oxo acids (2-oxoglutarate or pyruvate depending on the type of consortium). Growth of one type of the phototrophic consortium "*Chlorochromatium aggregatum*" in enrichment cultures critically depended on the addition of 2-oxoglutarate (Fröstl and Overmann 1998). In situ, "*Pelochromatium roseum*" incorporates 2-oxoglutarate as demonstrated by microautoradiography of natural water samples with 2-[^{14}C(U)]-oxoglutarate. Since the epibiont isolated from "*Chlorochromatium aggregatum*" does not utilize 2-oxoglutarate for growth in pure culture, and since epibionts of "*Pelochromatium roseum*" were shown to grow photoautotrophically in situ (see above), 2-oxoglutarate most likely is taken up and metabolized by the central bacterium. In situ, the half saturation constant of 2-oxoglutarate uptake was determined to be ≤10–40 nM (Glaeser and Overmann 2003b), indicating a high affinity of the uptake system(s) similar to the uptake systems for glucose, acetate or aspartate present in other planktonic bacteria.

It remains to be elucidated whether the same colorless motile bacterium occurs in different phototrophic consortia or whether each of the different epibionts is associated with another type of central bacterium. First evidence for at least physiological differences between central bacteria of different consortia comes from chemotaxis experiments with consortia from an Eastern German and a Spanish lake (Glaeser and Overmann 2004). Whereas "*Chlorochromatium aggregatum*" from Dagowsee (Eastern Germany) was attracted by 2-oxoglutarate, the same type of consortium exhibited chemotaxis toward pyruvate in Lake Sisó (Spain). As the chemotaxis toward these organic compounds is likely to be mediated by the central bacterium, not only the epibionts, but also the central bacteria may differ between the various types of phototrophic consortia. Future molecular studies are needed to gain further insight into the diversity of the central bacteria of phototrophic consortia.

4
Modes and Specificity of Interaction between the Bacterial Partners

Theoretically, phototrophic consortia could form randomly from bacterial cells which encounter each other just by chance. Under these conditions, morphologically identical consortia would be expected to harbor different types of green sulfur bacteria and, secondly, free-living epibiont cells should also be detected in the habitat of phototrophic consortia. However, when the epibionts associated with a particular phototrophic consortium were analyzed, all epibiont cells invariably contained the same 16S rRNA gene sequence. Evidently, phototrophic consortia with the same morphology that share the same habitat contain only one single type of epibiont (Glaeser and Overmann 2004). Furthermore, none of the 16S rRNA gene sequences of epibionts has so far been detected in free-living green sulfur bacteria, suggesting that the bacterial interaction within phototrophic consortia is highly specific, and that epibionts have adapted to a life in association.

In line with the results of the phylogenetic analyses, the epibiont numbers per consortium show a nonrandom frequency distribution in consortia from natural samples, as well as from enrichment cultures. Thus, most consortia of one particular type contain the same number of epibionts (Overmann et al. 1998). Consortia multiply as a whole. After a simultaneous doubling of epibionts and the central bacterium, the elongated consortium divides by a transverse constriction through the aggregate. Evidently, the cell division of all epibiont cells proceeds in a highly synchronized fashion and parallel to that of the central bacterium.

Several studies provide additional insight into the close interaction between the two partner bacteria. Evidence has accumulated that the two unrelated bacteria interact in a unique way with respect to signal transduction. When a microspectrum of light is projected onto a suspension of "*Chlorochromatium aggregatum*", the consortia accumulate at a wavelength of 740 nm. Since this value matches the position of the long wavelength absorption of bacteriochlorophyll c in vivo, the antenna pigments are most likely the photoreceptors of the scotophobic response (Fröstl and Overmann 1998). On the other hand, electron microscopy clearly demonstrated that the central bacterium is monopolarly monotrichously flagellated, whereas the green sulfur bacterial epibionts are nonmotile (Glaeser and Overmann 2003b). Taken together, these experimental results clearly suggest that a rapid interspecies signal transfer occurs between the nonmotile, phototrophic component and the motile, colorless central bacterium. So far, this type of signal transfer between unrelated bacteria is unique in the microbial world.

Intact consortia exhibit chemotactic behavior toward sulfide (Fröstl and Overmann 1998; Glaeser and Overmann 2003b, 2004). Since sulfide is utilized by the epibiont, a rapid signal exchange between the epibiont and the central motile bacterium could possibly occur also during chemotaxis, yet has to be confirmed experimentally.

Physiological experiments have provided additional information on a mutual signal exchange between the two bacterial partners. When a natural population of "*Pelochromatium roseum*" was incubated with radioactively labeled 2-oxoglutarate and the incorporation of radiolabel followed by microautoradiography of the samples, 87.5% of the consortia incorporated 2-oxoglutarate when both light and sulfide were present, whereas uptake was detected in less than 1.4% of the consortia if either light or sulfide were absent (Glaeser and Overmann 2003b). Because the epibionts in this population of "*Pelochromatium roseum*" had been shown to grow photoautotrophically, the metabolic state of the green sulfur bacterial epibiont seems to regulate the incorporation of 2-oxoglutarate by the central bacterium. Evidently, interaction between the two partner bacteria in phototrophic consortia is not limited to a rapid interspecies signal exchange during tactic behavior, but also occurs during the regulation of central metabolic processes. It may be speculated that such a mechanism serves to synchronize the growth and multiplication of the two partner bacteria.

It is to be expected that the tightly packed layer of epibionts impedes the diffusion of essential substrates and nutrients to the embedded central bacterium. The diffusion coefficient in cytoplasma is moderately decreased (by a mean factor of 0.3; Koch 1996) as compared to typical aqueous environments. Accordingly, the epibiont may not only regulate the uptake of exogenous carbon compounds by the central bacterium, but may actually supply the central bacterium directly with certain carbon substrates. Such an exchange of organic carbon compounds, vitamins and chelators has been suggested for other consortia, like the associations of heterotrophic bacteria with filamentous cyanobacteria (Paerl and Pinckney 1996). Free-living Chlorobiaceae have been shown to excrete up to 30% of the photosynthetically fixed carbon (Czeczuga and Gradski 1972), mainly as 2-oxo-3-methylvalerate and 2-oxoglutarate (Sirevåg and Ormerod 1970). The latter compound is a direct intermediate of the reverse tricarboxylic acid cycle which is employed by green sulfur bacteria for CO_2 fixation (Evans et al. 1966; Sirevåg and Ormerod 1970). Similar to other green sulfur bacteria, the epibionts may thus excrete 2-oxo acids, which could then be utilized as carbon substrates by the central bacterium.

A typical syntrophic sulfur cycle is less likely to occur within phototrophic consortia (see above). Theoretically, interspecies hydrogen transfer may occur between the central bacterium and the epibionts in a manner similar to the aggregates of fermentative bacteria with methanogenic archaea or sulfate-reducing bacteria in activated sludge. Based on typical metabolic rates for bacteria, it has been calculated that the single layer of epibionts is not sufficient to shield the central bacterium from high concentrations of exogenous, easily diffusible compounds, such as hydrogen gas, oxygen or sulfide (Overmann 2001a).

5
Selective Advantage of the Interaction

Phototrophic consortia have been reported to occur worldwide in numerous freshwater lakes and ponds where light reaches anoxic sulfide-containing water layers (Caldwell and Tiedje 1975; Croome and Tyler 1984; Gasol et al. 1995; Overmann et al. 1998; Overmann 2001a). The consortia appear to thrive exclusively in the chemocline of these aquatic environments where they prefer a very narrow regime of low light intensities (<5 µmol quanta m^{-2} s^{-1}) and sulfide concentrations (0–200 µM) (Overmann et al. 1998; Glaeser and Overmann 2004). As the doubling time of a natural population of phototrophic consortia was estimated to amount to 15.3 days (Glaeser and Overmann 2003a), the natural conditions must severely limit the growth of intact consortia in situ. Under these conditions, however, bacteria in phototrophic consortia seem to have a strong selective advantage over their free-living relatives, since epibionts can constitute up to 88% of all green sulfur bacteria present in the natural habitat (Glaeser and Overmann 2003a). A selective advantage of the bacteria in the associated state is also substantiated by the experimental result that low concentrations of sulfide (300 µM) act highly selective for the enrichment of "*Chlorochromatium aggregatum*" consortia, whereas higher sulfide concentrations favor free-living green sulfur bacteria (Fröstl and Overmann 1998).

Although they are physiologically rather similar, one major difference between the epibionts of phototrophic consortia and the free-living green sulfur bacteria is motility. Epibionts, like all other known Chlorobiaceae, do not synthesize flagella. The preferred association with motile bacteria suggests that motility and tactic responses must be of ecological significance. Phototrophic consortia are capable of detecting light intensities as low as 0.4–0.7 µmol quanta m^{-2} s^{-1} (Fröstl and Overmann 1998). Epibionts in association not only are adapted to even lower light intensities than most of the known free-living Chlorobiaceae, but the high light sensitivity of the consortia in combination with their motility provides a means to quickly move to water layers suitable for phototrophic growth.

Many planktonic bacteria contain gas vesicles that provide buoyancy and provide a means for maintaining the vertical position or conduct diurnal vertical migrations in a stratified water column. Energy costs for slow-growing bacterial cells (generation time >11 h) in suspension are substantially less for gas-vacuolated than for flagellated forms (Walsby 1994). If vertical gradients alone would exert a selective pressure to favor motile phototrophic consortia, not only flagellated but also gas vacuolated forms should consequently be selected for and hence should occur frequently in nature. Interestingly, only one of the 19 known types of phototrophic consortia, namely *Chloroplana vacuolata*, contains gas vesicles but does not bear flagella. Almost all (17) other types are flagellated. In the aquatic environment, particulate matter represents a major source of dissolved carbon substrates (Smith et al. 1992). Organic particles accumulate in the chemocline of stagnant water bodies (Culver and Brunskill 1969) where they can create a horizontally heterogeneous environment (Overmann et al. 1994). Motile chemotrophic aquatic bacteria have been shown to occur in patches extending over tens to hundred µm around point sources of nutrients (Blackburn et al. 1998; Krembs et al. 1998; Grossart et al. 2001). Hence, chemotaxis of intact phototrophic consortia not only provides a means to rapidly follow diurnal variations in the vertical position of sulfide-containing water layers but also may permit the consortia to rapidly detect and exploit local accumulations of sulfide and organic carbon compounds released by other microorganisms in horizontally inhomogeneous environments. Movements by means of flagella are much faster than the vertical migrations mediated by changes in the cellular gas vesicle content (Overmann et al. 1991).

Theoretically, one may suppose that the motility of phototrophic consortia enhances the transport of sulfide towards the cells when they are moving in an environment with limiting sulfide concentrations. The ratio of transport by convection to that by diffusion can be assessed by calculating the Péclet number (Jørgensen 2001) for phototrophic consortia (length $L = 10$ µm, velocity v 30 µm s^{-1}, diffusion coefficient D (H$_2$S) = $1.55 \cdot 10^{-5}$ cm^2 s^{-1}) according to:

$$P_e = \frac{v \cdot L}{D}$$

Swimming can outweigh diffusion as a means of providing substrate to phototrophic consortia only at P_e numbers >>1. Since the Péclet number in the case of phototrophic consortia is 0.19, the speed and size of phototrophic consortia is far too small to enhance the supply of sulfide for the cells. In fact, phototrophic consortia would need to swim >155 µm·s^{-1} to increase the availability of sulfide for the epibionts (calculated according to Berg and Purcell 1977).

6
"Symbiosis Genes" of the Epibiont

All epibionts analyzed to date represent 16S rRNA gene sequence types that could not be detected in any free-living green sulfur bacterium (see (2.)3.1). Although phylogenetically distinct, the epibiont of "*Chlorochromatium aggregatum*" did not exhibit unusual physiological properties with respect to the capability of assimilating organic carbon sources, the temperature dependence or the pH dependence of growth. The only phenotypic trait that clearly differed between epibionts and free-living green sulfur bacteria was the composition of photosynthetic pigments. It therefore appears possible that other, not tested traits, distinguish the epibionts of phototrophic consortia from their free-living relatives and/or that specific "symbiosis" genes are induced in the associated state.

Gene acquisition and deletion are viewed as the major events underlying the emergence and evolution of bacterial pathogens and symbionts (Ochman and Moran 2001). Recently, the first genome sequence of the green sulfur bacterium *Chlorobium tepidum* has become available (Eisen et al. 2002). The size of the genome is comparatively small (2.15 Mbp). Analyses of the *Chl. tepidum* genome revealed that only a few proteins for regulatory responses are present and that capabilities for the assimilation of organic compounds are very limited. This currently available information suggests that at least the genome of the phototrophic bacterial partner is sufficiently simple to search for candidate genes involved in the symbiotic interaction.

Accordingly, genetic differences between the epibiont and other Chlorobiaceae were assessed by a molecular biology approach with the aim of identifying particularly genes involved in mutual cell–cell recognition and in the exchange of signals between the partners. Subtractive hybridization of epibiont DNA against genomic DNA of 16 different strains of nonsymbiotic Chlorobiaceae resulted in the isolation of three different genes which appear to be highly unusual for green sulfur bacteria, including a large adhesion exoprotein, and a RTX-toxin homologue (Vogl and Overmann in prep.). The recently completed genome sequence of the epibiont from "*Chlorochromatium aggregatum*" (http://genome.jgi-psf.org/draft_microbes/chlag/chlag.home.html) will provide further insight into the molecular basis of the symbiotic interaction within phototrophic consortia.

7
Outlook

For decades, the morphologically less structured syntrophic cocultures (Chap. 1) have remained the only type of association that were amenable to experimental manipulation. In the case of phototrophic consortia, the highly ordered structure resembles that of other types of structured associations observed in nature. Therefore, the enrichment cultures of phototropic consortia now available offer the possibility to study novel aspects of this particularly tight interaction between non-related prokaryotes.

Possibly the most fascinating – and so far unique – aspect of symbiosis in phototrophic consortia is the signal exchange between the two partners. A detailed analysis of its structural and functional basis will help to reach a better understanding not only of the mechanisms of symbiosis, but possibly also of pathogenicity. Since they consist of two bacterial partners with a limited amount of genetic information, phototrophic consortia are certainly very well-suited model systems for the study of these fundamental aspects of symbiotic interactions between non-related organisms.

Theoretically, the highly specific association of bacteria in phototrophic consortia could have emerged only once during evolution. However, based on the phylogenetic analysis of a large data set of 41 independently sampled 16S rRNA gene sequences, the extant epibionts of phototrophic consortia are not monophyletic (Glaeser and Overmann 2004). Thus, the ability to form symbiotic associations either arose independently from different ancestors or, alternatively, the common ancestor of the extant green sulfur bacteria was symbiotic. In the latter case, several descendent lineages would have developed the ability to sustain an independent life style. In any event, our phylogenetic analysis implies that the switch between the symbiotic and the free-living state (or vice versa) occurred more than once during the radiation of the green sulfur bacteria. Thus, identification of additional "symbiosis" genes and analysis of the structure, function, and origin of these genes in the green sulfur bacterial epibiont is also relevant from an evolutionary perspective in order to understand the genetic basis of the transition between the symbiotic and the free-living state of these very special green sulfur bacteria.

Acknowledgements. Several motivated students have participated in our research on phototrophic consortia: Jürgen Fröstl, Jens Glaeser, Kajetan Vogl, Martina Schlickenrieder, Martina Müller, Birgit Kanzler and Kristina Pfannes. Their contributions were decisive for the success of the project. Support by the Deutsche Forschungsgemeinschaft (grants Ov20/3–1, Ov20/3–2, Ov20/3–3, Ov20/10–1) is gratefully acknowledged.

References

Abella CA, Cristina XP, Martinez A, Pibernat I, Vila X (1998) Two new motile phototrophic consortia: "*Chlorochromatium lunatum*" and "*Pelochromatium selenoides*". Arch Microbiol 169:452–459

Baas-Becking LGM (1934) Geobiologie of inleiding tot de milieukunde. Van Stockum and Zoon, The Hague, The Netherlands
Bassler BL (2002) Small talk: cell-to-cell communication in bacteria. Cell 109:421–424
Bassler BL, Greenberg EP, Stevens AM (1997) Cross-species induction of luminescence in the quorum sensing bacterium *Vibrio harveyi*. J Bacteriol 179:4043–4045
Beijerinck MW (1913) De infusies en de ontdekking der backteriën, Jaarboek van de Koninklijke Akademie v. Wetenschappen. Müller, Amsterdam, The Netherlands
Berg HC, Purcell EM (1977) Physics of chemoreception. Biophys J 20:193–219
Blackburn N, Fenchel T, Mitchell J (1998) Microscale nutrient patches in planktonic habitats shown by chemotactic bacteria. Science 282:2254–2256
Caldwell DE, Tiedje JM (1975) A morphological study of anaerobic bacteria from the hypolimnia of two Michigan lakes. Can J Microbiol 21:362–376
Conrad R, Phelps TJ, Zeikus JG (1985) Gas metabolism evidence in support of the juxtaposition of hydrogen-producing and methanogenic bacteria in sewage sludge and lake sediments. Appl Environ Microbiol 50:595–601
Croome RL, Tyler PA (1984) Microbial microstratification and crepuscular photosynthesis in meromictic Tasmanian lakes. Verh Int Verein Limnol 22:1216–1223
Culver DA, Brunskill GJ (1969) Fayetteville Green Lake, New York. V. Studies of primary production and zooplankton in a meromictic marl lake. Limnol Oceanogr 14:862–873
Czeczuga B, Gradski F (1972) Relationship between extracellular and cellular production in the sulphuric green bacterium *Chlorobium limicola* Nds. as compared to primary production of phytoplankton. Hydrobiologia 42:85–95
Dubinina GA, Kuznetsov SI (1976) The ecological and morphological characteristics of microorganisms in Lesnaya Lamba (Karelia). Int Rev Ges Hydrobiol 61:1–19
Eisen JA et al (2002) The complete genome sequence of *Chlorobium tepidum* TLS, a photosynthetic, anaerobic, green sulfur bacterium. Proc Natl Acad Sci USA 99:9509–9514
Evans MCW, Buchanan BB, Arnon DI (1966) A new ferredoxin-dependent carbon reduction cycle in a photosynthetic bacterium. Proc Natl Acad Sci USA 55:928–934
Finlay BJ (2002) Global dispersal of free-living microbial eukaryote species. Science 296:1061–1063
Finlay BJ, Clarke KJ (1999) Ubiquitous dispersal of microbial species. Nature 400:828
Fröstl J, Overmann J (1998) Physiology and tactic response of "*Chlorochromatium aggregatum*". Arch Microbiol 169:129–135
Fröstl J, Overmann J (2000) Phylogenetic affiliation of the bacteria that constitute phototrophic consortia. Arch Microbiol 174:50–58
Gasol JM, Jürgens K, Massana R, Calderón-Paz JI, Pedrós-Alió C (1995) Mass development of *Daphnia pulex* in a sulfide-rich pond (Lake Cisó). Arch Hydrobiol 132:279–296
Glaeser J, Overmann J (2003a) Characterization and in situ carbon metabolism of phototrophic consortia. Appl Environ Microbiol 69:3739–3750
Glaeser J, Overmann J (2003b) The significance of organic carbon compounds for in situ metabolism and chemotaxis of phototrophic consortia. Environ Microbiol 5:1053–1063
Glaeser J, Overmann J (2004) Biogeography, evolution and diversity of the epibionts in phototrophic consortia. Appl Environ Microbiol 70(8):4821–4830

Glaeser J, Baneras L, Rütters H, Overmann J (2002) Novel bacteriochlorophyll e structures and species-specific variability of pigment composition in green sulfur bacteria. Arch Microbiol 177:475–485

Gorlenko VM, Kuznetzov SI (1972) Vertical distribution of phototrophic bacteria in the Kononér Lake of the Mari ASSR. Microbiol 40:651–652

Grossart H-P, Riemann L, Azam F (2001) Bacterial motility in the sea and its ecological implications. Aquat Microb Ecol 25:247–258

Huber H, Hohn MJ, Rachel R, Fuchs T, Wimmer VC, Stetter KO (2002) A new phylum of archaea represented by a nanosized hyperthermophilic symbiont. Nature 417:63–67

Jacobi CA, Aßmus B, Reichenbach H, Stackebrandt E (1997) Molecular evidence for association between the *Sphingobacterium*-like organism "*Candidatus comitans*" and the Myxobacterium *Chondromyces crocatus*. Appl Environ Microbiol 63:719–723

Jørgensen BB (2001) Life in the diffusive boundary layer. In: Boudreau BP, Jørgensen BB (eds) The Benthic boundary layer: transport processes and biogeochemistry, chap 14. Oxford Univ Press, Oxford, pp 348–373

Kanzler B, Pfannes KR, Vogl K, Overmann J (2005) Molecular characterization of the non-photosynthetic partner bacterium in the consortium "*Chlorochromatium aggregatum*". Appl Environ Microbiol 71:7434–7441

Koch A (1996) What size should a bacterium be? A question of scale. Annu Rev Microbiol 50:317–348

Krembs C, Juhl AR, Long RA, Azam F (1998) Nanoscale patchiness of bacteria in lake water studied with the spatial information preservation method. Limnol Oceanogr 43:307–314

Lauterborn R (1906) Zur Kenntnis der sapropelischen Flora. Allg Bot 2:196–197

Ochmann H, Moran NA (2001) Genes lost and genes found: evolution of bacterial pathogenesis and symbiosis. Science 292:1096–1099

Overmann J (2001a) Green sulfur bacteria. In: Garrity GM (ed) Bergey's manual of systematic bacteriology, vol 1. Williams and Wilkins, Baltimore, pp 601–623

Overmann J (2001b) Phototrophic consortia: a tight cooperation between non-related eubacteria. In: Seckbach J (ed) Symbiosis. Mechanisms and model systems. Kluwer Academic Publ, Dordrecht, pp 239–255

Overmann J, Schubert K (2002) Phototrophic consortia: model systems for symbiotic interrelations between prokaryotes. Arch Microbiol 177:201–208

Overmann J, Lehmann S, Pfennig N (1991) Gas vesicle formation and buoyancy regulation in *Pelodictyon phaeoclathratiforme* (Green sulfur bacteria). Arch Microbiol 157:29–37

Overmann J, Cypionka H, Pfennig N (1992) An extremely low-light-adapted phototrophic sulfur bacterium from the Black Sea. Limnol Oceanogr 32:150–155

Overmann J, Beatty JT, Hall KJ (1994) Photosynthetic activity and population dynamics of Amoebobacter purpureus in a meromictic saline lake. FEMS Microbiol Ecol 15:309–320

Overmann J, Tuschak C, Fröstl J, Sass H (1998) The ecological niche of the consortium "*Pelochromatium roseum*" Arch Microbiol 169:120–128

Paerl HW, Pinckney JL (1996) A mini-review of microbial consortia: their roles in aquatic production and biogeochemical cycling. Microb Ecol 31:225–247

Perfiliev BV (1914) On the theory of symbiosis of *Chlorochromatium aggregatum* Lauterb. (*Chloronium mirabile* Buder) and *Cylindrogloea bacterifera* nov. gen., nov. spec. (in Russian). I Mikrobiol Petrogr 1:222–225

Pfennig N (1980) Syntrophic mixed cultures and symbiotic consortia with phototrophic bacteria: a review. In: Gottschalk G, Pfennig N, Werner (eds) Anaerobes and anaerobic infections. Fischer, Stuttgart, pp 127–131

Reichenbach H, Dworkin M (1992) The myxobacteria. In: Trüper HG, Balows A, Dworkin M, Harder W, Schleifer K-H (eds) The Prokaryotes. Springer, Berlin Heidelberg New York, pp 3416–3487

Rudolph C, Wanner G, Huber R (2001) Natural communities of novel archaea and bacteria growing in cold sulfurous springs with a string-of-pearl like morphology. Appl Environ Microbiol 67:2336–2344

Schink B (1991) Syntrophism among prokaryotes. In: Balows A, Trüper HG, Dworkin M, Harder W, Schleifer K-H (eds) The prokaryotes, 2nd edn. Springer, Berlin Heidelberg New York, pp 276–299

Sirevåg R, Ormerod J (1970) Carbon dioxide-fixation in photosynthettic green sulfur bacteria. Science 169:186–188

Skuja H (1957) Taxonomische und biologische Studien über das Phytoplankton schwedischer Binnengewässer. Nova Acta Reg Soc Sci Upsala Ser IV(16):1–404

Smith DC, Simon M, Alldredge AL, Azam F (1992) Intense hydrolytic enzyme activity on marine aggregates and implications for rapid particle dissolution. Nature 359:139–142

Staley JT (1999) Bacterial biodiversity: a time for place. ASM News 65:681–687

Tonolla M, Demarta A, Peduzzi S, Hahn D, Peduzzi R (2000) In situ analysis of sulfate-reducing bacteria related to *Desulfocapsa thiozymogenes* in the chemocline of meromictic Lake Cadagno. Appl Environ Microbiol 66:820–824

Trüper HG, Pfennig N (1971) Family of phototrophic Green Sulfur Bacteria: *Chlorobiaceae* Copeland, the correct family name; rejection of *Chlorobacterium* Lauterborn; and the taxonomic situation of the consortium-forming species. Int J Syst Bacteriol 21:8–10

Tuschak C, Glaeser J, Overmann J (1999) Specific detection of green sulfur bacteria by in situ hybridization with a fluorescently labeled oligonucleotide probe. Arch Microbiol 171:265–272

Vogl K, Glaeser J, Pfannes K, Wanner G, Overmann J (2005) *Chlorobium chlorochromatii* sp. nov., a symbiotic green sulfur bacterium isolated from the phototrophic consortium "*Chlorochromatium aggregatum*" (submitted)

Walsby AE (1994) Gas vesicles. Microbiol Rev 58:94–144

Prokaryotic Symbionts of Termite Gut Flagellates: Phylogenetic and Metabolic Implications of a Tripartite Symbiosis

Andreas Brune, Ulrich Stingl

1
Introduction

The symbiotic associations of termites with their gut microorganisms have been studied for almost a century. While the earlier work had focused on the intestinal protozoa of lower termites and their role in digestion, the more recent investigations have been directed mainly at the prokaryotic gut microorganisms, their metabolic activities, and the structure and function of the bacterial and archaeal populations. In addition, the application of microsensor techniques has revealed the unexpected dynamics of the physiochemical gut conditions and given the first indications of the spatial organization of the major microbial activities, which has led to the recognition that the gut provides a variety of different microhabitats for its microbiota.

The most important habitats for the prokaryotes in the hindgut of lower termites are provided by the intestinal protozoa. Although it is long known that the gut flagellates are intimately associated with prokaryotes, the significance of this phenomenon has not been fully appreciated. The difficulties surrounding cultivation of the protozoa and the complete absence of pure cultures of the prokaryotic symbionts so far allowed merely a morphological description of the different associations. Their exact nature and the possible benefits for the partners are mostly in the dark. However, with the advent of molecular biology tools and the resulting new possibilities, the interest in the symbiotic associations between prokaryotes and termite gut flagellates has been renewed.

A definitive classification of these associations into the different categories of symbiosis, such as mutualism, parasitism, or commensalism, would require a level of understanding that remains to be reached. This chapter will therefore use the term symbiosis in its broader definition, as it was originally coined by Anton de Bary (1878), which comprises all kinds of associations between different species.

A. Brune
Max Planck Institute for Terrestrial Microbiology, Marburg, Germany
U. Stingl
Department of Microbiology, Oregon State University, Corvallis, Or., USA

2
The Termite Gut Microecosystem

The symbiotic digestion of lignocellulose by termites is a complex series of events involving both the host and its intestinal microbiota (Brune 2003, 2005). While the digestive activities in the foregut and midgut seem to be mainly caused by the host, the processes in the hindgut are largely controlled by the symbionts. The recalcitrance of the lignocellulosic diet and the dynamics of physicochemical gut conditions contribute to the heterogeneity of ecological niches and are reflected in a diverse community of prokaryotic and eukaryotic microorganisms.

2.1
Lignocellulose degradation

There is a large body of literature on the decomposition of wood and cellulose by termite gut flagellates, which are essential for the digestion of lignocellulose in lower termite (Radek 1999). By contrast, the majority of prokaryotes in termite guts do not seem to contribute to polymer degradation. They appear to be involved mainly in the fermentation of soluble metabolites released into the gut, which are derived either directly from the food by the digestive enzymes or from the fermentative activity of the intestinal protozoa (Breznak 2000; Ohkuma 2003; Brune 2005).

Lignocellulose is not only difficult to degrade, but it is also an extremely nitrogen-poor substrate that lacks most of the essential nutrients required by the termite, such as amino acids, vitamins, and sterols. The capacity of the intestinal prokaryotes to fix dinitrogen, to assimilate nitrate and ammonia, and to synthesize those amino acids and vitamins essential for the host makes them also an important source of nutrition (Breznak 2000; Machida et al. 2001). The microbial biomass produced in the hindgut is accessed via proctodeal feeding, which serves both the transfer of symbionts among individual termites and nitrogen cycling by subsequent digestion of microbial biomass in the midgut (Fujita et al. 2001).

2.2
The Gut Microenvironment

Based on the oxygen sensitivity and the fermentative metabolism of the intestinal protozoa, the high concentrations of microbial fermentation products, the presence of anaerobic or oxygen-sensitive activities such as methanogenesis and nitrogen fixation, and the isolation of obligately anaerobic bacteria, it was initially assumed that the termite hindgut is an anoxic habitat (for literature, see Breznak 2000; Brune 2005).

However, the maintenance of anoxia in an environment of such minute dimensions is not a trivial issue (Brune 1998). Oxygen microsensor measurements in the gut of the termite *Reticulitermes flavipes* have demonstrated that the steep gradient in oxygen partial pressure between the oxic gut epithelium and the anoxic gut contents drives a continuous influx of oxygen into the hindgut (Brune et al. 1995; Ebert and Brune 1997). Also, in all other termites investigated, oxygen penetrates 50–200 µm into the periphery of the hindgut lumen, leaving only the central portion of the dilated compartments anoxic (for literature, see Brune 2005).

While the redox potential in the hindgut contents of *Reticulitermes flavipes* basically mirrors the oxygen gradients, the massive hydrogen production by the intestinal protozoa and its efficient consumption by hydrogenotrophic microorganisms gives rise to steep radial counter gradients of hydrogen and enormous dynamics of hydrogen partial pressure between the gut center and gut periphery (Ebert and Brune 1997).

The hydrogen partial pressure in the different gut regions seems to be controlled by the spatial organization of the hydrogen-producing and hydrogen-consuming populations (Tholen and Brune 2000). Radiotracer analysis of the in situ metabolism in the hindgut of *Reticulitermes flavipes* demonstrated that also the high oxygen fluxes significantly influence the fermentation processes in the hindgut (Tholen and Brune 2000). The oxygen-reduction potential of obligately anaerobic bacteria isolated from termite guts (see Sect. (8.)2.3) corroborates that the location of microbial populations relative to the oxygen gradient will affect intestinal carbon fluxes.

2.3
Prokaryotic Diversity

As in most other environments, phylogenetic analysis of the 16S rRNA genes of the prokaryotes in the hindgut of several termites has revealed an enormous diversity of the intestinal microbiota (see Brune 2005 for a review). The majority of clones in bacterial clone libraries of several species in the genus *Reticulitermes* bears less than 90% sequence similarity to those of known, cultivated microorganisms. Clones in the so-called Termite Group 1 (Ohkuma and Kudo 1996; Hongoh et al. 2003; Yang et al. 2005) even represent a new bacterial phylum, the 'Endomicrobia' (Stingl et al. 2005), which contains no cultivated representatives and apparently consists exclusively of endosymbionts of termite gut protozoa (see (8).4.2.2.). The fact that most clones recovered in these studies fall within lineages consisting exclusively of clones obtained from the hindgut of termites, with the closest relatives of a given clone usually stemming from the most closely related termite (Yang et al. 2005), indicates that most of the prokaryotes in termite guts are specific for and also restricted to this particular habitat.

A number of prokaryotes from termite guts have been isolated in pure culture. Many of the isolates are unique to the termite gut habitat and catalyze key activities in hindgut metabolism, e.g., methanogenesis, reductive acetogenesis, nitrogen fixation, or uric acid fermentation (e.g., Breznak 1994; Leadbetter and Breznak 1996; Leadbetter et al. 1998, 1999; Lilburn et al. 2001; Boga et al. 2003; Graber et al. 2004). Also, oxygen reduction by anaerobic bacteria, such as lactic acid bacteria (Tholen et al. 1997; Bauer et al. 2000), homoacetogenic bacteria (Boga and Brune 2003), and sulfate-reducing bacteria (Kuhnigk et al. 1996; Fröhlich et al. 1999), emerges as a potentially important activity (see above).

Although the characterization of the isolates has provided valuable information on metabolic properties and other physiological features relevant for the colonization of this particular habitat, there are still enormous discrepancies between the phylogenetic groups dominating the clone libraries and the species recovered by cultivation, indicating the presence of a strong cultivation bias (Breznak 2000; Brune 2005). To date, there is not a single isolate representing the prokaryotes associated with the gut flagellates. In general, the lack of knowledge on the individual components of the prokaryotic microbiota and their metabolic capacities and activities in situ still make it difficult to define the essential functions and understand the complex interactions.

3
Hindgut Protozoa

Symbiotic flagellates are found exclusively in the phylogenetically lower termites and the closely related cockroaches of the genus *Cryptocercus*; the higher termites harbor a largely prokaryotic microbiota. The first studies that revealed the beneficial nature of these peculiar symbionts – discovered in 1856 by Lespes (see Leidy 1881) and initially considered as parasites – were reported by Cleveland (1925, 1926), who demonstrated that termites do not survive for long after elimination of their gut flagellates. The importance of the symbionts for their termite host is impressively documented by their enormous abundance in the enlarged hindgut paunch; it has been estimated that the fresh weight of the protozoa may account for more than one half of the fresh weight of the termite (Katzin and Kirby 1939). Phylogenetically, the gut flagellates are extremely diverse: almost 450 distinct species have been described to occur within the approx. 200 termite species investigated (Yamin 1979).

3.1
Phylogeny

Termite gut flagellates belong to three distinct taxa: trichomonads, hypermastigids, and oxymonads (Yamin 1979). Originally, all were considered primitive, primarily amitochondriate eukaryotes, but recent molecular data proved that they represent two separate eukaryotic lineages.

Phylogenetic analyses of the termite gut flagellates by various groups (reviewed by Ohkuma 2003; Gerbod et al. 2004), mainly based on 18S rRNA gene sequence analysis, confirmed the presence of two classes of the phylum Parabasalia, the Trichomonadea and Hypermastigea (Cavalier-Smith 2002), the latter representing the most basal lineage.

The phylogeny of Oxymonadea was long in the dark. Comparative sequence analysis revealed that they are a sister taxon to the excavate protists (Moriya et al. 1998, 2001, 2003; Dacks et al. 2001; Stingl and Brune 2003); they are now classified in the phylum Loukozoa (Cavalier-Smith 2002).

While Hypermastigea consist exclusively of termite gut symbionts, the Oxymonadea and the Trichomonadea comprise also species occurring in other habitats, such as the intestinal tract or body cavities of other animals, including humans.

3.2
Physiology and Function

The first studies on the anaerobic nature and the fermentation of cellulose by termite gut flagellates were performed by Hungate in the 1940s (reviewed by Hungate 1955).

Hungate's work with mixed protozoan suspensions and the few pure culture studies of parabasalid flagellates obtained from the hindgut of *Zootermopsis* species (Trager 1934; Yamin and Trager 1979; Yamin 1980, 1981; Odelson and Breznak 1985a,b) led to the current hypothesis that the glycosyl units of cellulose and possibly other polysaccharides in the wood particles taken up by the protozoa are fermented to acetate, CO_2, and H_2 according to the following equation:

$$\langle C_6H_{12}O_6 \rangle + 2\,H_2O \rightarrow 2\,CH_3COO^- + 2\,H^+ + 2\,CO_2 + 4\,H_2$$

The concept is supported by the high hydrogen partial pressure and the dominance of acetate among the short-chain fatty acids in the hindgut of all lower termites investigated (reviewed by Radek 1999; Breznak 2000; Brune 2005).

The acetate produced in the hindgut is resorbed and serves as the main respiratory substrate of the host; this is probably the most straightforward explanation why the well-being and survival of lower termites on a normal, lignocellulosic diet depend on their gut protozoa (Breznak 2000; Brune 2003, 2005). Despite the demonstration of host cellulases in the secretions of midgut and salivary glands of lower termites (see Watanabe and Tokuda 2001 for a review), the cellulolytic activities in the hindgut seem to be largely of protozoan origin (Ohtoko et al. 2000; Nakashima et al. 2002).

However, data on the physiology of gut flagellates are scarce, and the current concept of cellulose metabolism by termite gut flagellates is largely based on assumptions and inferences from other systems. It is assumed that the soluble sugars resulting from polysaccharide degradation in the food vacuoles are oxidized to pyruvate via glycolysis in the cytoplasm. Pyruvate is then imported into the hydrogenosomes – special organelles that enable anaerobic protozoa and fungi to produce molecular hydrogen. Hydrogenosomes were originally discovered by Lindmark and Müller (1973) in the parabasalid flagellates *Trichomonas vaginalis* and *Tritrichomonas foetus*, which are parasites of humans or bovines. They contain several key enzymes: pyruvate-ferredoxin oxidoreductase and hydrogenase, which convert pyruvate to acetate, CO_2, and H_2; and phosphotransacetylase and acetate kinase, which allow the formation of ATP, which is subsequently exported from the hydrogenosomes into the cytoplasm (Müller 1993; Hackstein et al. 2002). The oxidation of glucose to acetate according to Eq. (1) requires that also the reducing equivalents produced during glycolysis are transported into the hydrogenosomes and released as H_2.

It should be emphasized that this concept is mainly based on results obtained with trichomonad flagellates that are parasites of mammals. There is no physiological data on the hydrogenosomes of termite gut flagellates. The presence of hydrogenosomes in hypermastigids is based only on microscopic evidence and the fermentation products of a few axenic cultures (see above).

Virtually nothing is known about the physiology of oxymonads, and not a single pure culture has been obtained so far. These protists do not seem to possess hydrogenosomes (Bloodgood et al. 1974; Radek 1994), which means that their metabolism must be different from that of parabasalids. Assuming that oxymonads cannot form hydrogen (which remains to be tested), other reduced fermentation products have to be expected. In *Reticulitermes flavipes*, where oxymonads are the most abundant flagellates (Cook and Gold 1998), lactate has been identified as a key intermediate in hindgut metabolism (Tholen and Brune 2000). The human pathogen, *Trichomonas vaginalis* switches to a lactic acid fermentation when pyruvate-ferredoxin oxidoreductase is inhibited with metronidazol (Cerkasova et al. 1986).

The gut flagellate communities of each termite species are stable and host-specific, but also quite diverse (Yamin 1979), which implies also a high degree of functional diversity. There is evidence for a resource partitioning among the different protozoan species colonizing the same gut, which seem to be specialized on cellulose or wood particles of different size classes

(Yoshimura et al. 1993; Inoue et al. 2000). Also a specialization on other wood components such as hemicellulose would create additional niches (Inoue et al. 1997). Certain flagellates do not seem to ingest wood particles (e.g., Yoshimura et al. 1996), and also pinocytosis of dissolved substrates (Hollande and Valentin 1969) or a bacteriovorous lifestyle (Huntenburg et al. 1986) would be alternative strategies of survival.

4
Symbiotic Associations with Prokaryotes

It is long known that most of the gut flagellates are associated with prokaryotes (Pierantoni 1936; Kirby 1941; Ball 1969; Bloodgood and Fitzharris 1976; Dolan 2001). In view of the enormous biovolume of protozoa in the hindgut of lower termites and the resulting cell surface area available for colonization, it is not astonishing to find that the vast majority of the prokaryotes in the dilated part of the hindgut – approx. 90% in the case of *Mastotermes darwiniensis* (Berchtold et al. 1999) – is associated with the protozoan fraction.

The associations between prokaryotes and protozoa seem to be quite specific. For *Reticulitermes santonensis*, it has been shown that the structure and composition of the bacterial community associated with the protozoan fraction is clearly different from that of the hindgut wall or that freely suspended in the hindgut fluid (Yang et al. 2005). In situ hybridization of hindgut contents with phylogenetic probes specific for the epibionts or endosymbionts of certain gut flagellates did not detect any unassociated target cells (Stingl et al. 2004, 2005).

4.1
Prokaryotic Epibionts

Many of the termite gut flagellates are colonized by epibiotic bacteria. Some of the associations are already evident by observation with a light microscope, but only the electron microscope reveals the enormous variety and complexity of these associations (e.g., Cleveland and Grimstone 1964; Leander and Keeling 2004; Figs. 1–3). The presence of special attachment sites on the cell envelope of the flagellates (e.g., Tamm 1980; Radek et al. 1992; Dolan and Margulis 1997; Stingl et al. 2004) indicates a tight association. In some cases, the epibionts seem to be responsible for the locomotion of the host (Cleveland and Grimstone 1964; Tamm 1982; see also Chapter by H. König, L. Li, M. Wenzel, and J. Fröhlich). Advances in molecular ecology made it possible to elucidate the identity of the microorganisms involved in these associations and the specificity of these symbioses.

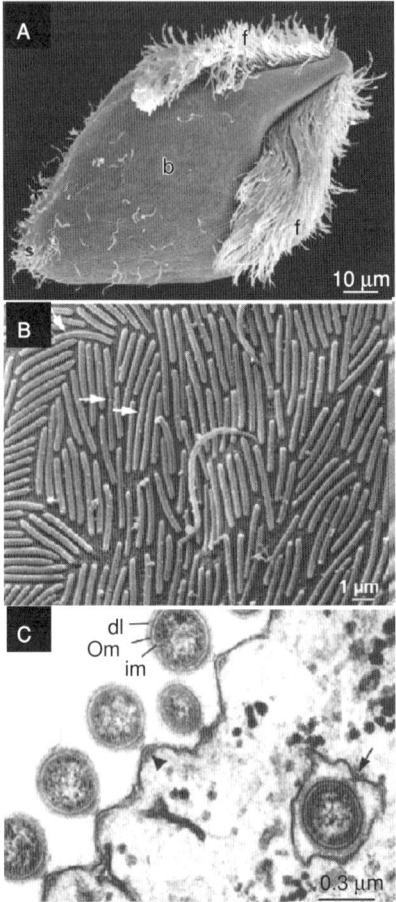

Fig. 1. '*Candidatus* Vestibaculum illigatum' colonizing *Staurojoenina* sp. from *Neotermes cubanus*. **A** Scanning electron micrograph illustrating the whole cell covered with rod-shaped and spirochetal epibionts (*b* bacteria, *f* flagella). **B** Scanning electron micrograph of the flagellate surface, showing the morphotype and the dense arrangement of 'Vestibaculum illigatum'. *Arrows* point to cells in division stages. **C** Transmission electron micrograph of a thin section of the flagellate cell, showing (1) the Gram-negative cell wall structure of 'Vestibaculum illigatum' (*dl* diffuse layer, *om* outer membrane, *im* inner membrane), (2) the presence of special attachment sites (*arrow*), and (3) the presence of 'Vestibaculum illigatum' in vacuoles within the cytoplasm of the host cell. (Figures from Stingl et al. 2004)

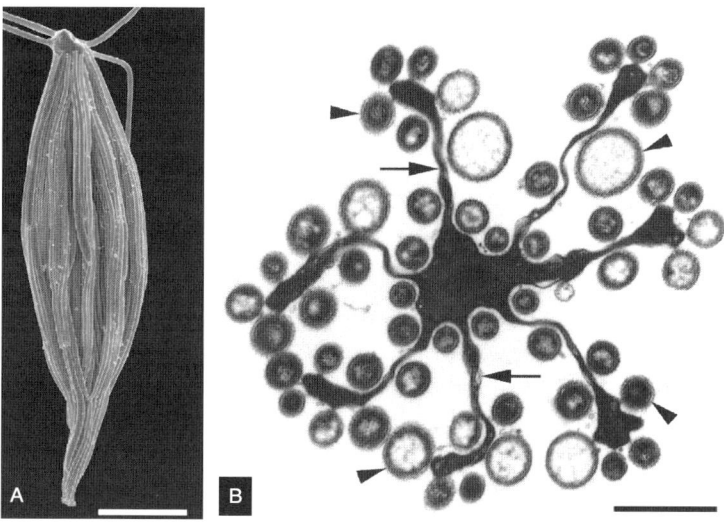

Fig. 2. Symbiotic bacteria attached to the oxymonad flagellate *Streblomastix strix* from *Zootermopsis angusticollis*. **A** Scanning electron micrograph showing a *S. strix* cell covered with epibiotic bacteria. *Bar* 5 µm. **B** Transmission electron micrograph of a thin section of a *S. strix* cell. The host is organized as a central core (*c*) with seven thin vanes radiating outward (*arrows*). Seven to ten epibiotic bacteria with distinctive morphotypes were clustered around each vane (*arrowheads*). *Bar* 1 µm. (Figure from Leander and Keeling 2004)

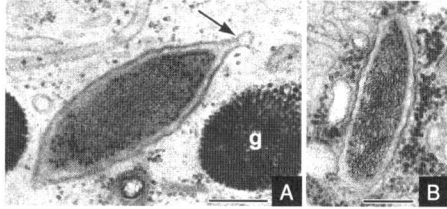

Fig. 3. Transmission-electron micrographs of ultra-thin sections of *Trichonympha agilis* (**A**) and *Pyrsonympha vertens* (**B**) showing the ultrastructure of Termite Group 1 (TG-1) bacteria, which are very abundant in the cytoplasm of these flagellate species. TG-1 cells are surrounded by two membranes; the outer-most membrane forms tube-like elongations at the tapered cell poles (*arrow* in **A**). *g* glycogen. *Bars* 0.2 µm. (Figure from Stingl et al. 2005)

4.1.1
Methanogens

Methanogenic archaea in termite guts are easily visualized by the characteristic autofluorescence of their coenzyme F_{420}. While the methanogens in the hindgut of *Reticulitermes flavipes* are mostly attached to the gut wall or to the surface of other prokaryotes (Leadbetter and Breznak 1996), they are also associated with protozoa in other termites, either as epibionts (Fig. 4) or as endosymbionts (see below) of certain smaller gut flagellates (Lee et al. 1987; Tokura et al. 2000). Most of the clones of methanoarchaea retrieved from lower termites cluster among the family Methanobacteriaceae (for references, see Ohkuma et al. 1999b; Tokura et al. 2000; Shinzato et al. 2001), and are closely related to the three *Methanobrevibacter* species that have been isolated from the gut of *Reticulitermes flavipes* (Leadbetter and Breznak 1996; Leadbetter et al. 1998). They are assumed to participate in the consumption of H_2 formed by the flagellates (see Sect. (8.)5).

4.1.2
Spirochaetes

Although spirochetes are extremely abundant in the intestinal tract of most wood-feeding termites, only a few species have been characterized in pure culture (Breznak and Leadbetter 2002; Graber et al. 2004). Termite gut spirochetes display an enormous morphological diversity (Margulis and Hinkle 1999; Breznak and Leadbetter 2002), but several cultivation-independent analyses have documented that they belong exclusively to the *Treponema* branch of the Spirochaetes, forming two distinct clusters (Lilburn et al. 1999; Ohkuma et al. 1999c).

In many instances, the spirochetal cells are attached to the surface of flagellates (for a review, see Breznak and Leadbetter 2002). Specific attachment

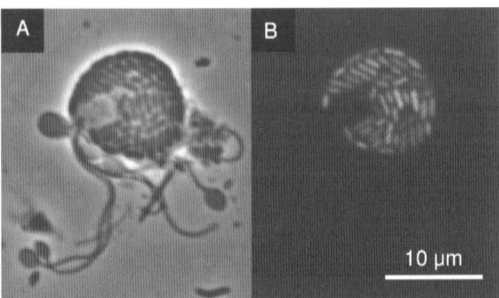

Fig. 4. F_{420}-fluorescent epibiotic methanogens colonizing a small trichomonad from *Schedorhinotermes lamanianus*. Phase-contrast (**A**) and epifluorescence (**B**) photomicrographs of the samoe field of view. (Figure from Brune 2004)

structures have been identified (e.g., Bloodgood et al. 1974; Bloodgood and Fitzharris 1976; Radek et al. 1992, 1996), but only in a few cases have the symbionts been assigned to particular genotypes (Iida et al. 2000; Noda et al. 2003; Wenzel et al. 2003). Attached spirochetes belong to both *Treponema* clusters; interestingly, up to five different genotypes can be attached to a single flagellate.

4.1.3
Bacteroidales

The surface of many of the larger gut flagellates is densely covered with morphologically similar rod-shaped bacteria (Bloodgood and Fitzharris 1976; Tamm 1982; Dolan and Margulis 1997; Stingl et al. 2004; Fig. 1). The epibionts of a *Staurojoenina* sp. from *Neotermes cubanus* were recently identified as members of a new, termite-associated lineage of Bacteroidales (Stingl et al. 2004) and have been assigned to the candidate taxon 'Vestibaculum illigatum' (Fig. 5), and also the epibionts of *Barbulanympha* and *Caduceia* seem to be members of the *Bacteroidales* (see Stingl et al. 2004).

Also the rod-shaped epibionts of *Mixotricha paradoxa* (Wenzel et al. 2003; see also Chapter by H. König, L. Li, M. Wenzel, and J. Fröhlich) fall into this cluster, and it is likely that the other clones in the cluster recovered from other termites (Ohkuma et al. 2002; Hongoh et al. 2003; Yang et al. 2005) represent

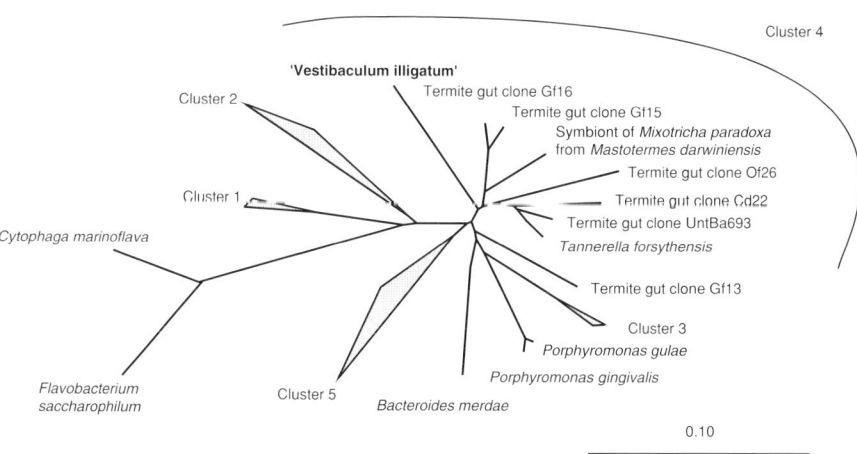

Fig. 5. Phylogenetic relationships between the 16S rRNA gene sequences of '*Candidatus* Vestibaculum illigatum' colonizing *Staurojoenina* flagellates in *Neotermes cubanus* and the rod-shaped epibiont of *Mixotricha paradoxa* in *Mastotermes darwiniensis* to clones obtained from various other termites and the most-closely related isolates bacteria among the Bacteroidales. (Figure from Stingl et al. 2004)

symbionts of gut flagellates. The rod-shaped epibionts of *Caduceia versatilis* (d'Ambrosio et al. 1999), whose attachment structures differ considerably from that of the other rod-shaped epibionts of devescovinid flagellates (Tamm 1980), are involved in an unusual motility symbiosis (Tamm 1982).

4.2
Prokaryotic Endosymbionts

Far less is known about the bacterial endosymbionts of termite flagellates, and both their phylogeny and their metabolic roles are often completely obscure.

4.2.1
Methanogens

The flagellate *Pentatrichomonoides scroa* from the gut of the termite *Mastotermes darwiniensis* (Fröhlich and König 1999) and certain *Dinenympha* and *Microjoenia* species in *Reticulitermes speratus* and *Hodotermopsis sjoestedti* (Tokura et al. 2000) are associated with methanogens that seemingly exist within the cell. As in the case of the epibiotic methanogens, the 16S rRNA gene sequences indicate that they belong to the genus *Methanobrevibacter*. The guts of each termite species usually harbor more than one genotype of methanogens, and there is evidence that the methanogens associated with the flagellates are different from those attached to the gut wall (Tokura et al. 2000). Although methanoarchaea are usually not encountered in the larger gut flagellates (Lee et al. 1987), they probably participate in the consumption of H_2 accumulating within the gut (see Sect. (8.)5).

4.2.2
The 'Endomicrobia'

For a long time, the nature and identity of the endosymbionts of the larger gut flagellates was completely obscure. Recently, the cytoplasmic symbionts of the hypermastigid *Trichonympha agilis* and the oxymonad *Pyrsonympha vertens* in the gut of *Reticulitermes santonensis* were identified as members of the so-called 'Termite Group 1' (TG-1) (Stingl et al. 2005). Based on their 16S rRNA gene sequences, whose origin was verified by fluorescence in situ hybridization with clone-specific probes, and their peculiar ultrastructure (Fig. 3), the endosymbionts of the two protozoa were provisionally classified as two species in the candidate genus 'Endomicrobium' (Stingl et al. 2005). The symbionts are specific for their respective flagellate host and do not occur on the surface of the flagellates or freely suspended in the gut fluid.

Up to that point, 16S rRNA gene sequences of TG-1 bacteria had been retrieved only from termites of the genus *Reticulitermes* (Ohkuma and Kudo 1996; Hongoh et al. 2003; Yang et al. 2005). A broad survey of termites from different families, using a PCR assay with TG-1-specific primers, documented that TG-1 bacteria are present in and also restricted to the guts of all lower termites and the wood-feeding cockroach *Cryptocercus punctulatus* – insects that are in an exclusive, obligately mutualistic association with parabasalid and oxymonad flagellates (Stingl et al. 2005).

Only a small fraction of termites has been investigated to date, but already the present dataset indicates an enormous phylogenetic diversity among TG-1 bacteria (Fig. 6). Based on their isolated phylogenetic position and their endosymbiotic lifestyle, the candidate phylum 'Endomicrobia' has been proposed (Stingl et al. 2005). Although there are indications for a horizontal transfer of TG-1 bacteria between hypermastigid and oxymonad flagellates, the bacterial symbionts seem to have evolved together with their flagellate hosts, which would also explain the apparently simultaneous loss of flagellates and 'Endomicrobia' in the higher termites (Stingl et al. 2005).

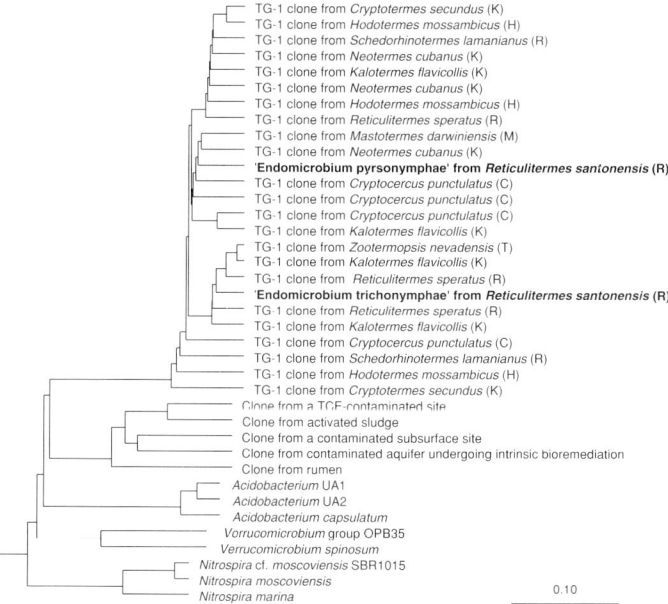

Fig. 6. Phylogenetic tree of the 'Endomicrobia', illustrating the relationship between the 16S rRNA gene sequences of the cytoplasmic symbionts of *Trichonympha agilis* and *Pyrsonympha vertens* to clones of 'Termite Group 1' (TG-1) bacteria obtained from various other termites and the wood-feeding cockroach *Cryptocercus punctulatus*. The candidate phylum 'Endomicrobia' is clearly separated from the most-closely related environmental clones and other bacterial phyla. The TG-1 clones from different termite families do not form separate clusters. (*K* Kalotermitidae, *H* Hodotermitidae, *R* Rhinotermitidae, *M* Mastotermitidae, *T* Termopsidae, *C* Blattodea:Cryptocercidae). *Bar* represents 0.1 substitutions per site. (Tree modified after Stingl et al. 2005)

4.2.3
Endonuclear Bacteria

Endonuclear bacteria are found in many termite gut flagellates, but only ultrastructural investigations have been carried out (for review, see Ball 1969). Nothing is known about their identity and the nature of the symbiosis. There are certain analogies to the endonuclear bacteria of ciliates (Görtz 2002; see also Chapter by J. Kusch and H.-D. Görtz).

5
Functional and Metabolic Implications

The most intriguing question concerns the physiology and metabolic function of the symbiotic bacteria of the termite gut flagellates. Assuming that the growth rate of gut bacteria is not necessarily sufficient to compensate for the high dilution rates of the hindgut contents, it would be advantageous for prokaryotes to associate with the protozoa to prevent wash-out from this substrate-rich and favorable environment. The intimacy of many associations, especially the elaborate attachment structures, however, indicate that there is a mutual advantage.

5.1
Hydrogen Metabolism

The symbiosis with methanogens is likely based on the interspecies transfer of molecular hydrogen. Little is known about the physiology of the protozoa, but from thermodynamic considerations, any fermentative metabolism releasing part of the reducing equivalents as H_2 will become more exergonic if this product is not allowed to accumulate (Schink 1997; see also Chapter by B. Schink). Virtually all 16S rRNA gene clones of methanogens obtained from lower termites fall into the phylogenetic radiation of the genus *Methanobrevibacter*. Since all members of this genus, including the three species isolated from the gut of *Reticulitermes flavipes*, perform methanogenesis from H_2 and CO_2, but use other substrates less efficiently, it is reasonable to assume that also the hitherto uncultivated *Methanobrevibacter* strains associated with the protozoa are potential hydrogen consumers (Lee et al. 1987; Tokura et al. 2000).

Based on the results of various treatments that selectively eliminate or affect certain members of the gut microbiota of *Zootermopsis angusticollis*, Messer and Lee (1989) concluded that the large protozoa of the genus *Trichonympha* are the most important hydrogen source in the hindgut, whereas the methanogenic symbionts of *Trichomitopsis termopsidis* produce most of the methane.

It is therefore possible that the bacterial symbionts of the larger protozoa are also involved in hydrogen metabolism. Reductive acetogenesis plays a major role as hydrogen sink in the guts of virtually all lower termites (Brauman et al. 1992; Breznak 1994). So far, only a few homoacetogens have been isolated from termite guts (Breznak 1994; Boga and Brune 2003; Boga et al. 2003). The recent demonstration of reductive acetogenesis in *Treponema primitia* and other termite gut isolates (Leadbetter et al. 1999) makes them likely candidates for this function, especially since spirochetes are often associated with certain gut flagellates (see Sect. (8.)4.1.2).

However, in view of the high H_2 partial pressures in the hindgut lumen of *Reticulitermes flavipes* (Ebert and Brune 1997), it is unlikely that – at least in this termite – an improved hydrogen transfer is the basis for the close association. The fact that spirochetes and methanogens are also attached to oxymonad flagellates, which do not possess hydrogenosomes (Bloodgood et al. 1974; Radek 1994), indicates that there may be other reasons for these associations.

5.2
Other Possible Functions

The fermentative metabolism of termite gut flagellates is poorly studied, and especially in the case of oxymonads, where hydrogenosomes are lacking, further work is sorely needed (see above). Reduced fermentation products other than hydrogen would also open the possibility of different metabolic links between the protozoa and their prokaryotic symbionts. Microinjection of radiolabeled metabolites into the hindgut of *Reticulitermes flavipes* has revealed that lactate is an important intermediate in the carbon flux from polysaccharides to acetate, but the microorganisms responsible for the production and consumption of lactate remain to be identified (Tholen and Brune 2000).

There are also other hypotheses concerning possible roles of the prokaryotic symbionts of gut flagellates. One of them would be the production of cellulases, as suggested for the endosymbionts of the larger flagellates (Bloodgood and Fitzharris 1976). However, after elimination of the endosymbionts by antibiotics, the hypermastigote *Trichonympha agilis* is still capable of degrading cellulose (Yamin 1981), and the hindgut protozoa seem to posses their own cellulase genes (Ohtoko et al. 2000).

Another possible function would be in the fixation of atmospheric nitrogen. Nitrogen fixation is an important process in wood-feeding termites (reviewed by Breznak 2000). It has been estimated that 30–60% of the total nitrogen in *Neotermes koshunensis* is derived from dinitrogen (Tayasu et al. 1994), and a few diazotrophs have been isolated from termite guts (for references, see Breznak 2000). Recently, also spirochetes – including the termite gut isolate *Treponema azonutricum* – have been shown to fix atmospheric nitrogen (Lilburn et al. 2001; Graber et al. 2004).

There is an enormous diversity in the *nifH* genes present in the guts of each termite species, and most of them cannot be assigned to described species of prokaryotes (Ohkuma et al. 1996, 1999a; Lilburn et al. 2001). In addition, the analysis of *nifH* gene expression in the gut of *Neotermes koshunensis* has shown the preferential expression of *nifH* from an alternative nitrogenase (*anf*), which could not be assigned to a known microorganism (Noda et al. 1999) and may belong to one of the epibiotic or endosymbiotic prokaryotes.

It has been postulated that the epibiotic bacteria of *Streblomastix strix* are chemotactic sensors for acetate (Dexter and Khalsa 1993), but in view of the high acetate concentrations in termite hindguts (see Breznak 2000 for review), it is not convincing that a chemotaxis towards acetate should be of any advantage for the flagellate. Leander and Keeling (2004), who investigated the ultrastructure of this fascinating association, could determine at least three different morphotypes among the epibionts that differed mainly in size. Elimination of epibionts by antibiotic treatment resulted in a modified morphology of the flagellate, which suggests an influence of the symbionts on the shape of the host.

6
Conclusions

The large number of symbioses among termite gut flagellates and prokaryotes, the high level of integration evidenced by the elaborate attachment structures or the intracellular location, and the large proportion of the prokaryotic gut microbiota that is associated with the protozoa suggest a major importance of such symbioses for the hindgut metabolism of lower termites.

Molecular tools allow the identification of the phylogeny of the partners involved in the symbioses, and although these investigations are still far from complete, it is apparent that the symbionts represent unusual and mostly unstudied phylogenetic groups. As a consequence of the unusual phylogenetic position and the complete lack of isolates, the metabolic capacities of the symbionts and their role in the symbiosis are still largely obscure.

In order to understand the hindgut metabolisms of lower termites, it will be essential to elucidate the metabolic relationship between the flagellates and their symbionts. It is reasonable to assume that the functional roles of the partners are less diverse than their phylogenetic diversity, and in view of the possible co-evolution of the partners, the symbioses between prokaryotes and gut flagellates are also excellent case studies in the microbial ecology and evolution.

References

Ball GH (1969) Organisms living on and in protozoa. In: Chen TT (ed) Research in protozoology, vol 3. Pergamon Press, New York, pp 565–718

Bauer S, Tholen A, Overmann J, Brune A (2000) Characterization of abundance and diversity of lactic acid bacteria in the hindgut of wood- and soil-feeding termites by molecular and culture-dependent techniques. Arch Microbiol 173:126–173

Berchtold M, Chatzinotas A, Schönhuber W, Brune A, Amann R, Hahn D, König H (1999) Differential enumeration and in situ localization of micro-organisms in the hindgut of the lower termite *Mastotermes darwiniensis* by hybridization with rRNA-targeted probes. Arch Microbiol 172:407–416

Bloodgood RA, Fitzharris TP (1976) Specific association of prokaryotes with symbiotic flagellate protozoa from the hindgut of the termite *Reticulitermes* and the wood-eating roach, *Cryptocercus*. Cytobios 17:103–122

Bloodgood RA, Miller KR, Fitzharris TP, McIntosh JR (1974) The ultrastructure of *Pyrsonympha* and its associated microorganisms. J Morphol 143:77–106

Boga H, Brune A (2003) Hydrogen-dependent oxygen reduction by homoacetogenic bacteria isolated from termite guts. Appl Environ Microbiol 69:779–786

Boga HI, Ludwig W, Brune A (2003) *Sporomusa aerivorans* sp. nov, an oxygen-reducing homoacetogenic bacterium from the gut of a soil-feeding termite. Int J Syst Evol Microbiol 53:1397–1404

Brauman A, Kane MD, Labat M, Breznak JA (1992) Genesis of acetate and methane by gut bacteria of nutritionally diverse termites. Science 257:1384–1387

Breznak JA (1994) Acetogenesis from carbon dioxide in termite guts. In: Drake HL (ed) Acetogenesis. Chapman and Hall, New York, pp 303–330

Breznak JA (2000) Ecology of prokaryotic microbes in the guts of wood- and litter-feeding termites. In: Abe T, Bignell DE, Higashi M (eds) Termites: evolution, sociality, symbiosis, ecology. Kluwer Academic Publ, Dordrecht, pp 209–231

Breznak JA, Leadbetter JR (2002) Termite gut spirochetes. In: Dworkin M, Falkow S, Rosenberg E, Schleifer, K-H, Stackebrandt E (eds) The prokaryotes: an online electronic resource for the microbiological community, 3rd edn, release 3.10, September 2002, Springer-SBM, New York, release 3.10. Springer, Berlin Heidelberg New York

Brune A (1998) Termite guts: the world's smallest bioreactors. Trends Biotechnol 16:16–21

Brune A (2003) Symbionts aiding digestion. In: Cardé RT, Resh VH (eds) Encyclopedia of insects. Academic Press, New York, pp 1102–1107

Brune A (2005) Symbiotic associations between termites and prokaryotes. In: Dworkin M, Falkow S, Rosenberg E, Schleifer K-H, and Stackebrandt E (eds) The prokaryotes: an online electronic resource for the microbiological community, 3rd ed, Springer-SBM, New York (in press)

Brune A, Emerson D, Breznak JA (1995) The termite gut microflora as an oxygen sink: microelectrode determination of oxygen and pH gradients in guts of lower and higher termites. Appl Environ Microbiol 61:2681–2687

Cavalier-Smith T (2002) The phagotrophic origin of eukaryotes and phylogenetic classification of Protozoa. Int J Syst Evol Microbiol 52:297–354

Cerkasova A, Cerkasov J, Kulda J (1986) Resistance of trichomonads to metronidazole. Acta Universitatis Carolinae Biol 30:485–503

Cleveland LR (1925) The effects of oxygenation and starvation on the symbiosis between the termite, *Termopsis*, and its intestinal flagellates. Biol Bull 48:309–327
Cleveland LR (1926) Symbiosis among animals with special reference to termites and their intestinal flagellates. Quart Rev Biol 1:51–64
Cleveland LR, Grimstone AV (1964) The fine structure of the flagellate *Mixotricha paradoxa* and its associated micro-organisms. Proc R Soc Lond Ser B Biol Sci 159:668–686
Cook TJ, Gold RE (1998) Organization of the symbiotic flagellate community in three castes of the Eastern Subterranean Termite, *Reticulitermes flavipes* (Isoptera: Rhinotermitidae). Sociobiology 31:25–39
Dacks JB, Silberman JD, Simpson AGB, Moriya S, Kudo T, Ohkuma M, Redfield RJ (2001) Oxymonads are closely related to the excavate taxon *Trimastix*. Mol Biol Evol 18:1034–1044
D'Ambrosio U, Dolan M, Wier AM, Margulis L (1999) Devescovinid trichomonad with axostyle-based rotary motor ("Rubberneckia"): Taxonomic assignment as *Caduceia versatilis* sp. nov. Eur J Protistol 35:327–337
De Bary A (1878) Über Symbiose. Ber Vers Deut Naturf Aerzte, Cassel, pp 121–126
Dexter Dyer B, Khalsa O (1993) Surface bacteria of *Streblomastix strix* are sensory symbionts. Biosystems 31:169–180
Dolan M (2001) Speciation of termite gut protists: the role of bacterial symbionts. Int Microbiol 4:203–208
Dolan M, Margulis L (1997) *Staurojoenina* and other symbionts in *Neotermes* from San Salvador Island, Bahamas. Symbiosis 22:229–239
Ebert A, Brune A (1997) Hydrogen concentration profiles at the oxic-anoxic interface: a microsensor study of the hindgut of the wood-feeding lower termite *Reticulitermes flavipes* (Kollar). Appl Environ Microbiol 63:4039–4046
Fröhlich J, König H (1999) Rapid isolation of single microbial cells from mixed natural and laboratory populations with the aid of a micromanipulator. Syst Appl Microbiol 22:249–257
Fröhlich J, Sass H, Babenzien H-D, Kuhnigk T, Varma A, Saxena S, Nalepa C, Pfeiffer P, König H (1999) Isolation of *Desulfovibrio intestinalis* sp. nov. from the hindgut of the lower termite *Mastotermes darwiniensis*. Can J Microbiol 45:145–152
Fujita A, Shimizu I, Abe T (2001) Distribution of lysozyme and protease, and amino acid concentration in the guts of a wood-feeding termite, *Reticulitermes speratus* (Kolbe): possible digestion of symbiont bacteria transferred by trophallaxis. Physiol Entomol 26:116–123
Gerbod D, Sanders E, Moriya S, Noel C, Takasu H, Fast NM, Delgado-Viscogliosi P, Ohkuma M, Kudo T, Capron M, Palmer JD, Keeling PJ, Viscogliosi E (2004) Molecular phylogenies of Parabasalia inferred from four protein genes and comparison with rRNA trees. Mol Phylogenet Evol 31:572–580
Görtz H-D (2002) Symbiotic associations between ciliates and prokaryotes. In: Dworkin M et al (eds) The prokaryotes: an evolving electronic resource for the microbiological community, 3rd edn, release 3.11. Springer, Berlin Heidelberg New York <http://141.150.157.117:8080/prokPUB/index.htm>
Graber JR, Leadbetter JR, Breznak JA (2004) Description of *Treponema azonutricum* sp. nov and *Treponema primitia* sp. nov., the first spirochetes isolated from termite guts. Appl Environ Microbiol 70:1307–1314

Hackstein JHP, Akhmanova A, Voncken F, van Hoek A, van Alen T, Boxma B, Moon-van der Staay SY, van der Staay G, Leunissen J, Huynen M, Rosenberg J, Veenhuis M (2002) Hydrogenosomes: convergent adaptations of mitochondria to anaerobic environments. Zoology 104:290–302

Hollande A, Valentin J (1969) Appareil de Golgi, pinocytose, lysosomes, mitochondries, bactéries symbiontiques, atractophores et pleuromitose chez les Hypermastigines du genre *Joenia*. Affinités entre Joenidae et Trichomonadines. Protistologica 5:39–86

Hongoh Y, Ohkuma M, Kudo T (2003) Molecular analysis of bacterial microbiota in the gut of the termite *Reticulitermes speratus* (Isoptera; Rhinotermitidae). FEMS Microbiol Ecol 44:231–242

Hungate RE (1955) Mutualistic intestinal protozoa. In: Hutner SH, Lwoff A (eds) Biochemistry and physiology of protozoa, vol 2. Academic Press, New York, pp 159–199

Huntenburg W, Stockert L, Smith-Somerville HE, Buhse HE Jr (1986) *Trichomitus trypanoides* (Trichomonadida) from the termite *Reticulitermes flavipes*. I. In vitro cultivation and cloning. Trans Am Microsc Soc 105:211–222

Iida T, Ohkuma M, Ohtoko K, Kudo T (2000) Symbiotic spirochetes in the termite hindgut: phylogenetic identification of ectosymbiotic spirochetes of oxymonad protists. FEMS Microbiol Ecol 34:17–26

Inoue T, Murashima K, Azuma J-I, Sugimoto A, Slaytor M (1997) Cellulose and xylan utilization in the lower termite *Reticulitermes speratus*. J Insect Physiol 43:235–242

Inoue T, Kitade O, Yoshimura T, Yamaoka I (2000) Symbiotic associations with protists. In: Abe T, Bignell DE, Higashi M (eds) Termites: evolution, sociality, symbiosis, ecology. Kluwer Academic Publ, Dordrecht, pp 275–288

Katzin LI, Kirby H (1939) The relative weight of termites and their protozoa. J Parasitol 25:444–445

Kirby H Jr (1941) Organisms living on and in protozoa. In: Calkins GN, Summers FM (eds) Protozoa in biological research. Columbia Univ Press, New York, pp 1009–1113

Kuhnigk T, Branke J, Krekeler D, Cypionka H, König H (1996) A feasible role of sulfate-reducing bacteria in the termite gut. System Appl Microbiol 19:139–149

Leadbetter JR, Breznak JA (1996) Physiological ecology of *Methanobrevibacter cuticularis* sp. nov. and *Methanobrevibacter curvatus* sp. nov., isolated from the hindgut of the termite *Reticulitermes flavipes*. Appl Environ Microbiol 62:3620–3631

Leadbetter JR, Crosby LD, Breznak JA (1998) *Methanobrevibacter filiformis* sp. nov., a filamentous methanogen from termite hindguts. Arch Microbiol 169:287–292

Leadbetter JR, Schmidt TM, Graber JR, Breznak JA (1999) Acetogenesis from H_2 plus CO_2 by spirochetes from termite guts. Science 283:686–689

Leander BS, Keeling PJ (2004) Symbiotic innovation in the oxymonad *Streblomastix strix*. J Euk Microbiol 51:291–300

Lee MJ, Schreurs PJ, Messer AC, Zinder SH (1987) Association of methanogenic bacteria with flagellated protozoa from a termite hindgut. Curr Microbiol 15:337–341

Leidy J (1881) The parasites of the termites. J Acad Nat Sci Philadelphia. 2nd Ser VIII:425–447

Lilburn TG, Schmidt TM, Breznak JA (1999) Phylogenetic diversity of termite gut spirochaetes. Environ Microbiol 1:331–345

Lilburn TG, Kim KS, Ostrom NE, Byzek KR, Leadbetter JR, Breznak JA (2001) Nitrogen fixation by symbiotic and free-living spirochetes. Science 292:2495–2498

Lindmark DG, Müller M (1973) Hydrogenosome, a cytoplasmic organelle of the anaerobic flagellate, *Tritrichomonas foetus*, and its role in pyruvate metabolism. J Biol Chem 248:7724–7728

Machida M, Kitade O, Miura T, Matsumoto T (2001) Nitrogen recycling through proctodeal trophallaxis in the Japanese damp-wood termite *Hodotermopsis japonica* (Isoptera, Termopsidae). Insect Soc 48:52–56

Margulis L, Hinkle G (1999) Large Symbiotic Spirochetes: Clevelandina, Cristispira, Diplocalyx, Hollandina, and Pillotina. In: Dworkin M, Falkow S, Rosenberg E, Schleifer K-H, and Stackebrandt E (eds) The prokaryotes: an online electronic resource for the microbiological community, 3rd edn, release 3.0, May 1999, Springer-SBM, New York

Messer M, Lee MJ (1989) Effect of chemical treatments on methane emission by the hindgut microbiota in the termite *Zootermopsis angusticollis*. Microb Ecol 18:275–284

Moriya S, Ohkuma M, Kudo T (1998) Phylogenetic position of symbiotic protist *Dinenympha exilis* in the hindgut of the termite *Reticulitermes speratus* inferred from the protein phylogeny of elongation factor 1 alpha. Gene 210:221–227

Moriya S, Tanaka K, Ohkuma M, Sugano S, Kundo T (2001) Diversification of the microtubule system in the early stage of eukaryote evolution: elongation factor 1alpha and alpha-tubulin protein phylogeny of termite symbiotic oxymonad and hypermastigote protists. J Mol Evol 52:6–16

Moriya S, Dacks JB, Takagi A, Noda S, Ohkuma M, Doolittle WF, Kudo T (2003) Molecular phylogeny of three oxymonad genera: *Pyrsonympha*, *Dinenympha* and *Oxymonas*. J Euk Microbiol 50:190–197

Müller M (1993) The hydrogenosome. J Gen Microbiol 139:2879–2889

Nakashima K, Watanabe H, Saitoh H, Tokuda G, Azuma J-I (2002) Dual cellulose-digesting system of the wood-feeding termite, *Coptotermes formosanus* Shiraki. Insect Biochem Mol Biol 32:777–784

Noda S, Ohkuma M, Usami R, Horikoshi K, Kudo T (1999) Culture-independent characterization of a gene responsible for nitrogen fixation in the symbiotic microbial community in the gut of the termite *Neotermes koshunensis*. Appl Environ Microbiol 65:4935–4942

Noda S, Ohkuma M, Yamada A, Hongoh Y, Kudo T (2003) Phylogenetic position and in situ identification of ectosymbiotic spirochetes on protists in the termite gut. Appl Environ Microbiol 69:625–633

Odelson DA, Breznak JA (1985a) Nutrition and growth characteristics of *Trichomitopsis termopsidis*, a cellulolytic protozoan from termites. Appl Environ Microbiol 49:614–621

Odelson DA, Breznak JA (1985b) Cellulase and other polymer-hydrolyzing activities of *Trichomitopsis termopsidis*, a symbiotic protozoan from termites. Appl Environ Microbiol 49:622–626

Ohkuma M (2003) Termite symbiotic systems: efficient bio-recycling of lignocellulose. Appl Microbiol Biotechnol 61:1–9

Ohkuma M, Kudo T (1996) Phylogenetic diversity of the intestinal bacterial community in the termite *Reticulitermes speratus*. Appl Environ Microbiol 62:461–468

Ohkuma M, Noda S, Usami R, Horikoshi K, Kudo T (1996) Diversity of nitrogen-fixation genes in the symbiotic intestinal microflora of the termite *Reticulitermes speratus*. Appl Environ Microbiol 62:2747–2752

Ohkuma M, Noda S, Kudo T (1999a) Phylogenetic diversity of nitrogen fixation genes in the symbiotic microbial community in the gut of diverse termites. Appl Environ Microbiol 65:4926–4934

Ohkuma M, Noda S, Kudo T (1999b) Phylogenetic relationships of symbiotic methanogens in diverse termites. FEMS Microbiol Lett 171:147–153

Ohkuma M, Iida T, Kudo T (1999c) Phylogenetic relationships of symbiotic spirochetes in the gut of diverse termites. FEMS Microbiol Lett 181:123–129

Ohtoko K, Ohkuma M, Moriya S, Inoue T, Usami R, Kudo T (2000) Diverse genes of cellulase homologues of glycosyl hydrolase family 45 from the symbiotic protists in the hindgut of the termite *Reticulitermes speratus*. Extremophiles 4:343–349

Ohkuma M, Noda S, Hongoh Y, Kudo T (2002) Diverse bacteria related to the Bacteroides subgroup of the CFB phylum within the gut symbiotic communities of various termites. Biosci Biotechnol Biochem 66:78–84

Pierantoni U (1936) La simbiosi fisiologica nei termitidi xilofagi e nei loro flagellati intestinali. Arch Zool Ital 22:135–173

Radek R (1994) *Monocercomonoides termitis* n sp., an Oxymonad from the lower termite *Kalotermes sinaicus*. Arch Protistenkd 144:373–382

Radek R (1999) Flagellates, bacteria, and fungi associated with termites: diversity and function in nutrition – a review. Ecotropica 5:183–196

Radek R, Hausmann K, Breunig A (1992) Ectobiotic and endocytobiotic bacteria associated with the termite flagellate *Joenia annectens*. Acta Protozool 31:93–107

Radek R, Rösel J, Hausmann K (1996) Light and electron microscopic study of the bacterial adhesion to termite flagellates applying lectin cytochemistry. Protoplasma 193:105–122

Schink B (1997) Energetics of syntrophic cooperation in methanogenic degradation. Microbiol Mol Biol Rev 61:262–280

Shinzato N, Matsumoto T, Yamaoka I, Oshima T, Yamagishi A (2001) Methanogenic symbionts and the locality of their host lower termites. Microbes Environ 16:43–47

Stingl U, Brune A (2003) Phylogenetic diversity and whole-cell hybridization of oxymonad flagellates from the hindgut of the wood-feeding lower termite *Reticulitermes flavipes*. Protist 154:147–155

Stingl U, Maass A, Radek R, Brune A (2004) Symbionts of the gut flagellate *Staurojoenina* sp. from *Neotermes cubanus* represent a novel, termite-associated lineage of bacteroidales: description of '*candidatus* vestibaculum illigatum'. Microbiology 150:2229–2235

Stingl U, Radek R, Yang H, Brune A (2005) "Endomicrobia": Cytoplasmic symbionts of termite gut protozoa form a separate phylum of prokaryotes. Appl Environ Microbiol 71:1473–1479

Tamm S (1980) The ultrastructure of prokaryotic-eukaryotic cell junctions. J Cell Sci 44:335–352

Tamm S (1982) Flagellated ectosymbiotic bacteria propel a eukaryotic cell. J Cell Biol 49:697–709

Tayasu I, Sugimoto A, Wada E, Abe T (1994) Xylophagous termites depending on atmospheric nitrogen. Naturwissenschaften 81:229–231

Tholen A, Brune A (2000) Impact of oxygen on metabolic fluxes and in situ rates of reductive acetogenesis in the hindgut of the wood-feeding termite *Reticulitermes flavipes*. Environ Microbiol 2:436–449

Tholen A, Schink B, Brune A (1997) The gut microflora of *Reticulitermes flavipes*, its relation to oxygen, and evidence for oxygen-dependent acetogenesis by the most abundant *Enterococcus* sp. FEMS Microbiol Ecol 24:137–149

Tokura M, Ohkuma M, Kudo T (2000) Molecular phylogeny of methanogens associated with flagellated protists in the gut and with the gut epithelium of termites. FEMS Microbiol Ecol 33:233–240

Trager W (1934) The cultivation of a cellulose-digesting flagellate, *Trichomonas termopsidis*, and of certain other termite protozoa. Biol Bull 66:182–190

Watanabe H, Tokuda G (2001) Animal cellulases. Cell Mol Life Sci 58:1167–1178

Wenzel M, Radek R, Brugerolle G, König H (2003) Identification of the ectosymbiotic bacteria of *Mixotricha paradoxa* involved in movement symbiosis. Eur J Protistol 39:11–24

Yamin MA (1979) Termite flagellates. Sociobiology 4:1–119

Yamin MA (1980) Cellulose metabolism by the termite flagellate *Trichomitopsis termopsidis*. Appl Environ Microbiol 39:859–863

Yamin MA (1981) Cellulose metabolism by the flagellate *Trichonympha* from a termite is independent of endosymbiotic bacteria. Science 211:58–59

Yamin MA, Trager W (1979) Cellulolytic activity of an axenically-cultivated termite flagellate, *Trichomitopsis termopsidis*. J Gen Microbiol 113:417–420

Yang H, Schmitt-Wagner D, Stingl U, Brune A (2005) Niche heterogeneity determines bacterial community structure in the termite gut (*Reticulitermes santonensis*). Environ Microbiol 7:916–932

Yoshimura T, Watanabe T, Tsunoda K, Takahashi M (1993) Distribution of the cellulolytic activities in the lower termite, *Coptotermes formosanus* Shiraki (Isoptera: Rhinotermitidae). Mater Org 27:273–284

Yoshimura T, Fujino T, Ito T, Tsunoda K, Takahashi M (1996) Ingestion and decomposition of wood and cellulose by the protozoa in the hindgut of *Coptotermes formosanus* Shiraki (Isoptera: Rhinotermitidae) as evidenced by polarizing and transmission electron microscopy. Holzforschung 50:99–104

Towards an Understanding of the Killer Trait: *Caedibacter* endocytobionts in *Paramecium*

Jürgen Kusch, Hans-Dieter Görtz

1
Introduction

Bacteria of the genus *Caedibacter* are endocytobionts of *Paramecium*. Mostly, the endocytobionts colonize the cytoplasm and sometimes the macronucleus of a host cell (Fig. 1). Infected paramecia may be recognized by the fact that they release toxins produced by *Caedibacter* that kill sensitive paramecia. Paramecia bearing *Caedibacter* are therefore called "killers". Host cells of *Caedibacter* are resistant against the toxins. This killer trait was first described by Sonneborn in 1938, who observed that sensitive paramecia exhibit distinct morphological symptoms upon ingestion of toxic particles (at that time designated kappa-particles) released by the killer strain and ultimately die (Sonneborn 1938). Evidences for a bacterial nature of the particles were found by Preer (1950), Hamilton and Gettner (1958) and Dippel (1958). After the initial discovery of killer bacteria, others were found that acted differently on sensitive cells. In the case of *Caedibacter,* the toxic effect is associated with refractile inclusion bodies (R bodies) found inside the bacteria (Schmidt et al. 1988). R bodies consist of highly insoluble protein ribbons (Fig. 2). They unwind under certain conditions and are associated with the toxicity of the bacteria. Besides *Caedibacter*, several free-living bacteria were found, too, that have the ability to produce R bodies.

J. Kusch
Department of Ecology, Faculty of Biology, TU Kaiserslautern, Erwin-Schroedinger-Str. 13/14, 67663 Kaiserslautern, Germany
H.-D. Görtz (e-mail: hans-dieter.goertz@bio.uni-stuttgart.de)
Department of Zoology, Biological Institute, University of Stuttgart, Pfaffenwaldring 57, 70569 Stuttgart, Germany

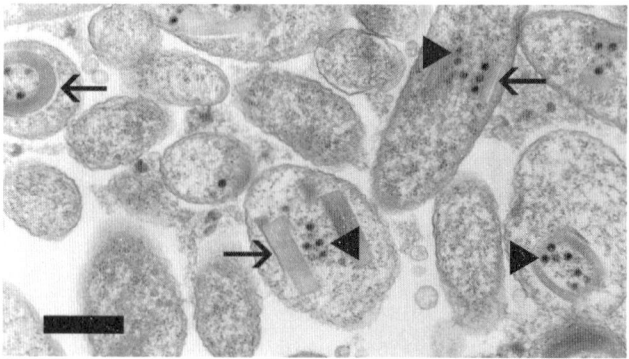

Fig. 1. Cluster of *Caedibacter caryophilus* in the macronucleus of *Paramecium caudatum*. Some bacteria contain R bodies (*arrows*) with phage capsids (*arrowheads*). Electron micrograph, courtesy of Michael Schweikert, University of Stuttgart. *Bar* 0.5 µm

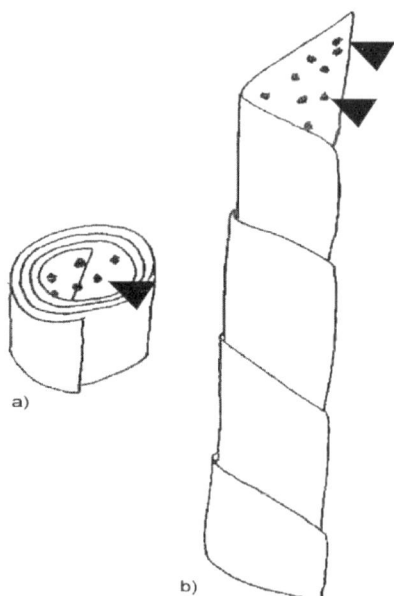

Fig. 2. R bodies are coiled protein ribbons that are refractile in phase contrast. **a** The unrolled ribbons have a length of up to 20 µm and a width of 0.4 to 0.8 µm. **b** By appropriated triggers, such as low pH, they are induced to unroll. Certain R bodies are associated with phage capsids (*arrowheads*)

All R body producing, obligate endosymbionts of paramecia were combined into the genus *Caedibacter* and classified based upon morphological, functional, and phenotypic properties (Preer and Preer 1982). Within the genus *Caedibacter*, six different species were identified in paramecia (*C. caryophilus*, *C. macronucleorum*, *C. paraconjugatus*, *C. pseudomutans*, *C. taeniospiralis*, *C. varicaedens*) (Preer et al. 1974; Preer and Peer 1982; Schmidt et al. 1987; Fokin and Görtz 1993, for review see Görtz and Schmidt 2004). In recent years, endocytobionts of small free-living amoeba were classified as *Caedibacter* based upon rDNA-similarities to *C. caryophilus* (Horn et al. 1999). However, none of these bacteria in amoeba was observed to express R bodies. R bodies of *C. caryophilus*, *C. macronucleorum*, *C. paraconjugatus*, *C. pseudomutans*, *C. varicaedens* are often associated with phage capsids, and the proteins of R bodies are encoded by phage genomes or plasmids (see for reviews Quackenbush 1988; Pond et al. 1989). The plasmids, in addition may contain transposons. Thus, killer paramecia are subject of multilevel infections. The presence of killer-symbionts was first recognized by Sonneborn as an example of extrachromosomal inheritance in *Paramecium* (Sonneborn 1938), and the complexity of this symbiosis is most impressively addressed in a title of a paper by Bob Quackenbush and his co-workers: Extrachromosomal elements of extrachromosomal elements of *Paramecium* and their extrachromosomal elements (Quackenbush et al. 1986a). In this title the genome of *Caedibacter* being an extrachromosomal element of *Paramecium*, the plasmids (or phage genomes) being extrachromosomal elements and the transposons found in plasmids are addressed. In earlier times many authors discussed the significance of *Caedibacter* for its host. Most authors have emphasized the advantage paramecia are getting from their killer symbionts (see Landis 1988). Yet, neither is the nature of the association clarified, nor has the origin and relationship of plasmids and phages been discovered. Do we have to expect a gene transfer from *Caedibacter* to the host genome, as has been suggested repeatedly? Is this a mutualistic symbiosis or rather a good example of parasitism, and why do we still find uninfected paramecia in nature? There are some new answers to these questions that shall be addressed in this chapter.

2
Evolutionary Ecology of the *Caedibacter* Symbiosis and R Body Production

Intracellular bacteria are favorable for the host and may be regarded as mutualists when uninfected competitors are killed due to these bacterial symbionts. This view, however, may be too simple, as ecological and molecular studies revealed fascinating details on the evolutionary ecology of this long known occurrence of "killer"-bacteria in paramecia. The bacteria are closely related to pathogens and exploit the host cells just as invasive agents in metazoans do. *Caedibacter* species are energy parasites that depend on the import of metabolites from the host cytoplasm (Linka et al. 2003), and therefore limit the reproduction of their host; under conditions that are unfavorable for paramecia *Caedibacter* may even overgrow the host cell, finally causing its death (Schmidt et al. 1987; Kusch et al. 2002). A carrier-mediated transport of nucleotides occurs for an acquisition of energy from the host cell. Recently, the evolutionary origin of the nucleotide transporters (NTT) of *Caedibacter caryophilus* that are specific ATP / ADP antiporters, has been elucidated. The phylogeny of bacterial NTT appears highly complex (Fig. 3) with NTT of *C. caryophilus* exhibiting significant sequence similarity with NTTs in intracellular pathogens of humans (*Rickettsia*). It was unexpected that NTTs of *C. caryophilus* and *Holospora obtusa* are more distantly related to one another, although these two endonucleobionts of *Paramecium* are close relatives in 16S rRNA trees. It is therefore suggested that earlier horizontal gene transfer is responsible for the distribution of NTT paralogs in bacterial endocytobionts of *Paramecium* (Linka et al. 2003).

For each parasite-host system there may be an optimal strategy of host exploitation maximizing the parasite's propagation (Poulin 1998). *Caedibacter* endocytobionts limit growth of their host due to energy theft. Too fast growth with subsequent virulence (host damage) of *Caedibacter* should kill

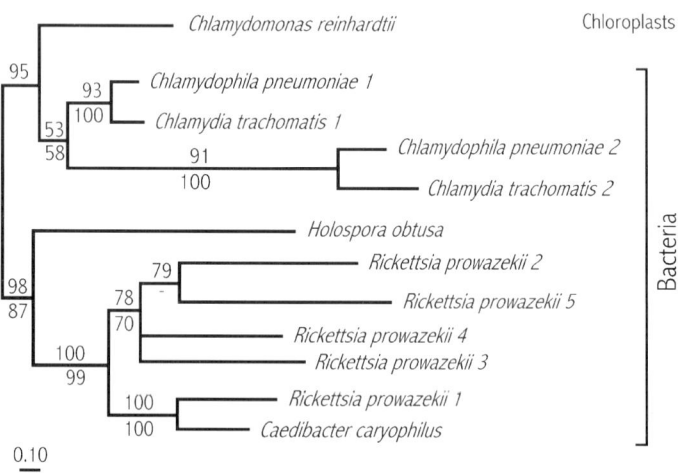

Fig. 3. Phylogeny of nucleotide transport proteins (NTT) in intracellular bacteria and *Chlamydomonas* chloroplasts. Bootstrap support (calculated from 1,000 replicates) is given as numbers at branches (%) for tree construction with maximum likelihood (TREE-PUZZLE; *upper numbers*) or distance (PHYLIP; *lower numbers*) approaches. Support of different tree topology at one branch is indicated by a '-'. The *scale bar* (0.10) represents mean number of substitutions per site. (After Linka et al. 2003, with changes)

host paramecia. The observation that infections of paramecia by *Caedibacter* are rare (Preer et al. 1974; Landis 1981, 1987; Kusch et al. 2002) may indicate that killing of host cells can result in the extinction of the responsible bacterial genotype. On the other hand, *Caedibacter* endocytobionts, besides limiting host reproduction, are giving competitive advantage to host paramecia under certain circumstances (Landis 1981, 1987; Kusch et al. 2002). Competition for food occurs within *Paramecium* species, when food is scarce, and interspecific competition, too, should occur in case of overlapping resources (Gause 1934; Maruyama et al. 2001).

It is the evolution of R body production by *Caedibacter* that has resulted in a mechanism for competition by allelopathic interference for paramecia bearing these bacterial endocytobionts. Individuals of both, the same species as well as closely related paramecia may be killed by toxins of *Caedibacter* when the paramecia are not bearing these bacteria. In *Caedibacter*, the toxins are associated with R body production. Parasites, in general, depend on the abundance of host organisms and on mechanisms for their successful infection. The rate at which an infected host is transmitting the parasite to susceptible hosts depends on host density. The increase in host abundance should result in an increase of parasite abundance in the next generation, and propagation of parasites is reciprocal to the mortality rate of parasite-free hosts (Anderson and May 1982, 1991; Poulin 1998). Toxic *Caedibacter* bacteria, instead, kill potential hosts. Infections and a subsequent growth of the bacteria are rare. Therefore, propagation of *Caedibacter* cells predominantly occurs via vertical transmission by vegetative reproduction of their hosts. Lethal effects on competitors increase the relative population growth rate of infected paramecia and their bacterial parasites. Indeed, infected strains could outcompete genetically identical but uninfected strains in mixed cultures of either of the species, *P. tetraurelia* or *P. novaurelia* (Kusch et al. 2002). Reproduction rates of *Caedibacter* and virulence effects on hosts should be well adapted to interference competition effects by R body production.

Only *Caedibacter* with R bodies are toxic, and R bodies are expressed only in part of the bacteria. R bodies are refractile in phase contrast microscopy. R body-bearing bacteria are therefore called "bright bodies" or simply "brights". *Caedibacter* without R bodies are called "non-brights". Since non-brights are not toxic but nevertheless may contain plasmids or phages genomes, it may happen that paramecia not bearing *Caedibacter* become infected by non-brights. Yet, this may be a rare case, since any brights that are co-ingested by non-infected paramecia will kill the cells. Sometimes, infection of formerly uninfected host paramecia may occur during conjugation, the type of sexual reproduction in ciliates. Paramecia undergoing conjugation do not feed and therefore do not take up the toxin released by an infected mating partner. Sensitive paramecia are therefore protected and can mate with infected "killer" paramecia. During conjugation of ciliates, a cytoplasmic bridge forms, that enables exchange of migrating pronuclei between the two partner cells.

Eventually, some cytoplasm and even cell organells or intracellular bacteria may be exchanged, too (Balsley 1967; Landis 1981). Infection during this cytoplasmic exchange at conjugation would increase the proportion of killers in a population (Landis 1987). For *Caedibacter paraconjugatus* this possibility generally does not exist, as it is a so-called mate killer symbiont. The toxic principle only acts on sensitive paramecia during mating with *C. paraconjugatus*-bearing ciliates.

Fokin et al. (2004) observed that non-infective bacteria infected the macronucleus of *Paramecium caudatum*, when infectious *Holospora obtusa* were present, too. If co-infections of *Paramecium* by *Caedibacter* were frequent in the presence of *Holospora* species, then the release of *Caedibacter* by its host paramecia could significantly increase the transmission rate and prevalence. Otherwise the reproduction rate of *Caedibacter* should be as low as to ensure stably infected host populations (so, growth of *Caedibacter* should equal its dilution by growth and reproduction of its hosts). Thereby, virulence effects should be minimized. If infection of new host paramecia would be a predominant way for reproduction and spreading of *Caedibacter* endocytobionts, selection should favor *Caedibacter* genotypes without toxin production, as toxic *Caedibacter* kill their potential hosts. Competitive advantages of host paramecia due to bacterial toxin production, in combination with vegetative reproduction of the host, renders infection of new hosts by the bacterial parasites unnecessary for their propagation, and should be responsible for the persistence of "killer paramecia" in nature.

Concerning the ecological significance of *Caedibacter* infections, it is especially interesting that the genetic condition of potential hosts is crucial for the maintenance of *Caedibacter* in *Paramecium*. In genetic crossings, Sonneborn (1943) observed that only paramecia with certain genotypes may harbor *Caedibacter*. Namely, for one special gene it turned out that an allele K is crucial for the persistence of the bacteria in host cells, while *Caedibacter* could not live in cells with k in the homozygous state. It is not determined, whether K is actively supporting growth of the endocytobionts or k is suppressing growth. In this context, the question arises whether the parasites affect the rate of conjugation or autogamie in host paramecia, and whether genetical recombination of hosts affects the prevalence of *Caedibacter* in a host population.

In various cases of host-specific intracellular microorganisms, evidences for gene transfer from endocytobionts to host cell have been obtained, e.g., in certain *Euplotes* and in *Amoeba proteus* (Heckmann 1983; Jeon and Jeon 1997). It is a first indication that many intracellular bacteria possess small genomes. However, *Euplotes* and *A. proteus* became dependent upon their symbionts, but this is not the case for *Paramecium* bearing *Caedibacter*. The demands of the bacteria, on the other hand, may not be very specific. The most important need may be ATP, which is not a specific compound. So, at the time being, there may be no real indication for a gene transfer between *Caedibacter* and *Paramecium*.

Depending on the environmental conditions given, the infection of *Paramecium* by *Caedibacter* may be regarded as parasitic or even mutualistic. Infected hosts should have a selective advantage over uninfected and, thereby, sensitive paramecia. In interspecific competition of *Caedibacter*-bearing cells with other bacterial feeders like tiny rotifers, nematods, or unrelated ciliates the toxins may not help. In these cases the bacterial load with *Caedibacter* being energy parasites may be a severe burden and of disadvantage. This conflict of intraspecific versus interspecific competition may be the reason why some natural populations of *Paramecium* are infected by *Caedibacter* while others are not. In the case of very low abundance of *Caedibacter*-bearing paramecia, cells that have lost their bacteria and thereby have become sensitive to the toxin may not be harmed, because only low amounts of toxin are present. Uninfected paramecia grow faster and namely under nutrient-poor conditions may outcompete infected ones.

We do not yet understand the mechanisms, and how resistance against the toxins is achieved. *Caedibacter*-bearing paramecia are resistant against the toxins, and resistance is even mediated by *Caedibacter* that have lost the ability for R body expression (Schmidt et al. 1987). Schmidt et al. observed that host cells are less affected by *Caedibacter* that are no longer able to express R bodies, these paramecia bearing only non-brights grow faster than paramecia with brights. One might therefore expect that *Caedibacter*-lacking phages or plasmids are frequent in nature, but this doesn't seem to be the case. While R body-free *Caedibacter* were found in small free-living amoeba, in nature they have not been detected in *Paramecium*. However, there are good evidences that we only know a few of the bacterial endocytobionts in ciliates, and perhaps R body-free *Caedibacter* have simply been overlooked. It is therefore of interest to look systematically for bacterial infections in paramecia and other ciliates, the more so as we may expect potential pathogens among the bacterial endocytobionts of protozoa (Görtz and Michel 2003).

3
Host Specificity of *Caedibacter*

Parasites are believed to have a major impact on host evolution. On the other hand, it is necessary for a parasite to adapt its growth dynamics to the host species or even to its genotype because the host must not be killed before the parasite has developed infectious stages. Optimal virulence should result in high parasite fitness in a host, whereas maladapted virulence could drive a parasite or the host extinct ('Suicide King' hypothesis; e.g., Dybdahl and Storfer 2003). Co-evolution between parasites and hosts in some cases resulted in very specific (genotype-specific) interactions. Certain host genotypes can persist only together with certain parasite genotypes (towards which they are well adapted), but not with others (towards which they are not adapted).

Various *Paramecium* species in general are infected by different *Caedibacter* species (Pond et al. 1989), but certain *Caedibacter* species live in different host species. *C. taeniospiralis* or *C. pseudomutans*, both, occur in *P. tetraurelia*, and *C. varicaedens* or *C. paraconjugatus* inhabit *P. biaurelia*. Furthermore, *C. caryophilus* of different genotypes inhabit different host species and here live inside different compartments of the cell (Kusch et al. 2000). *C. caryophilus* inhabits the macronucleus of *P. caudatum*, and lives inside the cytoplasm of *P. novaurelia*, too. A 16S-rDNA sequence of 1695 base pairs length differed in a total of only one base pair in *C. caryophilus* of the two *Paramecium* species. This difference, though only small, may indicate that the bacteria have different adaptations. The infection of specific host cell compartments as a result of interspecific communication may depend on certain host genes, but not on basically different traits of the infecting endocytobiont species. Yet, so far we do not know about the extent of genetic differences of the total genomes of *Caedibacter* strains; they may be larger than the differences of their rDNA. Although this example hints to a quite specific adaptation of *Caedibacter* genotypes and host species, we do not know whether different genotypes of one *Caedibacter* species occur in different genotypes of one *Paramecium* species. Alternatively, the infection of different compartments of a cell in different host species could result from different infection mechanisms (e.g., cytoplasmic exchange during conjugation versus co-infection with nuclear specific *Holospora* species).

Recent work shows that *Caedibacter* bacteria are not limited to the genus *Paramecium* as hosts. Predators of paramecia, e.g., *Didinium nasutum* (Ciliophora) or *Amoeba proteus* (Amoebozoa, Gymnamoebia) were frequently infected when the predators were fed on infected paramecia (Reisser et al. 1985; Kusch et al. 2002). Most of the newly infected predators lost their parasites during a few cycles of vegetative reproduction, but some did not. The genus *Acanthamoeba* comprises numerous species of free-living amoeba that are increasingly being recognized as hosts for obligate bacterial endocytobionts. Acanthamoebae serve as natural vehicles for the human pathogen *Legionella pneumophila* and for a variety of other clinically relevant bacteria (Barker and Brown 1994; Atlas 1999). In contrast to these transient interrelationships, which result in lysis of the infected amoebas, the presence of stably infecting, obligate bacterial endocytobionts has been reported for about 25% of all *Acanthamoeba* isolates (Fritsche et al. 1993). Comparative analyses of 16S rDNA sequences revealed that the endocytobionts of *Acanthamoeba* strains are related to *Caedibacter caryophilus*, *Holospora elegans* and *Holospora obtusa*, all of which live in the nuclei of *Paramecium caudatum* (Horn et al. 1999). With overall 16S rRNA sequence similarities to their closest relative, *C. caryophilus*, of between 87 and 93%, these endocytobionts of amoebas represent three distinct new species that were proposed to be classified as *Caedibacter acanthamoeba, Paracaedibacter acanthamoebae* and *Paracaedibacter symbiosus*. The *Caedibacter*-specific R body production was not observed in these endocytobionts from acanthamoebae. These findings suggest a specific association of *Caedibacter*-species with different host species of unrelated higher taxa and a relatively high diversity of hosts. The specializations and ecological adaptations that enable a certain *Caedibacter* species (or genotype) to colonize a special host, but not others, should be of great interest for future research.

4
Molecular Evolution of R Body Production

Caedibacter endocytobionts of paramecia are infected by phages or plasmids, either of which is responsible for the production of R bodies and probably the toxin (Preer et al. 1972, 1974; Pond et al. 1989; Kusch et al. 2000). Bacteriophages are evidently non-infective, and bacterial lysis does not seem to occur (Pond et al. 1989), with the exception of *Caedibacter caryophilus* in the host *Paramecium caudatum* (Schmidt et al. 1987). The R body coding region of the plasmid of *C. taeniospiralis* is of 1.8 kb and consists of the three genes *rebA*, *rebB*, and *rebC* coding for proteins of 18, 13 and 10 kDa, respectively. A putative fourth gene, *rebD*, may be involved in R body production and expression of the toxin (Heruth et al. 1994). Extrachromosomal elements coding for R bodies in different *Caedibacter* species seem to be more closely related than their hosts (Quackenbush et al. 1986a), that are polyphyletic, with e.g., *C. caryophilus* belonging to the α-Proteobacteria and *C. taeniospiralis* being a member of the γ-Proteobacteria (Beier et al. 2002). R body production occurs in non-endobiotic bacteria, too, including pseudomonads (Pond et al. 1989). It appears that different bacterial genera gained (or were infected by) identical or functionally similar extrachromosomal elements during their evolution.

Among species of the genus *Caedibacter*, *C. taeniospiralis* is the only one bearing a plasmid (pKAP) as an R body coding extrachromosomal element (Dilts 1976; Quackenbush 1983; Quackenbush and Burbach 1983). Plasmid pKAP is present in all strains of *C. taeniospiralis*, but varying in size (41.5 to 50.5 kb). The plasmids of *C. taeniospiralis* strains differ by the insertion of transposons at different plasmid locations (Quackenbush et al. 1986b). The transposons are present in the genome of the bacterial host, too. This was evident because of spontaneous insertions of those transposons into pKAP plasmids and from hybridizations of the transposons with genomic DNA (Dilts and Quackenbush 1986; Quackenbush et al. 1986b).

The transcriptional activity of the isolated plasmid pKAP298 of *C. taeniospiralis* strain 298 and its complete sequence (49112 bp) were analyzed (Fig. 4) to gain insight into the constitution of R body coding extrachromosomal elements (Jeblick and Kusch 2005). A significant homology of the total pKAP298 to other plasmids or phages was not detected, indicating a so far unique molecular structure of pKAP298. Yet, the occurrence of sequences with homology to several phage proteins indicate that pKAP298 is derived from a former phage, as was hypothesized by Preer et al. (1974). Stress conditions induce the transcription of R body coding genes. This is why R bodies occur predominantly in the stationary phase of paramecia resulting from a lack of food (Pond et al. 1989). Additional transcription activity of pKAP298 indicates functions besides R body synthesis.

Regions active in transcription contain some of the genes with homology to phage proteins (Jeblick and Kusch 2005). Because of this observation one may assume a synthesis of phages or incomplete phages in *Caedibacter taeniospiralis*, too. Long ago, Preer, Preer and Jurand already discussed the possibility

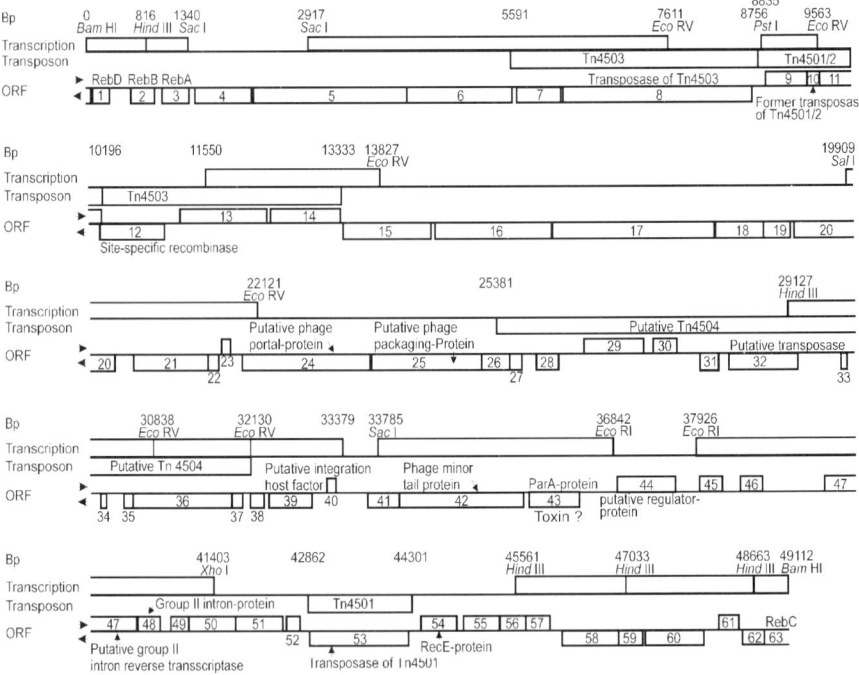

Fig. 4. Map of the R-body coding plasmid pKAP298 from *Caedibacter taeniospiralis* as deduced from analysis of its sequence and transcription activity. The only Bam HI restriction site was arbitrarily defined as bp 0. Each of the five double segments gives 10 kbp, except the last one with 9.112 kbp. The upper part of each double segment shows regions with detected transcription activity above the line and transposons below the line. Base pair (*Bp*) numbers belong to vertical lines below the indicated Bp values. Some restriction sites are also given (below bp numbers). The lower parts of each double segments show the detected open reading frames (*ORF*), with the direction of transcription indicated by *arrows* on the left side of the figure. Putative functions are given for some ORFs. (After Jeblick and Kusch 2005, changed)

that R body proteins are related to the proteins of the defective phages of *Caedibacter* (Preer et al. 1974). Prophages of some temperate phages, e.g., of coliphages P1 (Ikeda and Tomizawa 1968; Walker and Walker 1975) and N15 (Ravin and Shulga 1970) also are autonomous plasmids. Furthermore, Casjens et al. (2004) reported that the *Klebsiella oxytoca* linear plasmid pKO2 is a prophage. Sequences with homology to proteins that are involved in the phage-induced lysis of a host cell were not detected on pKAP298. Schmidt et al. (1987) presumed from the occurrence of R bodies in the macronucleus of *C. caryophilus*-infected *Paramecium caudatum* that bacteriophages of *C. caryophilus* induced host cell lysis. In *C. taeniospiralis* induced lysis was never observed. Plasmid pKAP298 may stem from a bacteriophage of the order *Caudovirales*, to which all bacteriophages that have sequence homology to pKAP298 belong (Jeblick and Kusch 2005). Some of the sequences that were identified to originate from a phage, have homologies to proteins of the order *Rhizobiales* (Kaneko et al. 2002; Galibert et al. 2001), to which the original host may have belonged. The plasmid pKAP298 may have evolved from a bacteriophage to a broad host range-plasmid, since host strains classified as *C. taeniospiralis* have only low DNA homologies (Quackenbush 1977, 1978).

Apparently, pKAP298 and its R body genes are related to a bacteriophage that does not have a complete set of phage head and tail genes. In several other bacteria, sequence analysis and electron microscopy revealed a common ancestry for some bacteriocins, bacterial products with specific bactericidal activity, and bacteriophages (e.g., Strauch et al. 2001; Jabrane et al. 2002). These bacteriocins have been regarded as defective bacteriophages, which might have arisen from temperate phages by several successive mutations (Bradley 1967; Daw and Falkiner 1996). The bacterial bacteriocin production obviously evolved from phage structural genes, and it seems that *Caedibacter* species similarly derived R-body genes from a phage. The gene transfer agent (GTA) of *Rhodobacter capsulatus* is another example, where bacterial functions may have evolved from phage genes (Marrs 1974; Yen et al. 1979). Therefore, the present genes in pKAP298 should come from an adaptive selection for toxicity that preserved this genetic element during evolution.

Toxicity of R bodies to paramecia ingesting *Caedibacter* species may be associated with a protein that has homology to the Soj-/ParA-family (ORF 43, Fig. 4) and also has homologies to a membrane associated ATPase (Jeblick and Kusch 2005). This ATPase is involved in eukaryotic ATPase dependent ion carriers (Zhou et al. 2000) and may cause toxic effects on paramecia ingesting this protein. As an ATPase the protein may primarily function in the distribution of pKAP298 on daughter cells after host divisions (Motallebi-Vesharesh et al. 1990). It also may interfere with the distribution of the host genome due to misrelations of ParA and ParB proteins (Easter and Gober 2002), leading to the observed low reproduction rates observed for *Caedibacter* species (Preer et al. 1972). The toxin might disturb the osmoregulatory abilities of *Paramecium* via effects on the cell membrane (Butzel and Pagliara

1962) and therefore leads to hump-killing or vacuolization as typical killing mechanisms (Preer et al. 1974).

5
Conclusions

In conclusion of old and new observations, the old killer symbionts of *Paramecium* may rather be parasites than mutualists. However, this view does not completely do justice to the system. By coding for R bodies and presumably also for toxins, plasmids and phage genomes ensure the persistence of *Caedibacter* and of the plasmids / phages themselves in paramecia populations. *Caedibacter*-free populations of *Paramecium* may develop under unfavorable conditions, e.g., in case of poor nutrients being the reason for low abundance of the ciliates. Only then paramecia without endocytobionts would survive, not being threatened by R body-bearing *Caedibacter* that are released by host cells. *Caedibacter*-free paramecia need less food and grow faster. Since, however, in many habitats the conditions are favorable for bacterial feeders like *Paramecium*, R body-coding plasmids or phages ensure persistence of the killer symbioses, no matter whether *Caedibacter* is a mutualist or parasite. As always in microbially infected protozoa, host and endocytobiont are forming an entirely new unit facing selection.

References

Anderson RM, May RM (1982) Coevolution of hosts and parasites. Parasitology 85:411–426
Anderson RM, May RM (1991) Infectious diseases of humans. Oxford Univ Press, Oxford
Atlas RM (1999) *Legionella*: from environmental habitats to disease pathology, detection and control. Environ Microbiol 1:283–293
Balsley M (1967) Dependence of the kappa particles of stock 7 of *Paramecium aurelia* on a single gene. Genetics 56:125–131
Barker J, Brown MRW (1994) Trojan Horses of the microbial world: protozoa and the survival of bacterial pathogens in the environment. Microbiology 140:1253–1259
Beier CL, Horn M, Michel, R, Schweikert, M, Görtz H-D, Wagner M (2002) The genus *Caedibacter* comprises endobionts of *Paramecium* spp. related to the *Rickettsiales* (*Alphaproteobacteria*) and to *Francisella tularensis* (Gammaproteobacteria). Appl Environ Microbiol 68:6043–6050
Bradley DE (1967) Ultrastructure of bacteriophage and bacteriocins. Bacteriol Rev 31:230–314
Butzel HM, Pagliara A (1962) The effect of biochemicals inhibitors upon the killer sensitive system in *Paramecium aurelia.* Exp Cell Res 27:382–395

Casjens SR, Gilcrease EB, Huang WM, Bunny KL, Pedulla ML, Ford ME, Houtz JM, Hatfull GF, Hendrix RW (2004) The PKO2 linear plasmid prophage of *Klebsiella oxytoca*. J Bacteriol 186:1818–1832

Daw MA, Falkiner FR (1996) Bacteriocins: nature, function and structure. Micron 27:467–479

Dilts JA (1976) Covalently closed, circular DNA in kappa endobionts of *Paramecium*. Genet Res 27:161–170

Dilts JA, Quackenbush RL (1986) A mutation in the R-body-coding sequences destroys expression of the killer trait in *P. tetraurelia*. Science 232:641–643

Dippel RV (1958) The fine structure of kappa in killer stock 51 of *Paramecium aurelia*. Preliminary observations. J Biophys Biochem Cytol 4:125–128

Dybdahl MF, Storfer A (2003) Parasite local adaptation: Red Queen versus Suicide King. TREE 18:523–530

Easter J, Gober JW (2002) ParB-stimulated nucleotide exchange regulates a switch in functionally distinct ParA activities. Mol Cell 10:427–434

Fokin SI, Görtz H-D (1993) *Caedibacter macronucleorum* sp. nov., a bacterium inhabiting the macronucleus of *Paramecium duboscqui*. Arch Protistenkd 143:319–324

Fokin SI, Skovorodkin IN, Schweikert M, Görtz H-D (2004) Co-infection of the macronucleus of *Paramecium caudatum* by free-living bacteria together with the infectious *Holospora obtusa*. J Eukaryot Microbiol 51:417–424

Fritsche TR, Horn M, Seyedirashti S, Gautom RK, Schleifer K-H, Wagner M (1993) In situ detection of novel bacterial endosymbionts of *Acanthamoeba* spp. phylogenetically related to members of the order *Rickettsiales*. Appl Environ Microbiol 65:206–212

Galibert F, Finan TM, Long SR, Puhler A, Abola P, Ampe F, Barloy-Hubler F, Barnett MJ, Becker A, Boistard P, Bothe G, Boutry M, Bowser L, Buhrmester J, Cadieu E et al (2001) The composite genome of the legume symbiont *Sinorhizobium meliloti*. Science 293:668–672

Gause GF (1934) The struggle for existence. Hafner Publ, New York

Görtz H-D (2002) Bacterial symbionts of protozoa in aqueous environments – potential pathogens? In: Greenblatt C, Spigelman M (eds) Emerging pathogens. Oxford Univ Press, Oxford, pp 25–37

Görtz H-D, Michel R (2003) Bacterial symbionts of protozoa in aqueous environments – potential pathogens? In: Greenblatt C, Spigelman M (eds) Emerging pathogens. Oxford Univ Press, Oxford, pp 25–37

Görtz H-D, Schmidt HJ (2004) *Caedibacter, Holosporaceae, Lyticum, Paracaedibacter, Pseudocaedibacter, Pseudolyticum, Tectibacter* and *Polynucleobacter*. In: Garrity GM, Brenner DJ, Krieg NR, Staley JT (eds) Bergey's manual of systematic bacteriology, vol 2. Springer, Berlin Heidelberg New York (in press)

Hamilton LD, Gettner ME (1958) Fine structure of kappa in *Paramecium aurelia*. J Biophys Biochem Cytol 4:122–123

Heckmann K (1983) Endosymbionts of *Euplotes*. In: Jeon KW (ed) Intracellular symbiosis. Academic Press, New York, pp 111–144

Heruth DC, Pond FR, Dilts JA, Quackenbush RL (1994) Characterization of genetic determinants for R body synthesis and assembly in *Caedibacter taeniospiralis* 47 and 166. J Bacteriol 176:3559–3567

Horn M, Fritsche TR, Gautom RK, Schleifer K-H, Wagner M (1999) Novel bacterial endosymbionts of *Acanthamoeba* spp. Related to the *Paramecium caudatum* symbiont *Caedibacter caryophilus*. Environ Microbiol 1:357–367

Ikeda H, Tomizawa J (1968) Phage P1, an extrachromosomal replicator unit. Cold Spring Harbor Symp Quant Biol 33:791–798

Jabrane A, Sabri A, Compère P, Jacques P, Vandenberghe I, van Beeumen J, Thonart P (2002) Characterization of Serracin P, a phage-tail-like bacteriocin, and its activity against *Erwinia amylovora*, the fire blight pathogen. Appl Environ Microbiol 68:5704–5710

Jeblick J, Kusch J (2005) Sequence, transcription activity and evolutionary origin of the R-body coding plasmid pKAP298 from the intracellular parasitic bacterium *Caedibacter taeniospiralis*. J Mol Evol 60:164–173

Jeon WP, Jeon KW (1997) A symbiont-produced protein and bacterial symbiosis in *Amoeba proteus*. J Euk Microbiol 44:614–619

Kaneko T, Nakamura Y, Sato S, Minamisawa K, Uchiumi T, Sasamoto S, Watanabe A, Idesawa K, Iriguchi M, Kawashima K, Kohara M, Matsumoto M, Shimpo S, Tsuruoka H, Wada T, Yamada M, Tabata S (2002) Complete genomic sequence of nitrogen-fixing symbiotic bacterium *Bradyrhizobium japonicum* USDA110. DNA Res 9:189–197

Kusch J, Stremmel M, Schweikert M, Adams V, Schmidt HJ (2000) The toxic symbiont *Caedibacter caryophila* in the cytoplasm of *Paramecium novaurelia*. Microb Ecol 40:330–335

Kusch J, Czubatinski L, Wegmann S, Hübner M, Alter M, Albrecht P (2002) Competitive advantages of *Caedibacter*-infected Paramecia. Protist 153:47–58

Landis WG (1981) The ecology, role of the killer trait, and interactions of five species of the *Paramecium aurelia* complex inhabiting the littoral zone. Can J Zool 59:1734–1743

Landis WG (1987) Factors determining the frequency of the killer trait within populations of the *Paramecium aurelia* complex. Genetics 115:197–205

Landis W (1988) Ecology. In: Görtz H-D (ed) Paramecium. Springer, Berlin Heidelberg New York, pp 419–436

Linka N, Hurka H, Lang BF, Burger G, Winkler HH, Stamme C, Urbany C, Seil I, Kusch J, Neuhaus HE (2003) Phylogenetic relationships of non-mitochondrial nucleotide transport proteins in bacteria and eukaryotes. Gene 306:27–35

Marrs BL (1974) Genetic recombination in *Rhodopseudomonas capsulata*. Proc Natl Acad Sci USA 71:971–973

Maruyama C, Fujisawa H, Takagi Y (2001) Age-associated survival and extinction in mixed cultures of *Paramecium*. Eur J Protistol 37:303–312

Motallebi-Veshareh M, Rouch DA, Thomas CM (1990) A family of ATPases involved in active partitioning of diverse bacterial plasmids. Mol Microbiol 4:1455–1463

Pond F, Gibson I, Lalucat J, Quackenbush RL (1989) R-body-producing bacteria. Bacteriol Rev 53:25–67

Poulin R (1998) Evolutionary ecology of parasites – from individuals to communities. Chapman and Hall, London

Preer JR (1950) Microscopically visible bodies in the cytoplasm of the 'killer' strains of Paramecium aurelia. Genetics 35:344–362

Preer JRJr, Preer LB (1982) Revival of names of protozoan endosymbionts and proposal of *Holospora caryophila* nom. nov. Int J Syst Bacteriol 32:140–141

Preer LB, Jurand A, Preer JR, Rudman BM (1972) The classes of kappa in *Paramecium aurelia*. J Cell Sci 11:581–600

Preer JR Jr, Preer LB, Jurand A (1974) Kappa and other endobionts in *Paramecium aurelia*. Bacteriol Rev 38:113–163

Quackenbush RL (1977) Phylogenetic relationships of bacterial endobionts of *Paramecium aurelia*: deoxyribonucleotide sequence relationship of 51 kappa and its mutants. J Bacteriol 129:895–900

Quackenbush RL (1978) Genetic relationships among bacterial endobionts of *Paramecium aurelia*: polynucleotide sequence relationships among members of *Caedibacter*. J Gen Microbiol 108:181–187

Quackenbush RL (1988) Endosymbionts of killer paramecia. In: Görtz H-D (ed) Paramecium. Springer, Berlin Heidelberg New York

Quackenbush RL (1983) Plasmids from bacterial endobionts of hump-killer paramecia. Plasmid 9:298–306

Quackenbush RL, Burbach JA (1983) Cloning and expression of DNA sequences associated with the killer trait of *Paramecium tetraurelia* stock 47. Proc Natl Acad Sci USA 80:250–254

Quackenbush RL, Cox BJ, Kanabrocki JA (1986a) Extrachromosomal elements of extrachromosomal elements of paramecia, and their extrachromosomal elements. In: Wickner RB, Hinnebusch A, Lambowitz AM, Gunsalus IC, Hollaender A (eds) Extrachromosomal elements in lower eukaryotes. Plenum Press, New York, pp 265–278

Quackenbush RL, Dilts JA, Cox BJ (1986b) Transposonlike elements in *Caedibacter taeniospiralis*. J Bacteriol 166:349–352

Ravin VK, Shulga MG (1970) Evidence for extrachromosomal location of prophage N15. Virology 40:800–807

Reisser W, Meier R, Görtz H-D, Kwang WJ (1985) Establishment, maintenance, and integration mechanisms of endosymbionts in protozoa. J Protozool 32:383–390

Schmidt HJ, Pond FR, Görtz H-D (1987) Refractile bodies (R bodies) from the macronuclear killer particle *Caedibacter caryophila*. J Cell Sci 88:177–184

Schmidt HJ, Görtz H-D, Pond FR, Quackenbush RL (1988) Characterization of *Caedibacter* endonucleosymbionts from the macronucleus of *Paramecium caudatum* and the identification of a mutant with blocked R body synthesis. Exp Cell Res 174:49 57

Sonneborn TM (1938) Mating types in *P. aurelia*: diverse conditions for mating in different stocks; occurrence, number and interrelations of the type. Proc Am Phil Soc 79:411–434

Sonneborn TM (1943) Gene and cytoplasm. I. The determination and inheritance of the killer character in variety 4 of *P. aurelia*. II. The bearing of determination and inheritance of characters in *P. aurelia* on problems of cytoplasmic inheritance, pneumococcus transformations, mutations and development. Proc Nat Acad Sci USA 29:329–343

Strauch E, Kaspar H, Schaudinn C, Dersch P, Madela K, Gewinner C, Hertwig S, Wecke J, Appel B (2001) Characterization of Enterocoliticin, a phage tail-like bacteriocin, and its effect on pathogenic *Yersinia enterocolitica* strains. Appl Environ Microbiol 67:5634–5642

Walker DH Jr, Walker JT (1975) Genetic studies of coliphage P1. I. Mapping by use of prophage deletions. J Virol 16:525–534

Yen HC, Hu NT, Marrs BL (1979) Characterization of the gene transfer agent made by an overproducer mutant of *Rhodopseudomonas capsulata*. J Mol Biol 131:157–168

Zhou T, Radaev S, Rosen BP, Gatti DL (2000) Structure of the ArsA ATPase: the catalytic subunit of a heavy metal resistance pump. EMBO J 19:4838–4845

Bacterial Ectosymbionts which Confer Motility: *Mixotricha paradoxa* from the Intestine of the Australian Termite *Mastotermes darwiniensis*

Helmut König, Li Li, Marika Wenzel, Jürgen Fröhlich

Dedicated to Prof. Dr. Karl Otto Stetter on the occasion of his 65th birthday

1
Introduction

Because of their intestinal flora, termites are among the most important wood- and litter-feeding insects (Wood and Sands 1978; Abe et al. 2000; König and Varma 2005). The gut microbes play an indispensable role in the digestion of food. The dense gut microbiota can include a variety of prokaryotic and eukaryotic microorganisms from different systematic positions (König et al. 2002; Chap. 3). Due to the interesting microbial symbionts and their ecological importance in the global carbon cycle, termites have attracted the interest of many scientists from different disciplines.

Flagellates from the intestine of termites have branched off very early in the evolution of the eukaryotes. These species seem not to occur elsewhere in nature except in wood-eating roaches of the genus *Cryptocercus*. All protozoa of the termite gut belong to the oxymonads, trichomonads, and the hypermastigotes (Fig. 5.4; Honigberg 1970; Yamin 1979; Radek and Hausmann 1993; Viscogliosi et al. 1993; Berchtold and König, 1995; Brugerolle and König 1997; Dacks and Redfield 1998; Keeling et al. 1998; Kitade and Matsumoto 1998; Moriya et al. 1998; Ohkuma et al. 1998; Fröhlich and König 1999b; Brugerolle 2000; Brugerolle and Lee 2000a,b; Delgado-Viscogliosi et al. 2000). From 205 examined termite species, 434 species of flagellates have been described until 1979. The flagellates occur in high number in the paunch (10^3-10^7) and they can occupy more than 90% of the paunch volume (Berchtold et al. 1999). Only three species of the flagellate flora have been obtained in culture: *Trichomitopsis termopsidis* (Yamin 1978, 1980), *Trichonympha sphaerica* (Yamin 1981) from *Zootermopsis* sp. and *Trichomitus trypanoides* from

H. König (e-mail: hkoenig@uni-mainz.de), L. Li, M. Wenzel, J. Fröhlich
Institute of Microbiology and Wine Research, Johannes Gutenberg-University,
Becherweg 15, 55099 Mainz, Germany

Reticulitermes santonensis (Berchtold et al. 1995). Whether lower termites obligatory depend on the flagellates is a matter of debate. We observed several colonies of *Zootermopsis angusticollis* and one colony of *Kalotermes flavicollis*, which lived without flagellates.

The microbiota is not distributed randomly in the gut, but play certain roles in the degradation of lignocellulose and occupy distinct microhabitats. Electron microscopy studies of termite guts have shown that prokaryotes occur either suspended in the contents, located within or on the surface of flagellates, or they are attached to the gut wall (Breznak and Pankratz 1977; Czolij et al. 1984, 1985, 1986; König er al. 2002).

2
The Intestinal Microbiota of *Mastotermes darwiniensis*

The lower wood-feeding termite *Mastotermes darwiniensis* Froggatt (Fig. 5.1.) is the only living member of the family Mastotermitidae. Today, this species is restricted to Northern Australia, but mastotermitid fossil specimen from the Eocene and Miocene have been found in Central America, the Caribic region, Europe, and Australia (Thorne et al. 2000). Termites are assigned to 13 families and 282 genera (Table 5.1; Myles 1999). *Mastotermes darwiniensis* is believed to be the most primitive existing termite species (Gay and Calaby 1970). *Mastotermes darwiniensis* developed a complex symbiotic hindgut flora, which consists of protozoa (formerly named Archaezoa; Cleveland and Grimstone 1964; Brugerolle et al. 1994; Berchtold and König 1995; Fröhlich and König 1999a,b), Bacteria (Berchtold and König 1996; Berchtold et al. 1999), Archaea (Fröhlich and König 1999a,b) and yeast (Prillinger et al. 1996; Schäfer et al. 1996). Defaunation experiments showed that the protozoa appeared to be essential for the termites survival (Veivers et al. 1983). In amber containing the Miocene termite *Mastotermes electrodominicus*, a 20-million-year-old fossil microbial community consisting of protists, spirochetes, and other bacteria has been observed (Wier et al. 2002).

Despite their small volumes of about 0.5–10 µl, the hindguts of termites are morphologically complex systems. The digestive system of termites consists of the foregut with the crop and the gizzard, the midgut and the hindgut (Noirot and Noirot-Timotheé 1969; Noirot 1995). The hindgut consists of five segments (P1–P5): the proctodeal segment, the enteric valve, the paunch, the colon and the rectum. The paunch is the microbial fermentation chamber, but the colon also contains microorganisms. Some higher termites possess a

Bacterial Ectosymbionts which Confer Motility

Fig. 5.1. Colony of *Mastotermes darwiniensis*. Soldiers and workers are present

Table 5.1. Proposed taxonomy of the order Isoptera. (Myles 1999; number of genera and species in parenthesis)

LOWER TERMITES

Family: Termopsidae (rottenwood termites; 5, 20)

Family: Hodotermitidae (harvester termites; 3, 19)

Family Mastotermitidae (primitive termites; 1, 1)

Familiy: Kalotermitidae (dampwood and drywood termites; 22, 419)

Family: Rhinotermitidae (subterranean termites; 14, 343)

Family: Serritermitidae (1, 1)

HIGHER TERMITES

Termitidae

Macrotermitinae (14, 349)

Apicotermitinae (43, 202)

Amitermitinae (17, 295)

Nasutitermitinae (91, 663)

Cubitermitinae (28, 161)

Termitinae (43, 288)

prolonged mesenteron resulting in a transitional mixed segment between the midgut and hindgut. In the lower, phylogenetically most primitive termite, *Mastotermes darwiniensis*, P1 and P2 are very small. P3, the paunch, is subdivided into a dilated, thin-walled region (P3a) and a thick-walled, more tubular region (P3b), which is followed by a thick-walled, tubular colon (P4) and the rectum (P5) (Fig. 5.2).

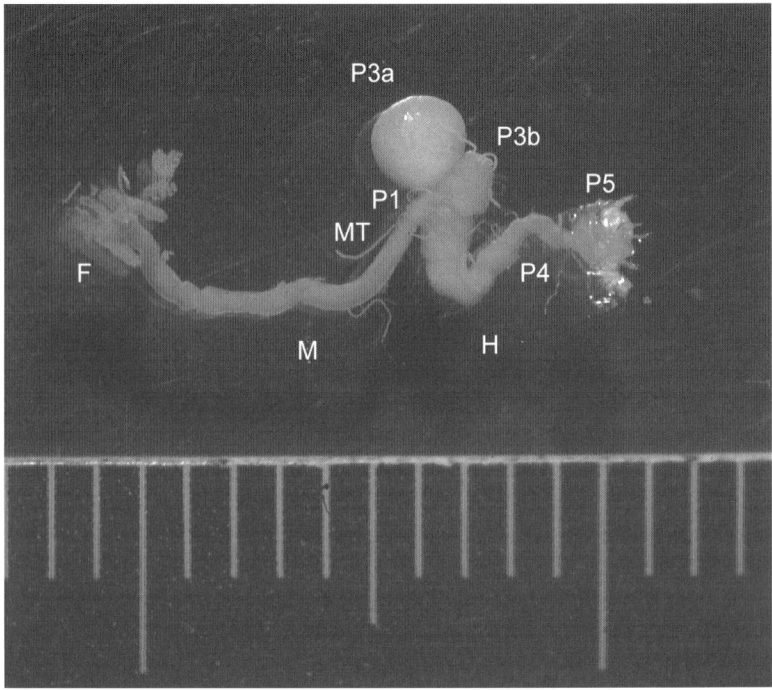

Fig. 5.2. Micrograph of the gut of *Mastotermes darwiniensis*. *F* foregut with salivary glands *M* midgut, *H* hindgut. *MT* malpighian tubules, *P1* proctodeal segment, followed by the enteric valve (P1, not marked), *P3a* thin-walled part of the paunch, *P3b* thick-walled part of the paunch, *P4* colon, *P5* rectum. Graduation: mm

In the case of *Mastotermes darwiniensis*, oxygen diffusion gradients could be detected up to 100 µm below the epithelium (Berchtold et al. 1999).

Combining the rRNA approach with confocal laser scanning microscopy and oxygen microelectrode measurements, the gut microbial community within the hindgut *Mastotermes darwiniensis* was examined (Berchtold et al. 1999; cf. König et al. 2002). The anterior part of the paunch (P3a region) of *Mastotermes darwiniensis* is tightly packed with large flagellates (1285 ± 244 cells of *Deltotrichonympha operculata*, *Deltotrichonympha nana*, *Koruga bonita*, and *Mixotricha paradoxa* per termite; Table 5.2) (Berchtold et al. 1999). From the combined volume of the larger flagellates it can be estimated that 95% of the anterior part of the paunch (P3a region) is occupied by flagellate protozoa. In *Mastotermes darwiniensis*, approximately 90% of the DAPI-stained cells were associated with the protozoa in the P3a region, only 2% were attached to the gut wall, and the rest were found in the residual liquid volume of the lumen fraction. In contrast, the flagellate population in the P3b/P4 region was much smaller. The flagellate cells represented only 10% of the total volume. The potentially colonizable surface area provided by the flagellates in the P3a region exceeds that of the wall by a factor of 18. In contrast to the P3a region, about 85% of the prokaryotes of the P3b region are attached to the wall. The prokaryotic cell density on the P3b/P4 epithelium (2×10^6 per mm^2) is considerably higher than that on the P3a surface (3×10^4 per mm^2). The concentration of non-attached cells in the residual volume is higher in the P3a region (7×10^9 cells per ml) than in the P3b/P4 region (1×10^9 cells per ml). The flagellates preferentially colonize the paunch, while low numbers are found in the colon (P4 region) (Berchtold et al. 1999).

3
Symbiotic Interactions between Spirochetes and Flagellates

One of the first detected symbioses between flagellates (*Pseudodevescovina uniflagellata*) and spirochetes was described by Kirby (1936). *Pseudodevescovina uniflagellata* lives in the gut of the Australian dry wood termite

Table 5.2. Intestinal protozoa flora of *Mastotermes darwiniensis*. (cf. König et al. 2002)

Species	Length (μm)	Titer (ml^{-1})
1. Hypermastigotes	100–500	10^4–10^5
(*Deltotrichonympha nana*,		
Deltotrichonympha operculata,		
Koruga bonita)		
2. Trichomonads		
Mixotricha paradoxa	300–500	10^3–10^4
Metadevescovina extranea	15–20	ca. 10^7
Pentatrichomonoides scroa	25	ca. 5×10^6

Neotermes insularis. Only three years earlier Sutherland (1933) published an article about *Mixotricha paradoxa* where the attached spirochetes were misconceived as cilia. A detailed description of the fine structure of *Mixotricha paradoxa* and the role of the ectosymbiotic bacteria in cell locomotion was provided by Cleveland and Grimstone (1964). Over the years, more and more examples of surface symbiosis between protists and prokaryotes from the termite gut appeared (Bloodgood and Fitzharris 1976; Smith et al. 1975; To et al. 1980; Dyer and Khalsa 1993), but examples of motility symbiosis in the termite gut could be rarely detected (Tamm 1982). Locomotory mechanisms of two larger flagellates from *Mastotermes darwiniensis* have been studied (Cleveland and Cleveland 1966; Tamm 1999).

Ectosymbiotic bacteria of flagellates can easily be detected by electron microscopy (Radek et al. 1992, 1996 ; Radek and Tischendorf 1999; Dyer and Khalsa 1993) or after staining the cells with ethidium bromide (Fröhlich and König 1999a). Ectosymbiotic spirochetes have been identified on the surface of flagellates (Iida et al. 2000; Noda et al. 2003; Wenzel et al. 2003). *Mixotricha paradoxa,* a large trichomonad from the hindgut of the Australian termite *Mastotermes darwiniensis* Froggatt, is a rare example of a movement symbiosis between eukaryotic and prokaryotic microorganisms (Wenzel et al. 2003).

Cleveland and Grimstone (1964) found two spirochete morphotypes on the surface of *Mixotricha paradoxa*, a small one, which covered the surface of the flagellate as a dense carpet and a longer spirochete, which only appeared sporadically. This longer spirochete could also be seen on the anterior part of the cell (Fig. 5.3). It is only loosely bound to the spirochete carpet. Cleveland and Grimstone (1964) described the regular arrangement of the spirochetes and a rod-shaped bacterium attached to the so-called brackets on the cell surface. These brackets seem to be significant for the locomotory function of the spirochetes. They form a regularly posteriorly oriented attachment site for the spirochetes, which allows the spirochetes to propel the flagellate cells forward. The rod-shaped bacteria, which are attached to the anterior site of the brackets, have no part in the locomotion of *Mixotricha paradoxa* (Cleveland and Grimstone 1964; König and Breunig 1997).

4
Morphology of *Mixotrixa paradoxa*

One of the larger wood-ingesting flagellates in the hindgut of *Mastotermes darwiniensis* is *Mixotricha paradoxa*, a member of the order Trichomonadida (Table 5.2). The pear-shaped cells are about 500 µm long and 250 µm in diameter (Fig. 5.3). The surface of *Mixotricha paradoxa* shows a highly ordered pattern of rod-shaped bacteria and in addition it is covered by a dense carpet

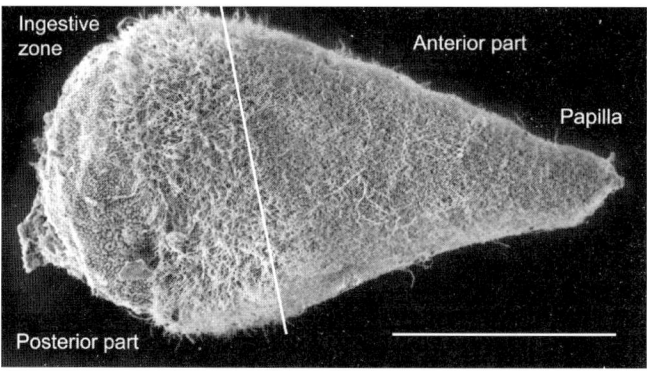

Fig. 5.3. Scanning electron micrograph of *Mixotricha paradoxa*. The *white line* indicates the borderline between the anterior and posterior part of the cell, which are colonized by different spirochete clones. The larger spirochete cells of the anterior part are only loosely associated and not tightly bound to the cell surface (from Wenzel et al. 2003; with permission). *Bar* 100 μm

of spirochetes with the exception of the posterior ingestive zone (Cleveland and Grimstone 1964). The rod-shaped bacteria and the spirochetes are attached to regularly arranged protrusions of the cell surface. Interestingly, Cleveland and Grimstone (1964) found that the spirochetes (not the relatively small four flagella) propel the cells. It is still unknown how the flagellates and the spirochetes communicate and coordinate the direction of movement. For hydro-mechanical reasons, it seems that cilia, flagella, sperm tails and spirochetes should automatically synchronize their movement when undulating in close proximity (Machin 1963). So far, it is not possible to cultivate either *Mixotricha paradoxa* or its ectosymbiotic spirochetes and rod-shaped bacteria. Although spirochetes are always a dominant part of the microflora of all termites (Margulis and Hinkle 1992), only four species have been obtained in pure culture (Leadbetter et al. 1999; Dröge 2006).

Spirochetes possess a cellular ultrastructure that is unique among eubacteria (Holt 1978; Canale-Parola 1991). The helical protoplasmic cylinder is encased by an outer envelope, which has some features analogous to the outer membrane of gram-negative bacteria. The spirochetes possess internal organelles of motility called periplasmic flagella, which are located between the protoplasmic cylinder and the outer envelope (Paster et al. 1996). Previously, several spirochetal 16S rDNA sequences originating from the hindgut of *Mixotricha darwiniensis* were published (Berchtold et al. 1994; Berchtold and König 1996; Wenzel et al. 2003). These clones also represent species from a side branch of the *Treponema* cluster (Lilburn et al. 1999; Ohkuma et al. 1999). Two additional spirochete clusters related to the genus *Spirochaeta* have been found recently (Dröge et al. 2006).

5
Phylogenetic Position of *Mixotricha paradoxa*

Figure 5.4 shows a phylogenetic tree of parabasalids. The SSU rDNA sequences were added to an existing database of 36 parabasalid sequences. This data set included 1,210 nucleotides which were analyzed with distance matrix, maximum parsomony, and maximum likelihood methods (Li 2003). The phylogenetic tree shows that four hindgut protozoa of *Mastotermes darwiniensis* form one subdivision: the small flagellate *Metadevescovina extranea* and the three large flagellates *Koruga bonita*, *Deltotrichonympha nana* and *Deltotrichonympha operculata*. They form a monophyletic group with a quite high bootstrap support. Parallel to this subdivision, *Mixotricha paradoxa* exhibits an earlier emergence, indicating its primitive position among the intestinal flagellates of *Mastotermes darwiniensis* and among the families Devescovinidae and Calonymphidae. *Mixotricha paradoxa* emerges earlier than *Foaina* species. *Foaina* has been proposed to be a primitive genus among the Devescovinidae and Calonymphidae (Gerbod et al. 2002). *Mixitricha paradoxa* should be clustered in the Devescovinidae/Calonymphidae group. *Koruga bonita*, *Deltotrichonympha nana* and *Deltotrichonympha operculata* classified in the Hypermastigea by morphological characters might be assigned to the Devescovinidae/ Calonymphidae group of Trichomonodea. This indicates that the hypermastigid flagellates are not a monophyletic branch (Fröhlich and König 1999b; Li 2003). *Pentatrichomonoides scroa*, the second small intestinal flagellate of *Mastotermes darwiniensis* is not closely related to the other five symbiotic flagellates, because it clusters with the Trichomonadinae.

6
Glycolytic Activities of *Mastotermes darwiniensis* and its Flagellates

Up to now, it has been believed that the hindgut flagellates produce nutrients using their own cellulolytic enzymes for the benefit of their termite host. The glycolytic activities found in separated cells of *Mixotricha paradoxa* are compiled in Table 5.3 (Berchtold and König, unpubl. res.). Surprisingly, not all of these activities seemed to be produced by the flagellates themselves, but rather taken up with the gut contents. Two endoglucanases, Cel I and Cel II,

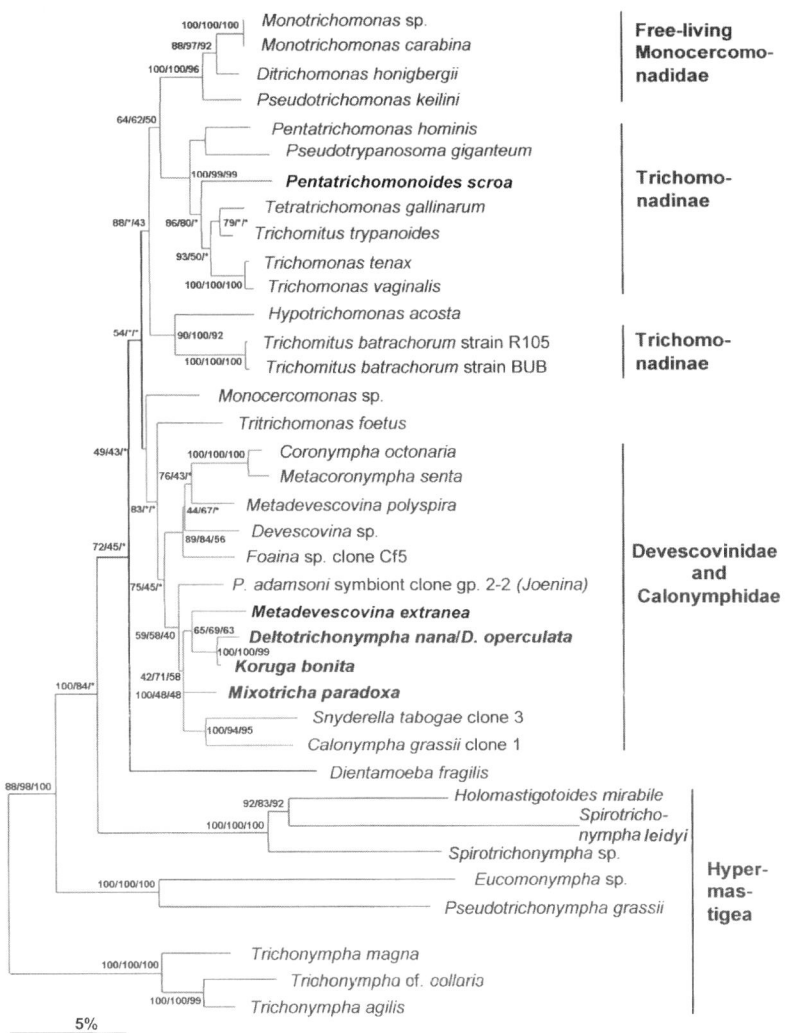

Fig. 5.4. Unrooted phylogenetic tree of parabasalids. Neighbor-joining analysis of SSU rDNA sequences. *Bar* represents five substitutions per 100 nucleotides. The bootstrap values are computed by three different reconstruction methods: distance matrix, maximum parsony and maximum likelihood. *Asterisks* designate nodes with bootstrap values below 40% (Li 2003).

with the molecular mass of approximately 48 kD, were isolated from the not yet culturable symbiotic flagellates living in the hindgut of the most primitive Australian termite *Mastotermes darwiniensis* (Li et al. 2003). The N-terminal sequences of these cellulases exhibited significant homology to cellulases of termite origin, which belong to glycosyl hydrolase family 9. The corresponding genes were detected not in the mRNA pool of the flagellates, but in the salivary glands of *Mastotermes darwiniensis*. A protein with the molecular mass of approximately 48 kD was also detected in crude extract of these flagellates, by western blot analysis using a polyclonal antiserum against the cellulase of the termite *Mastotermes darwiniensis*. The results gave evidence that cellulases isolated from the nutritive vacuole of the flagellates originated from the termite host. Probably, the cellulases are secreted from the salivary glands of *Mastotermes darwiniensis*. During the mechanical grinding of the wood particles by the termites, the cellulases are attached to wood particles or mixed with them, then the attached cellulases or the mixture move to the hindgut where they are most probably endocytosed by the flagellates.

Table 5.3. Determination of glycolytic and laccase activities of *Mixotricha paradoxa*

Enzyme	Activity [µU/cell]
α-L-Arabinosidase	0.3
β-L-Arabinosidase	–
β-D-Cellobiosidase	1.2
α-D-Galactosidase	0.2
β-D-Galactosidase	–
β-D-Glucosidase	4.0
β-D-Glucuronidase	–
α-D-Mannosidase	–
β-D-Mannosidase	0.2
β-D-Xylosidase	0.5
Cellulase	169.0
Xylanase	135.0
Laccase (ABTS)[1]	0.001

– = not found
[1] 2,2-azinobis-3-ethylbenzthiazolinesulfonic acid

It has been shown for *Coptotermes formosanus* that the endoglucanases of this termite are restricted to the salivary glands, the foregut, and the midgut (Nakashima et al. 2002). According to our work, the main endoglucanase activity found in cells of the hindgut flagellates of *Mastotermes darwiniensis* is likely to originate from the termite's cellulases. It has also been found that 40% of the endoglucanase activity of *Mastotermes darwiniensis* is present in the hindgut and most (ca. 84%) of the cellulase activity of the whole hindgut is present in the flagellate extract (Veivers et al. 1982). This implies that a certain amount of termite cellulases, secreted from the salivary glands, moves into the hindgut and enters the flagellate cells. They may be involved in the digestion of cellulose in the flagellate cells.

Using a PCR-based approach, DNA encoding cellulases belonging to the glycosyl hydrolase family 45 were obtained from micromanipulated nuclei of the flagellates *Koruga bonita*, *Deltotrichonympha nana* and *D. operculata*. The cellulase sequences of the termite symbiotic protists were phylogenetically monophyletic, showing more than 84% amino acid identity with each other. The deduced cellulase sequences of termite origin and flagellate origin consist of a single catalytic domain, lacking a cellulose-binding domain (CBD) and a spacer sequence found in most microbial cellulases (Li et al. 2003).

Although flagellate endoglucanase genes were even expressed in vivo, significant cellulase activity of flagellate origin was not found in the nutritive vacuole by SDS-PAGE. The flagellate cellulase proteins were not detected by western blot analysis. This implied that the native endoglucanase of flagellate origin has very low or even no CMC activity and the translation efficiency could also be very low. In the case of *Mixotricha paradoxa*, no cellulase gene could be detected.

It is conceivable that in the course of 200–300 million years, the symbiosis between termite and initially free-living intestinal flagellates mutually affected the enzymatic equipment of the other, such that the excess of termite cellulases led to a disuse of the flagellates' own enzymes. The lack of a selection pressure in the hindgut possibly directed low level translation, mutation, and inactivation of cellulolytic enzymes, from which the corresponding genes are still expressed. The production of an inactive enzyme may result in a complete loss of the corresponding genes. This means that the symbiotic flagellates are progressing to a state void of their own cellulolytic activities as was probably the case before the existence of cellulose-containing plants (though some cellulose-producing microorganisms, e.g., *Acetobacter*, appeared earlier than plants, the amount of the microbial cellulose should be much less than that of the plants). Presently, the symbiotic gut flagellates of the primitive Australian termite *Mastotermes darwiniensis* owe their endocellulolytic activity to their host.

7
Identification of the Ectosymbiotic Spirochetes of *Mixotricha paradoxa*

Berchtold and König (1996) and Wenzel (1998) found about 13 different spirochetal 16S rRNA gene clones in the intestinal tract of *Mastotermes darwiniensis*.

Although spirochetes constitute a main part of the gut flora of termites, the microhabitats and function of the identified spirochetes (Berchtold et al. 1994; Berchtold and König 1996; Wenzel 1998) in the gut of *Mastotermes darwiniensis* remained unclear. Therefore, the bacteria associated with the cell surface of *Mixotricha paradoxa* were identified. Six spirochetal 16S rDNA clones (mpsp 15, sp 40–7, mp1, mp3, mp4, mp5) were obtained from the bacteria on the cell envelope of *Mixotricha paradoxa* (Wenzel et al. 2003). Two clones (mpsp15, sp 40–7) were nearly identical (99%) to already described clones (Berchtold et al. 1994; Berchtold and König 1996), while the 16S rRNaA gene sequences of clones mp1, mp3, mp4 and mp5 have not been found previously.

The "spirochete" tree including the 16S rRNA gene sequences of the six spirochete clones mpsp15, sp 40–7, mp1, mp3, mp4, and mp5 obtained from the ectosymbiotic bacteria together with some representative spirochetes from the EMBL-database (Berchtold and König 1995) was constructed. The phylogenetic tree (Fig. 5.5) shows that the ectosymbiotic spirochetes of *Mixotricha paradoxa* belong all to the *Treponema* branch (Berchtold et al. 1994; Berchtold and König 1996; Paster et al. 1996; Lilburn et al. 1999; Ohkuma et al. 1999; Iida et al. 2000).

8
Identification of the Ectosymbiotic Rod-shaped Bacterium of *Mixotricha paradoxa*

Wenzel (1998) obtained 16S rDNA amplificates (e.g., clone B6; ca. 400 bp) from *Mixotricha paradoxa*, which were related to *Bacteroides* sp. With the specific primer B6.1 derived from the obtained sequence of clone B6 1338 bp were amplified.

The *Bacteroides* sp.-related clone B6 and some of its phylogenetic relatives were used to construct the "*Bacteroides*" tree. Figure 5.6 shows the phylogenetic relationship of clone B6 to representatives of *Bacteroides*-related species. This indicates that the epibiotic rod-shaped bacterium on the surface of *Mixotricha paradoxa* is a member of the *Bacteroides*-branch with probably *Bacteroides forsythus* and *B. merdae* as closest described species. The obtained sequence showed similarity to the uncultured clones Gf15 [AB055715; 92% in 857 aligned bp] and Gf16 [AB055716; 92% in 855 aligned bp] from *Glyptotermes fuscus* and *Bacteroides* cf. *forsythus* orale clone BUO63 [AY008308; 94% in 495 bp]. The low phylogenetic relationship indicated that clone B6 originated from a new species (Wenzel et al. 2003).

9
Assignment of 16S rDNA Sequences to the Corresponding Ectosymbiotic Bacterial Morphotypes

By using cell envelope preparations for in situ fluorescence hybridization, an interference with the fluorescence of wood particles was avoided. In addition, the fluorescence signals of the specific Cy3-labelled probes could be enhanced by applying helper oligonucleotides (Fuchs et al. 2000) and the binding of the probes was facilitated by performing a denaturing step.

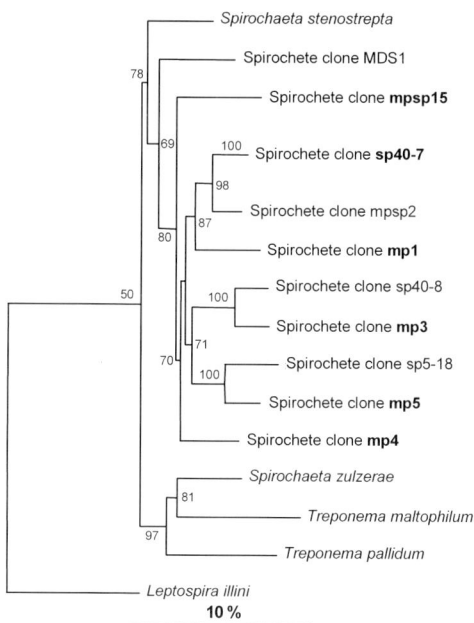

Fig. 5.5. Phylogenetic relationship of *Mixotricha paradoxa*-associated spirochete clones (*bold letters*). The relationship was determined by neighbor joining analysis. The data set contained 15 alignment positions and *Leptospira illini* as the outgroup. The bootstrap values (100 runs), obtained by using the program SEQBOOT from the PHYLIP program package (Felsenstein 1985, 1993), are inserted at the respective branching points. Bootstrap values under 50% are not shown. The *bar* represents 10 changes per 100 nucleotide positions (from Wenzel et al. 2003; with permission)

Fluorescence in situ hybridization was performed with specific Cy3-labelled probes derived from 16S rDNA amplificates obtained from the ectosymbiotic spirochetes and rod-shaped bacterium clone B6. Three spirochetal clones could be localized on the cell surface, clone mpsp15 at the anterior and clones mp1 and mp3 at the posterior part. Clone mp1 occurred in a lower number than clone mp3 (Fig. 5.7). The spirochetal clones that could not be localized on the cell surface form probably no dense population on the cell surface and their weak fluorescence signals might have been overlooked. Cleveland and Grimstone (1964) described two morphotypes of spirochetes, shorter ones which were attached to the brackets and longer ones which only appeared sporadically and were not linked to the brackets. In the light microscope we also observed the long spirochetes on living flagellate cells. They were not found on the isolated cell envelopes, indicating that they are not tightly bound

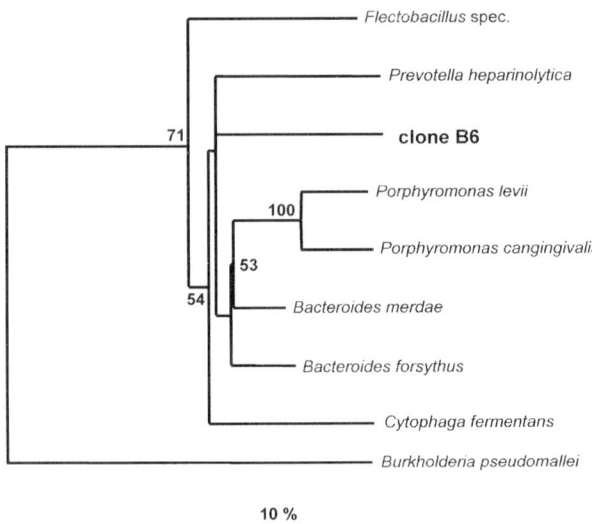

Fig. 5.6. Phylogenetic relationship of rod-shaped clone B6. 16S rDNA sequences were obtained from the EMBL-database (Stoesser et al. 2001). The relationship was determined by neighbor joining analysis. The data set contained nine alignment positions and *Burkholderia pseudomallei* as the outgroup. The bootstrap values (100 runs), obtained by using the program SEQBOOT, are inserted at the respective branching points. Bootstrap values under 50% are not shown. The *bar* represents 10 changes per 100 nucleotide position (from Wenzel et al. 2003; with permission)

to the cell surface. In contrast, the smaller morphotypes remained at the surface of the envelopes. Spirochete clone mpsp15 possesses one flagellum at each cell pole (Wenzel et al. 2003). After Margulis and Hinkle (1992) the characteristic flagella array is 1:2:1 corresponding to one flagellum at the end of the cell and two overlapping flagella in the middle of the cell. The other 16S rDNA clones localized at the posterior part of *Mixotricha paradoxa* have to be assigned to certain morphotypes in a future work.

Rod-shaped bacteria (length: 0.8–1.1 µm; width: 0.3 µm) are attached to cell surface brackets in a regular pattern (Cleveland and Grimstone 1964; König and Breunig 1997) perpendicular to the cell of *Mixotricha paradoxa*. The distance between the cells in a row is about 0.9 µm, and between two adjacent rows is 0.5 µm. The individual rods in the rows are staggered. The fluorescent probes B6.1 and B6.2 were specific for the *Bacteroides* sp-related clone B6. Positive hybridization results showed that clone B6 is spread all over the surface of *Mixotricha paradoxa* in a similar regular pattern as found in electron micrographs.

When the cell envelopes were incubated with the probes B6.1 and B6.2 or with the Cy3-labelled universal probe Eubac 338 (Amann et al. 1990) the same regular pattern of a rod-shaped bacterium was obtained indicating that clone B6 was the only rod associated with the cell surface (Fig. 5.7)

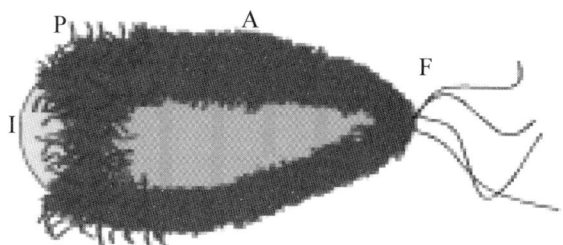

Fig. 5.7. Schematic drawing showing the proposed distribution of the bacterial ectosymbionts on the cell surface of *Mixotricha paradoxa* (from Wenzel et al. 2003; with permission). *A* anterior part, *P* posterior part, *F* flagella. *I* ingestive zone

10
Conclusions

Mixotricha paradoxa, a trichomonad from the hindgut of the Australian termite *Mastotermes darwiniensis* Froggatt, is a rare example of a movement symbiosis between eukaryotic and prokaryotic microorganisms (Cleveland and Grimstone 1964). It is known that a lot of symbiotic relationships of protozoa and spirochetes play no role in the locomotion of the protist cell (Bloodgood and Fitzharris 1976; Breznak 1984; Iida et al. 2000). This indicates that spirochetes must fulfill also other functions. Leadbetter et al. (1999) discussed the possibility that the spirochetes consume H_2 and CO_2 that is produced by the flagellates. They form acetate as end product, which is consumed by the termites, while other isolates live heterotrophic (Graber et al. 2004 a,b; Dröge et al. 2006).

From the movement symbiosis between spirochetes and the flagellate *Mixotricha paradoxa* the hypothesis was derived that eukaryotic locomotory organelles such as flagella and cilia originated from spirochetes (Bermudes et al. 1987; Margulis 1993), while the cytoplasm is assumed to be of archaeal origin. Our studies showed that several spirochete species synergistically contrive the movement of *Mixotricha paradoxa*. Since *Mixotricha paradoxa* belongs to the early branching flagellates the movement symbiosis should be an early invention during the evolution of the eukaryotic cell. If the above-mentioned hypothesis is correct, then the ancestor of locomotory organelles should have been a relative of *Treponema*. Since the other large symbiotic flagellates of *Mastotermes darwiniensis* (e.g. *Koruga bonita*) are relatively closely related to *Mixotricha paradoxa* it seems not likely that their flagella originated from spirochetes. The symbiosis between *Mixotricha paradoxa* and its spirochetes seems to be a unique example of movement symbiosis.

Acknowledgements We thank H. Hertel (BAM, Berlin) for termite cultures.

References

Abe T. Bignell DE, Higashi M (eds.) (2000) Termites: Evolution, Sociality, Symbioses, Ecology, Kluwer Academic Publ. Dordrecht.

Amann RI, Krumholz L, Stahl DA (1990) Fluorescent-oligonucleotide probing of whole cells for determinative, phylogenetic, and environmental studies in microbiology. J Bacteriol 172:762–770

Berchtold M, König H (1995) Phylogenetic position of two uncultivated trichomonads *Pentatrichomonoides scroa* Kirby and *Metadevescovina extranea* Kirby from the hindgut of the termite *Mastotermes darwiniensis* Froggatt. System Appl Microbiol 18:567–573

Berchtold M, König H (1996) Phylogenetic analysis and in situ identification of uncultivated spirochetes from the hindgut of the termite *Mastotermes darwiniensis*. System Appl Microbiol 19:66–73

Berchtold M, Ludwig W, König H (1994) 16S rDNA sequence and phylogenetic position of an uncultivated spirochete from the hindgut of the termite *Mastotermes darwiniensis* Froggatt. FEMS Microbiol Lett 123:269–274

Berchtold M, Breunig A, König H (1995) Culture and phylogenetic characterization of *Trichomitus trypanoides* Duboscque & Grassè 1924, n. comb. a trichomonad flagellate isolated from the hindgut of the termite *Reticulitermes santonensis* Feytaud. J Eukar Microbiol. 42: 388-391

Berchtold M, Chatzinotas A, Schönhuber W, Brune A, Amann R, Hahn D, König H (1999) Differential enumeration and in situ localization of microorganisms in the hindgut of the lower termite *Mastotermes darwiniensis*. Arch Microbiol 172: 407–416

Bermudes D, Margulis I, Tzertzinis G (1987) Prokaryotic origin of undulipodia. In: Lee JJ, Jerome FF (eds) Endocytobiology III. New York Academy of Sciences, New York, pp 187–197

Bloodgood RA, Fitzharris TP (1976) Specific associations of prokaryotes with symbiotic flagellate protozoa from the hindgut of the termite *Reticulitermes* and the wood-eating roach *Cryptocercus*. Cytobios 17:103–122

Breznak JA, Pankratz HS (1977) In situ morphology of the gut microbiota of wood-eating termites [*Reticulitermes flavipes* (Kollar) and *Coptotermes formosanus* (Shiraki)]. Appl Environ Microbiol 33:406–426

Breznak JA (1984) Biochemical aspects of symbiosis between termites and their intestinal microbiota. In: Anderson JM, Rainer ADM, Walton DWH (eds) Invertebrate-microbial interactions. Cambridge Univ Press, Cambridge, pp 173–203

Brugerolle G (2000) A microscopic investigation of the genus *Foaina*, a parabasalid protist symbiotic in termites and phylogenetic considerations. Eur J Protistol 36:20–28

Brugerolle G, König H (1997) Ultrastructure and organisation of the cytoskeleton in *Oxymonas*, an intestinal flagellate of termites. J Eukaryot Microbiol 44:305–313

Brugerolle G, Lee JJ (2000a) *Phylum Parabasalia*. In: Lee JJ, Leedale GF, Bradbury P (eds) The illustrated guide to the protozoa, 2nd edn, vol 2. Society of Protozoologists, Lawrence, Kansas, pp 1196–1250

Brugerolle G, Lee JJ (2000b) *Order Oxymonadida*. In: Lee JJ, Leedale GF, Bradbury P (eds) The illustrated guide to the protozoa, vol 2. Society of Protozoologists, Lawrence, Kansas, pp 1186–1195

Brugerolle G, Breunig A, König H (1994) Ultrastructural study of *Pentatrichomonoides* sp., a trichomonad flagellate from *Mastotermes darwiniensis*. Eur J Protistol 30:372–378

Canale-Parola E (1991) Free-living saccharolytic spirochetes: the genus *Spirochaeta*. In: Balows A, Trüper HG, Dworkin M, Harder W, Schleifer K (eds) The Prokaryotes, vol 4. Springer, Berlin Heidelberg New York, pp 3593–3607

Cleveland LR, Cleveland BT (1966) The locomotory waves of *Koruga*, *Deltotrichonympha* and *Mixotricha*. Arch Protistenk 109:39–63

Cleveland LR, Grimstone AV (1964) The fine structure of the flagellate *Mixotricha paradoxa* and its associated microorganisms. Proc R Soc Lond Ser B 159:668–686

Czolij R, Slaytor M, Veivers PC, O'Brien RW (1984) Gut morphology of *Mastotermes darwiniensis* Froggatt (Isoptera: Mastotermitidae). Int J Insect Morphol Embryol 13:337–355

Czolij R, Slaytor M, O'Brien RW (1985) Bacterial flora of the mixed segment and the hindgut of the higher termite *Nasutitermes exitiosus* Hill. (Termitidae: Nasutitermitinae). Appl Environ Microbiol 49:1226–1236

Czolij R, Slaytor M, O'Brien RW (1986) Bacterial flora of the mixed segment and the hindgut of the higher termite *Nasutitermes exitiosus* Hill (Termitidae, Nasutermitinae). Appl Environ Microbiol 49:1226–1236

Dacks JB, Redfield RJ (1998) Phylogenetic placement of *Trichonympha*. J Eukaryot Microbiol 45:445–447

Delgado-Viscogliosi P, Viscogliosi E, Gerbod D, Kulda J, Sogin ML, Edgcomb VP (2000) Molecular phylogeny of parabasalids based on small subunit rRNA sequences, with emphasis on the Trichomonadinae subfamily. J Eukaryot Microbiol 47:70–75

Dröge S, Fröhlich J, Radek R, König H (2005) *Spirochaeta coccoides* sp. nov., a novel coccoid spirochete from the hindgut of the termite *Neotermes castaneus*. Appl Environ Microbiol. In press

Dyer BD, Khalsa O (1993) Surface bacteria of *Streblomastix strix* are sensory symbionts. Biosystems 31:169–180

Felsenstein J (1985) Confidence limits on phylogenies: an approach using the bootstrap. Evolution. 39:783-791.

Felsenstein J (1993) Phylip (Phylogeny Inference Package) version 3.5. Department of genetics, University of Washington, Seattle.

Fröhlich J, König H (1999a) Ethidium bromide: a fast fluorescent staining procedure for the detection of symbiotic partnership of flagellates and prokaryotes. J Microbiol Meth 35:121–127

Fröhlich J, König H (1999b) Rapid isolation of single microbial cells from mixed natural and laboratory populations with the aid of a micromanipulator. System Appl Microbiol 22:249–257

Fuchs BM, Glöckner FO, Wulf J, Amann R (2000) Unlabeled helper oligonucleotides increase the in situ accessibility to 16S rRNA of fluorescently labeled oligonucleotide probes. Appl Environ Microbiol 66:3603–3607

Gay FJ, Calaby JH (1970) Termites of the Australian region. In: Krishna K, Weesner FM (eds) Biology of termites, vol 2. Academic Press, New York, pp 393–448

Gerbod D, Nöel C, Dolan MF, Edgcomb VP, Kitade O, Noda S, Dufernez F, Ohkuma M, Kudo T, Capron M, Sogin ML, Viscogliosi E (2002) Molecular phylogeny of parabasalids inferred from small subunit rRNA sequences, with emphasis on the Devescovinidae and Calonymphidae (Trichomonadea). Mol Phylogenet Evol 25:545–556

Graber JR, Breznak JA (2004a) Physiology and nutrition of *Treponema primitia*, an H_2/CO_2-acetogenic spirochete from termite hindguts. Appl Environ Microbiol 70:1307-1314

Graber JR, Leadbetter JR, Breznak JA (2004b) Description of *Treponema azotonutricium* sp. nov. and *Treponema primitia* sp. nov., the first spirochetes isolated from termite guts. Appl. Environ. Microbiol. 70:13154-1320

Holt SC (1978) Anatomy and chemistry of spirochetes. Microbiol Rev 42:114–160

Honigberg BM (1970) Protozoa associated with termites and their role in digestion. In: Krishna K, Weesner FM (eds) Biology of termites, vol 2. Academic Press, New York, pp 1–36

Iida T, Ohkuma M, Ohtoko K, Kudo T (2000) Symbiotic spirochetes in the termite hindgut: phylogenetic identification of ectosymbiotic spirochetes of oxymonad protists. FEMS Microbiol Ecol 34:17–26

Kirby H Jr (1936) Two polymastigote flagellates of the genera *Pseudodevescovina* and *Caduceia*. Quart J Microscop Sci 79:309–335

Keeling P, Poulsen N, McFadden GI (1998) Phylogenetic diversity of parabasalian symbionts from termites, including the phylogenetic position of *Pseudotrypanosoma* and *Trichonympha*. J Eukaryot Microbiol 45:643–650

Kitade O, Matsumoto T (1998) Characteristics of the symbiotic flagellate composition within the termite family Rhinotermitidae. Symbiosis 25:271–278

König H, Breunig A (1997) Ökosystem Termitendarm. Spektrum der Wissenschaft 68–76

König H, Varma A (eds.) (2005) Intestinal Microorganisms of Termites and Other Invertebrates. Springer Verlag, Heidelberg.

König H, Fröhlich J, Berchtold M, Wenzel M (2002) Diversity and microhabitats of the hindgut flora of termites. Rec Res Dev Microbiol 6:125–156

Leadbetter JR, Schmidt TM, Graber JR, Breznak JA (1999) Acetogenesis from H_2 plus CO_2 by spirochetes from termite guts. Science 283:686–689

Li L (2003) Cellulases and cellulase genes of the Australian termite *Mastotermes darwiniensis* and its hindgut Archaezoa. Thesis, Johannes Gutenberg-University, Mainz

Li L, Fröhlich J, Pfeiffer P, König H (2003) Termite's symbiotic gut Archaezoa are becoming living metabolic fossils. Eukaryotic Cell 2:1091–1098

Lilburn TG, Schmidt TM, Breznak JA (1999) Phylogenetic diversity of termite gut spirochaetes. Environ Microbiol 1:331–345

Machin KE (1963) The control and synchronization of flagellar movements. Proc R Soc Lond B Biol Sci 158:88–104

Margulis L (1993) Symbiosis in cell evolution, 2nd edn. Freeman, New York

Margulis L, Hinkle G (1992) Large symbiotic spirochetes: *Clevelandina, Cristispira, Diplocalyx, Hollandina,* and *Pillotina*. In: Balows A, Trüper HG, Dworkin M, Harder W, Schleifer K-H (eds) The prokaryotes, 2nd edn. Springer, Berlin Heidelberg New York, pp 3965–3978

Moriya S, Ohkuma M, Kudo T (1998) Phylogenetic position of symbiotic protist *Dinenympha exilis* in the hindgut of the termite *Reticulitermes speratus* inferred from the protein phylogeny of elongation factor 1 alpha. Gene 210:221–227

Myles TG (1999) Phylogeny and Taxonomy of the Isoptera. XIIIth international congress of the International Union for the Study of Social Insects 29, Adelaide, Australia

Nakashima K, Watanabe H, Saitoh H, Tokuda G, Azuma JI (2002) Dual cellulose-digesting system of the wood-feeding termite, *Coptotermes formosanus* Shiraki. Insect Biochem Mol Bio 32:777–784

Noda S, Ohkuma M, Yamada A, Hongoh Y, Kudo T (2003) Phylogenetic position and in situ identification of ectosymbiotic spirochetes on protists in the termite gut. Appl Environ Microbiol 69:625–633

Noirot C (1995) The gut of termites (Isoptera). Comparative anatomy, systematics, phylogeny. I. Lower termites. Ann Soc Entomol Fr 31:197–226

Noirot C, Noirot-Timotheé C (1969) The digestive system. In: Krishna K, Weesner FM (eds) Biology of termites, vol 2. Academic Press, New York, pp 49–88

Ohkuma M, Ohtoko K, Grunau C, Moriya S, Kudo T (1998) Phylogenetic identification of the symbiotic hypermastigote *Trichonympha agilis* in the hindgut of the termite *Reticulitermes speratus* based on small-subunit rRNA sequence. J Eukaryot Microbiol 45:439–444

Ohkuma M, Iida T, Kudo T (1999) Phylogenetic relationships of symbiotic spirochetes in the gut of diverse termites. FEMS Microbiol Lett 181:123–129

Paster BJ, Dewhirst FE, Cooke SM, Fussing V, Poulsen LK, Breznak JA (1996) Phylogeny of not-yet-cultured spirochetes from termite guts. Appl Environ Microbiol 2:347–352

Prillinger H, Messner R, König H, Bauer R, Lopandic K, Molnar O, Dangel P, Weigang F, Kirisitis T, Nakase T, Sigler L (1996) Yeasts associated with termites: a phenotypic and genotypic characterization and use of coevolution for dating evolutionary radiations in asco- and basidiomycetes. System Appl Microbiol 19:265–283

Radek R, Hausmann K (1993) Symbiontische Flagellaten im Termitendarm. In: Hausmann K, Kremer BP (eds) Extremophile Mikroorganismen in ausgefallenen Lebensräumen. VCH, Weinheim, pp 325–339

Radek R, Tischendorf G (1999) Bacterial adhesion to different termite flagellates: ultrastructural and functional evidence for distinct molecular attachment modes. Protoplasma 207:43–53

Radek R, Hausmann K, Breunig A (1992) Ectobiotic and endocytobiotic bacteria associated with the termite flagellate *Joenia annectens*. Acta Protozool 31:93–107

Radek R, Roesel J, Hausmann K (1996) Light and electron microscopic study of the bacterial adhesion to termite flagellates applying lectin cytochemistry. Protoplasma 193:105–122

Schäfer A, Konrad R, Kuhnigk T, Kämpfer P, Hertel H, König H (1996) Hemicellulose-degrading bacteria and yeasts from the termite gut. J Appl Bacteriol 80:471–478

Smith HE, Buhse HE, Stamler SJ (1975) Possible formation and development of spirochaete attachment sites found on the surface of symbiotic polymastigote flagellates of the termite *Reticulitermes flavipes*. BioSystems 7:374–379

Stoesser G, Baker W, van den Broek A, Camon E, Garcia-Pastor M, Kanz C, Kulikova T, Lombard V, Lopez R, Parkinson H, Redaschi N, Sterk P, Stoehr P, Tuli MA (2001) The EMBL nucleotide sequence database. Nucleic Acids Res 29:17–21

Sutherland JL (1933) Protozoa from Australian termites. Quart J Microscop Sci 76:145–173

Tamm SL (1982) Flagellated ectosymbiotic bacteria propel a eucaryotic cell. J Cell Biol 94:697–709

Tamm SL (1999) Locomotory waves of *Koruga* and *Deltotrichonympha*: flagella wag the cell. Cell Motil Cytoskeleton 43:145–158

Thorne BL, Grimaldi DA, Krishna K (2000) Early fossil history of termites. In: Abe T, Bignell DE, Higashi M (eds) Termites: evolution, sociality, symbioses, ecology. Kluwer Academic Publ, Dordrecht, pp 77–93

To LP, Margulis L, Chase D, Nuttung WL (1980) The symbiotic microbial community of the sonoran desert termite: *Pterotermes occidentis*. BioSystems 13:109–137

Veivers PC, Musca AM, O'Brien RW, Slaytor M (1982) Digestive enzymes of the salivary glands and gut of *Mastotermes darwiniensis*. Insect Biochem 12:35–40

Veivers PC, O'Brien RW, Slaytor M (1983) Selective defaunation of *Mastotermes darwiniensis* and its effect on cellulose and starch metabolism. Insect Biochem 13:95–101

Viscogliosi E, Philippe H, Baroin A, Perasso R, Brugerolle G (1993) Phylogeny of trichomonads based on partial sequences of large subunit rRNA and on cladistic analysis of morphological data. J Eukaryot Microbiol 40:411–421

Wenzel M (1998) Untersuchungen der symbiotischen Bakterien von *Mixotricha paradoxa*, eines Flagellaten aus der Termite *Mastotermes darwiniensis*. Diploma Thesis, Johannes Gutenberg-University Mainz, Germany

Wenzel M, Radek R, Brugerolle G, König H (2003) Identification of the ectosymbiotic bacteria of *Mixotricha paradoxa* involved in movement symbiosis. Eur J Protistol 39:11–23

Wier A, Dolan M, Grimaldi D, Guerrero R, Wagensberg J, Margulis L (2002) Spirochete and protist symbionts of a termite (*Mastotermes electrodominicus*) in Miocene amber. Proc Natl Acad Sci USA 99:1410–1413

Wood TG, Sands WA (1978) The role of termites in ecosystems. In: Brian JV (ed) Production ecology of ants and termites. Cambridge Univ Press, Cambridge, pp 245–292

Yamin MA (1978) Axenic cultivation of the cellulolytic flagellate *Trichomitopsis termopsidis* (Cleveland) from the termite, *Zootermopsis*. J Protozool 25:535–538

Yamin MA (1979) Flagellates of the orders Trichomondida Kirby, Oxymonadida Grassé, and Hypermastigida Grassi & Foà reported from lower termites (Isoptera families Mastotermitidae, Kalotermitidae, Hodotermitidae, Termopsidae, Rhinotermitidae, and Serritermitidae) and from the wood-feeding roach *Cryptocercus* (Dictyoptera: Cryptocercidae). Sociobiology 4:4–119

Yamin MA (1980) Cellulose metabolism by the termite flagellate *Trichomitopsis termopsidis*. Appl Environ Microbiol 39:859–863

Yamin MA (1981) Cellulose metabolism by the flagellate *Trichonympha* from the termite is independent of endosymbiotic bacteria. Science 211:58–59

Extrusive Bacterial Ectosymbiosis of Ciliates

Giovanna Rosati

1 Introduction

The number of well-established associations between ciliates and bacteria, reported in the literature, increases steadily. The majority of these studies refer to endosymbiotic associations; yet, ectosymbiotic bacteria have been reported to occur in different ciliate species. Although these associations appear to be well established and are potentially important for both the bacteria and the ciliate host, an interaction has actually been demonstrated only for a few cases.

Ectosymbiotic sulfur-reducing bacteria occur in a number of marine anaerobic ciliates, some of which also harbour endosymbiotic methanogens. The ectosymbiotic bacteria have different morphologies even in ciliate species of the same genus (Fenchel et al. 1977; Fenchel and Finlay 1991). Ectosymbiotic bacteria have also been reported for two anaerobic freshwater ciliate species, but only for those specimens retrieved from sulphate-rich environments (Finlay et al. 1991; Esteban and Finlay 1994). The production of H_2 in hydrogenosomes of the ciliates is probably significant in maintaining the symbiotic associations. The ciliates, very likely, could exist without the symbionts but the symbionts enhance their growth rate and yield, maintaining a low level of hydrogen tension (Fenchel 1991).

The association with chemolithoautotrophic sulfide-oxidizing bacteria has proven to be vital for some ciliates that inhabit marine sulfidic environments. The case of the genus *Kentrophoros*, originally described by Raikov (1971), is now a classic: these ciliates are known to carry a layer of bacteria on their dorsal side. The bacteria are able to divide on the ciliate surface and the ciliate phagocytizes the bacteria through the entire dorsal side, thus depending on its ectosymbionts for food. More recently, a second association has been described, which consists of sulfur-oxidising bacteria and the peritrich ciliate *Zoothamnium niveum* (Bauer-Nebelsick et al. 1996). The latter is a sessile, colonial species, which is invariably covered by bacteria. Preliminary studies indicate that nutrition of the host occurs by utilization of low molecular weight

G. Rosati (e-mail: rosatig@deee.unipi.it)
Dipartimento Etologia, Ecologia, Evoluzione, Università di Pisa, Via A. Volta 4–6, 56126 Pisa, Italy

organic compounds produced by the symbionts, and also by digestion of the symbionts in trophic cells of the host colony (Rinke et al. 2003). Both *Kentrophorus* and *Zoothamnium* adopted behavioural traits that are used to optimise the physiological activity of their symbionts. Chemolithotrophic sulfide-oxidising bacteria require both sulphide and oxygen that coexist at only low concentrations in the natural environment. *Kentrophoros* is a microaerophile and shows a chemosensory response toward oxygen; in the natural environment, it moves vertically in sediments, always positioned just at the interface between oxidised and anaerobic sulfide-reducing layers (Fenchel and Finlay 1995). The feather-shaped colonies of *Zoothamnium* rhythmically expand and contract, thus exposing the ectosymbiotic bacteria alternatively to oxygenated water above and sulphidic water below the boundary layer on the surface of highly sulphidic peat (Ott et al. 1998).

The present study deals with an ectosymbiotic association which does not involve chemolithoautotrophic bacteria. Based on the current knowledge, the association appears unique with respect to at least two characteristics: 1) the peculiarity of the bacteria involved (referred to as epixenosomes) and 2) the kind of ecological advantage provided to the ciliate host.

2
The Ciliate Host

The eukaryotic partners of the consortium are marine, sand-dwelling, hypotrich ciliates. They have been assigned to the *Euplotidium* genus. Since Noland (1937) erected this genus, six species have been described. *Euplotidium itoi* and *E. arenarium* are the only two species characterised at the electron microscopical level.

A lateral view of *E. itoi* at SEM is shown in Fig. 1. The organism is oval in shape, 60–90 μm long and 40–52 μm wide. The somatic ciliature on the flat ventral surface consists of 12 frontoventral cirri, 6 transversal cirri and 1 left marginal cirrus. The huge buccal cavity is bordered by a well-developed series of membranelles. A prominent oral plate is present. On the dorsal convex surface are five rows of short cilia (dorsal bristles). In a depression of the body surface lies the band in which epixenosomes (from the ancient Greek *epi* = on, *xenos* = alien, *soma* = body) are inserted. The band extends along both sides and at the anterior end of the dorsal surface. *E. itoi* and *E. arenarium* are morphologically very similar; they only differ in the shape of the macronucleus (two elongated pieces in the former species and moniliform, i.e., like a string of beads in the latter) and the absence of a peristomial plate in *E. arenarium*.

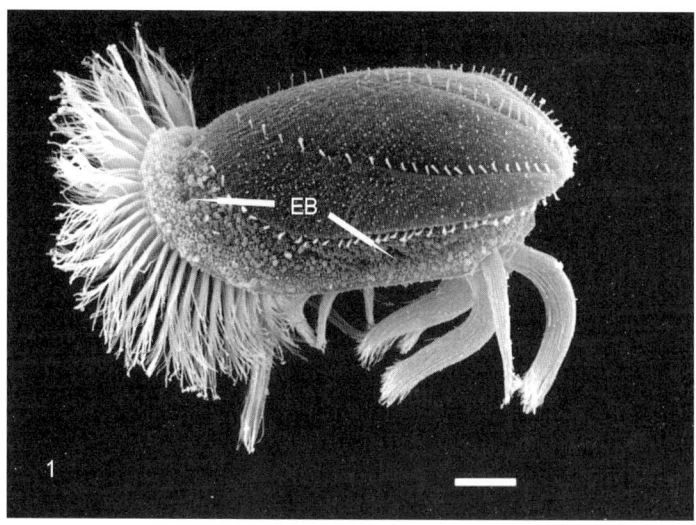

Fig. 1. SEM picture of *Euplotidium itoi*. Lateral view. The epixenosomal band (*EB*) is evident. *Bar* 10 µm

The specimens studied were repeatedly collected from a rocky shore near Leghorn (Ligurian Sea) in large pools, whose bottom was covered by a medium to coarse sand. Most of the pools are connected and linked to the open sea. In this restricted, well-characterised ecological niche, the two species coexist. Nevertheless, *E. itoi,* was more frequently found: in most collections, it was the only *Euplotidium* species present. Both species were subsequently grown over years in the laboratory in artificial seawater that was enriched with the flagellate *Dunaliella tertiolecta* and the diatom *Pheodactilum tricornutum* as a food source.

3
The Epixenososomal Band

The cortical depression in which the epixenosomes are inserted shows identical characteristics in *E. itoi* and *E. arenarium*. Their shape and localization correspond exactly to those of the structures reported by Ito (1958) in *E. itoi* and *Gastrocyrrus*, and by Tuffreau (1985) in *E. prosaltans*. At that time, the epixenosomes were interpreted as extrusomes. The latter studies were based only on optical microscopic observation; so it is possible that all of the above-mentioned structures are actually the same entity and that epixenosomes are present in different *Euplotidium* and *Gastrocirrus* species. This suggests that the association of ciliates with epixenosomes represents an ancient and well-established interaction.

4
Epixenosomes

4.1
Nature and Phylogenetic Affiliation

Analyses with an electron microscope clearly showed that epixenosomes are always external to the *Euplotidium* cortex. Since the epixenosome number is quite constant throughout the ciliate's life cycle, their number must increase prior to the binary fission of *Euplotidium*. These observations led us to interpret epixenosomes not as extrusomes but, rather, as epibionts. Initial evidence for a cellular nature of the epixenosomes was obtained by DAPI staining that showed the presence of DNA within epixenosomes (Verni and Rosati 1990). Subsequent in situ hybridization yielded inconsistent results (Rosati et al. 1998): epixenosomes were specifically labelled with three different eukaryotic probes while no signal was observed with three different prokaryotic probes, including the oligonucleotide EUB338, the sequence of which was considered to be complementary to all bacterial 16S rRNA sequences known at that time (Amann et al. 1991).

A detailed molecular study was then carried out in order to reconstruct the phylogenetic affiliation of *E. arenarium* epixenosomes (Petroni et al. 2000) based on comparative sequence analysis of genes coding for the small subunit of rRNA (SSU rRNA). Due to the unexpected, ambiguous results mentioned above, different oligonucleotide primers were used for in vitro amplification of the extracted DNA. A distinct amplification product that was suitable for cloning and direct sequencing was obtained employing one certain combination of bacterial with archeal primers. An oligonucleotide probe was designed to specifically target the 16S rRNA sequence from the *E. arenarium* culture. In situ hybridization with this probe proved that the retrieved 16S rRNA gene sequence in fact originated from epixenosomes (Fig. 3). Comparison of the sequence with the 16SrRNA sequences available in the ARB database (Ludwig and Strunk 1997) revealed that epixenosomes are bacteria phylogenetically related to Verrucomicrobia (Petroni et al. 2000). The latter represents a recently defined bacterial division (Hedlund et al. 1996) that comprises only a few cultivated, diverse species. Among them, members of the genus *Xiphinemobacter* are endosymbionts of nematodes (Vandekerckhove et al. 2000). However, the majority of the members of this division are so far only recognised by their 16rRNA gene sequences that were recovered as cultivation-independent clones from a variety of habitats. The closest relatives of epixenosomes appear to be two uncultivated marine clones (Fig. 2) and, among the cultivated species, free living bacteria of the anaerobic genus *Opitutus* (Chin et al. 2001).

Extrusive Bacterial Ectosymbiosis of Ciliates

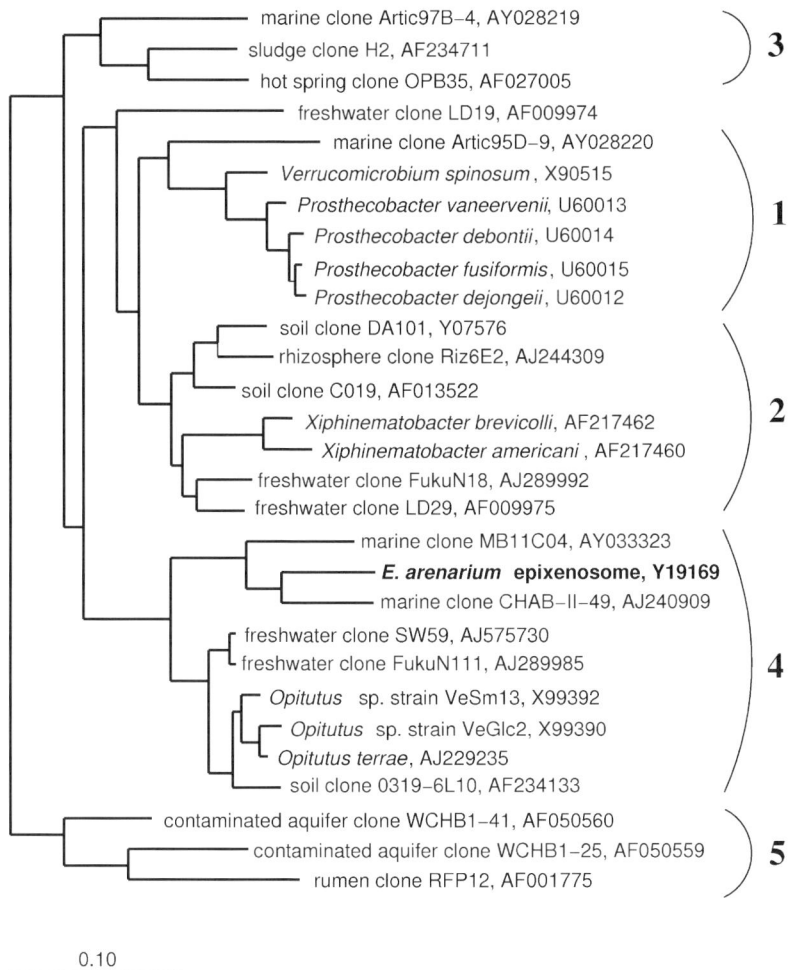

Fig. 2. Maximum likelihood tree showing the phylogenetic position of epixenosomes inferred from a selection of almost full 16SrRNA gene sequences of verrucomicrobia. A filter that retains only 1,349 positions conserved in at least 50% verrucomicrobia was used. The phylogenetic position of epixenosomes is shown. The habitat source of the environmental sequences is indicated before the clone names and the sequence accession numbers. Subdivisions are indicated at the right of the tree. *Bar* 10 nucleotide substitutions per 100 nucleotides

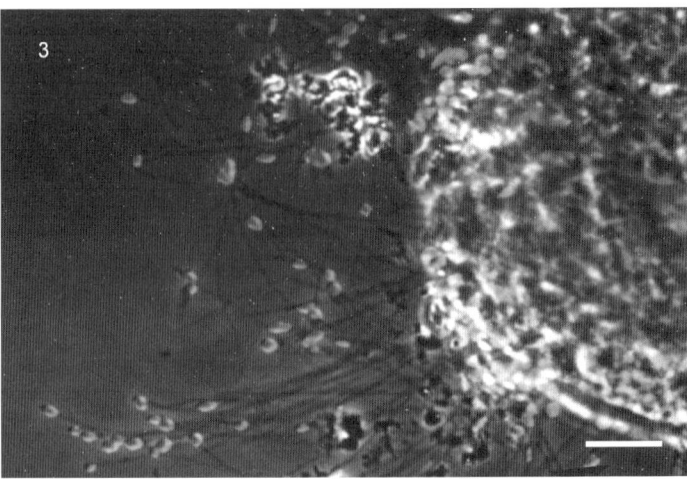

Fig. 3. Detail of ejected extrusive apparatuses in phase contrast superimposed with the fluorescent signal of epixenosome specific probe

Like other Verrucomicrobiales (Daims et al. 1999), the 16S rRNA gene sequence of epixenosomes has two base exchanges as compared to 16S rRNA gene sequences of other Bacteria that prevent the binding of the bacterial probe EUB338. This could also explain why, in previous experiments of in situ hybridization, oligonucleotide probes, usually used for bacterial detection, did not react with epixenosomes. Based on the data obtained with the specific probes, the positive results obtained with eukaryotic probes appear to represent artefacts, probably caused by unspecific binding.

4.2
Morphological Characteristics

Epixenosomes are present in two different forms. Form I, mainly localised in the central region of the epixenosomal band, are spherical, 0.5 μm in diameter and are surrounded by two membranes enclosing a granular matrix in which a clear central zone, containing thin filaments, can be distinguished. They are able to divide by binary fission (Fig. 4A). Form II epixenosomes (Fig. 4B) are mainly localised in the peripheral regions of the band, egg-shaped (2.2 μm in length and 1 μm in width), unable to divide and have a highly complex cytoplasmic organization compared to other prokaryotes. The following well-defined structures are present (Fig. 4B): 1) a dense, apical dome-shaped zone, 2) an inclusion body, 3) an extrusive apparatus, 4) a basket of tubules. This complex structure is gradually acquired following in a well-defined order (Rosati et al. 1993a) during the transformation of form I to form II. The appearance of the *EA* in the form of a few concentric layers immersed in the cytoplasm is the first sign of the transformation of form I epixenosomes to form II. Already at this early stage the cytoplasm surrounding the lamellar material is to some extent differentiated as preferential digestion with protease can be observed at its level. Successively, the structures become more complex: the concentric layers, which are tightly wound around a central core, increase in number and randomly dispersed tubule-like structures can be seen in the cytoplasm. In a more advanced step, dense material in the apical region of the organism can be observed in addition to the above-mentioned structures. The tubule-like structures become ordered in bundles. Finally, the epixenosomes become oval in shape and acquire the typical structure of form II. Cytochemical studies have revealed a precise localization of some enzymes. For example, active acid phosphatase, an enzyme cytosolic in prokaryotes but contained in membranous organelles in eukaryotes, in epixenosomes appears to be localised only in the space between the two external membranes, while endogenous peroxidase is only present in the inclusion body and at the top of extrusive apparatus. These data indicate that a functional cell compartmentalization corresponds to the structural complexity, in spite of the absence of true internal membranes (Rosati et al. 1996).

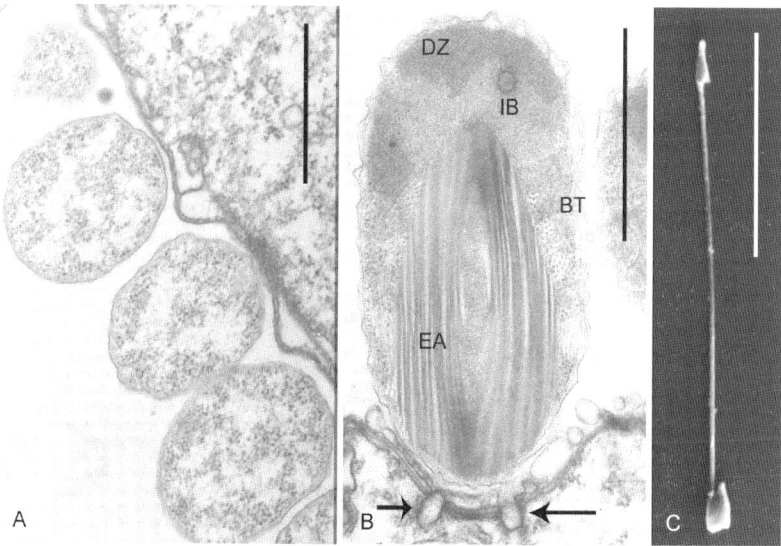

Fig. 4A–C. Epixenosomes at different stages. **A** Type I epixenosomes (from Verni and Rosati 1990); **B** Type II epixenosome (from Rosati et al. 1996). *BT* basket tubules; *DZ* dome-shaped zone; *EA* extrusive apparatus; *IB* inclusion body. *Arrow* points to a parasomal-like vesicle in the *Euplotidium* cortex. *Bars* 1 µm. **C** An ejecting epixenosome at SEM. *Bar* 5 µm

4.3
The Dome-Shaped Zone and Inclusion Body

The upper region of epixenosomes form II, just beneath the membrane, is occupied by a granular dense zone, referred to as dome-shaped zone (*DZ*) for its shape. It is about 0.9 µm thick with an apical hole of 150 nm. It contains DNA and basic proteins (Verni and Rosati 1990; Rosati et al. 1996). Its appearance is reminiscent of that of eukaryotic heterochromatin (Fig. 4B).

A round inclusion body (*RB*) with a rather constant diameter (200 nm) is localised under the *DZ* (Fig. 4B). The density is variable, the inclusion body appears empty after in vivo treatments which interfere with polysaccharide metabolism or after inhibition of protein synthesis. Both effects are reversible. These results indicate that *RB* contains polysaccharides (presumably storage polysaccharides) and enzymes, in particular endogenous peroxidases. After inhibition, the round inclusion body maintains its shape and size and appears surrounded by a thin layer. The latter can be visualised by staining procedures targeting glycosylated substances. These properties distinguish the structure from the storage granules common in prokaryotes and eukaryotes (Rosati et al. 1996).

4.4
The Extrusive Apparatus

The extrusive apparatus (*EA*) occupies most of the basal region of the epixenosome and extends toward the upper region (Fig. 4B). It has been compared to R-bodies described in a few bacterial species of different eubacterial genera. R-bodies are highly insoluble protein ribbons, typically coiled into cylindrical structures within the cells. R-bodies have been studied most intensively in bacteria of the genus *Caedibacter*, which are obligate symbionts of paramecia to which they confer the killer trait (Preer and Preer 1982). The genus has now been recognised as a polyphyletic group (Beier et al. 2002). Among free-living bacteria, R-bodies have been reported in the hydrogen-oxidising soil species *Pseudomonas teniospiralis* (Lalucat et al. 1982), the plant pathogen *P. avenae* (Wells and Horne 1983), the melanogenic marine bacterium *Marinomonas mediterranea* (Hernández-Romero et al. 2003), and the anoxygenic phototroph *Rhodocista centenaria* (Favinger et al. 1989). Several classes of R-bodies have been recognised on the basis of physical dimensions, morphology, and the effect of specific physical and chemical treatments. For example, lowered pH, to below 6.5, is apparently required for types 51 (derived from *Caedibacter taeniospiralis* symbiont of *Paramecium tetraurelia*) and Pa (synthesised in the free-living bacterium *P. avenae*) R bodies to unroll. Type 7 and Pt R bodies (synthesised by a symbiont and a free-living bacterium respectively) do not respond to changes in pH but can made to unroll by incubation for about 10 min at 70°C (Pond et al. 1989). They also differ in the way of uncoiling. *Marinomonas* R-bodies show an outer envelope, which has not been observed in other bacteria (Hernádez-Romero et al. 2003). In most cases, it has been demonstrated that the synthesis of the R-body is determined by extrachromosomal elements that are either phages or plasmids.

Many data point to the conclusion that although different classes of R-bodies exist, there is evolutionary homology among *Caedibacter* R-bodies. The relationship between the two *Pseudomonas* R-bodies and between *Pseudomonas* and *Caedibacter* R-bodies is less certain (Pond et al. 1989). Quackenbush and Burbach (1983) and Heruth et al. (1994) determined the nucleotide sequence of the genetic determinants for type 51 R-body synthesis and assembly for different strains of *C. teniospiralis*. Three independently transcribed genes were characterised. The R body-encoding sequences from both strains are identical. This, according to the authors, suggests that it is possible that R-bodies' structure and function requires an extremely high level of sequence conservation. It is now clear that the presence of R-bodies alone is not a phylogenetically meaningful taxonomic marker. R-body production may either be the result of convergent evolution (within different evolutionary lineages of the *Proteobacteria*) or of a single evolutionary event. In the latter case, this trait was passed to by horizontal gene transfer. Other structures have been described that resemble R-bodies: the cyanobacterial inclusions referred to as scroll-like membrane system (Jensen 1984) and ejectisomes, the extrusive organelles found in two flagellate families (Hausmann 1978; Kugrens et al. 1994), i.e., true cellular organelles in eukaryotic organisms.

The *EA* of epixenosomes (Fig. 4B) displays a far more complex ultrastructure in comparison with R-bodies. It consists of a ribbon rolled up around a granular central core (150 nm in diameter) containing a bundle of fibrils (20 nm thick and 0.8 µm long) in its upper region. The height of the ribbon decreases, both apically and distally, starting from the innermost to the outermost layer, thereby resulting in an oval shape of the *EA*. The matrix in which the *EA* is immersed can be selectively digested with proteases, which causes an unordered appearance of the *EA* layers. This result indicates that the matrix is proteinaceous and somehow different from the surrounding cytoplasm (Rosati et al. 1993a). Moreover it is separated from both the cytoplasm and the central core by thin layers in which active adenylate cyclase, a typical membrane bound enzyme in both prokaryotes and eukaryotes, has been detected (Rosati et al. 1996). The chemical nature of the *EA* ribbon itself is still unclear. It has a specific affinity to phosphotungstic acid, a substance used to stain acidic proteins or mucoproteins in cytochemical tests. Bacteriophages or capsomere-like structures have never been observed.

Further investigations are required to determine whether the synthesis of the whole structure is related to extrachromosomal elements and to evaluate the genetic relationships with the different R-bodies and/or the eukaryotic ejectisomes.

4.5
The Basket Tubules

Bundles of regularly arranged tubules, differently oriented with respect to each other, form a basket-like structure around the EA (Fig. 4B). The inner and outer diameter of the tubules is 20–24 nm and 12–14 nm, respectively, and their wall consists of globular structures. These dimensions and structural characteristics, revealed by different preparative techniques, are very similar to that of tubulin-based microtubules. *BT* share additional characteristics with tubulin microtubules: they are particularly well preserved by fixation with tannic acid, a substance generally used for the maintenance of tubulin microtubules. Furthermore, they are sensible to cold temperature (4°C) and antitubulin drugs such as nocodazole, and a positive immunoreaction was obtained at the *BT* level with three different antibodies all specific for tubulin. The immunocytochemical analysis was carried out by both fluorescence and electron microscopy (Rosati et al. 1993b).

Tubulin microtubules occur in most eukaryotes. Their acquisition very likely represented an important step in the evolution of the eukaryotic cell by facilitating the engulfment of bacterial endosymbionts, the ancestors of chloroplasts and mitochondria. Cytoplasmic tubules and fibrous structures within the size range of tubulin tubules and tubulin protofilament have been detected in prokaryotes. Whether they are homologous to eukaryotic tubulin microtubules has yet to be established (Bermudes et al. 1994). The leading candidate for an evolutionary precursor of tubulin in the bacterial and archeal domains is the cell division protein FtsZ (Erickson 1995). However, although FtsZ is able to form tubules in vitro (Bramhill and Thompson 1994), the formation of stable tubules has never been described in vivo. Moreover, tubulins and FtsZ share only very few sequence identity (de Boer et al. 1992).

Interestingly enough, two tubulin genes, bacterial tubulin a(*btuba*) and bacterial tubulin b (*btubb*) have been detected and sequenced in bacteria of the genus *Prosthecobacter* (Jenkins et al. 2002). Like epixenosomes, bacteria of this genus belong to the division Verrucomicrobia, but are assigned to a different subdivision than the former (Fig. 2). Despite extensive examination of cell sections, the authors were unable to locate any microtubule-like structures in *Prosthecobacter*. On the basis of comparative modelling data, the *Prosthecobacter* tubulins are predicted to be monomeric, unlike eukaryotic α and β tubulins. The latter are known to interact and to form dimers, which then polymerise during microtubule formation. Based on their monomeric nature, the bacterial tubulins are unlikely to form microtubule-like structures. Phylogenetic analyses indicate that *Prosthecobacter* tubulins are quite divergent and were not recently transferred horizontally from a eukaryote (Jenkins et al. 2002).

In contrast to *Prosthecobacter*, epixenosomes have well-organised cytoplasmic tubules that resemble eukaryotic tubulin microtubules more than any other bacterial tubular structure described so far. This is supported by the results of an in situ hybridization when the gene encoding for β tubulin in the ciliate *Euplotes crassus* was used as a probe. Two clusters of gold-particles, revealing that hybridisation took place, were repeatedly found at the *DZ* where the epixenosome genetic material is localised. The only other structure that was decorated by this method was some chromatin in the *Euplotidium* macronucleus (Rosati et al. 1998). Future sequencing of the alleged tubulin genes will reveal whether epixenosomes have true tubulins and determine whether these molecules are more closely related to the divergent forms of *Prosthecobacter* or to eukaryotic forms. The existence of tubulin genes in epixenosomes is presently under investigation but is complicated by the fact that epixenosomes cannot easily be separated from the host and the latter, being a ciliated protozoon, possesses several tubulin genes not yet sequenced.

5
Can *Euplotidium* and Epixenosomes Survive on their Own?

During the 15 years of studying the association, epixenosomes have been found in every specimen of *Euplotidium* when examined directly after collection. Secondly, the association can be indefinitely maintained in well-fed lab cultures. Thirdly, the host cell cycle is perfectly coordinated with the multiplication and the differentiation of epixenosomes from form I to form II (Giambelluca and Rosati 1996). This cumulative evidence strongly indicates that the association between *Euplotidium* and epixenosomes is obligatory at least in the natural environment. However, when *Euplotidium* stops dividing (for example during starvation or when the protein synthesis is artificially inhibited) epixenosomes stop division, too; under these conditions all form I epixenosomes change into form II and are gradually lost. Once conditions for growth are re-established, most of epixenosome-free euplotidia recover their reproductive capacity. In this way, strains without epixenosomes were obtained experimentally for both *E. itoi* and *E. arenarium*. The resulting strains reproduce and behave similar to the epixenosome-bearing strains. Thus, epixenosomes appear not to be vital for the survival of their host under laboratory conditions. On the contrary, the association may be vital for epixenosomes since attempts to grow the episymbionts in artificial media for different bacteria failed. Although the failure to cultivate epixenosomes may be attributed to other factors, like a peculiar metabolism, preliminary results from molecular studies indicate that epixenosomes could indeed be obligate symbionts. For example, differently from free-living Verrucomicrobia, they have only one ribosomal operon with 16S and 23S rRNA genes unlinked. This trait is characteristic of obligate symbiotic prokaryotes like members of the genus *Rickettsia* (Anderson et al. 1999) and is indicative of a reduced efficiency of ribosome synthesis.

6
The Ejection Mechanism and its Significance

The *EA* is a very efficiently engineered structure. In its resting position, its organisation is compact, fitting well into the intact epixenosome. During the ejection, it unrolls starting from inside, whereby the layers slip relative to each other. Thus, a tube is formed which fits exactly through the hole of the *DZ*. The tube penetrates the hole, and reaches the opening of the cell membrane formed as the first step of the ejecting process. As a result, the apical portion of the epixenosome cytoplasm containing the DZ and the IB is translocated away from the ciliate host. At the end of the process, the tube is 40 µm long, and 150 nm in diameter with a spear-shaped head of 2 µm length (Fig. 4C). Since the tube consists of overlapping layers, it remains continuous. The head is devoid of a membrane and encloses the terminal open portion of the tube. The internal fibrils represent the very tip of the whole extruded structure. Some *BT* are often found associated with the ejected tube (Rosati et al. 1993a). To date, no information is available concerning the chemical nature of the tube wall. However, the tubular shape, together with the ratio between length and diameter, render the tube much more resistant than a completely unwound flat ribbon, independent of its chemical nature. The ejectisomes of the chloromonadine *Pyramimonas* are the most similar stuctures as they also "form a tube by a spiral rolling of the ribbon" (Kugrens et al. 1994).

Ejectisomes routinely discharge when the water is disturbed in the vicinity of the cells. It has been experimentally demonstrated (Rosati et al. 1997) that the ejecting process in epixenosomes is activated by the detection of external signals through membrane receptors and involve the activation of the adenylate cyclase-cyclic AMP system as a transducing mechanism. The membrane receptors are localised at the top of the epixenosomes, just where the membrane interruption appears as a first step of the whole extrusion process. Membrane receptors have been detected based on their affinity for soybean agglutinin, a lectin that inhibits the ejection in live epixenosomes. On the contrary, ejection was stimulated (and the binding of soybean agglutinin prevented) in the presence of adrenalin. The latter is known to bind to the receptors with a high affinity for the lectin. The presence of receptors for mammalian hormones has already been reported in prokaryotes (Lenard

1992). In addition, the ejection of the epixenosomes is also inhibited by treatments with antitubulin drugs like nocodazole. This has led to the hypothesis that *BT* are involved in the ejection process (Rosati et al. 1993b).

Which stimuli trigger the ejection in the natural environment? What is the significance of a process in which the genetic material of bacterial epibionts is flung out of the remnant of their cell body and away from the host surface? As a massive ejection was always observed when the ciliate host is dying, it may be hypothesised that under these conditions *Euplotidium* produces the molecule(s) which stimulate the activation of symbiont ejection. In this case, the ejection would have a dispersive function similar to that of the polar filament in microsporidia spores. However, the reinfection of the epixenosome-free cultures of *Euplotidium* was unsuccessful. Under natural conditions, ejection may also serve the purpose of regulating the number of epixenosomes during division of the ciliate host. In this case, the ejection would be activated by substances produced by the host during a specific phase of its cell cycle. On the other hand, the apical position of the specific receptors, as well as the observation that the extrusive process can be stimulated or inhibited by the addition of different compounds to the *Euplotidium* culture medium, may indicate that stimuli are detected from the environment. A third hypothesis, still largely unexplored, is that the synthesis of the extrusive apparatus, like that of some R-bodies, could be determined by viral elements. In this case, the ejection would function as a dispersive tool for the latter.

7
Why Does *Euplotidium* Maintain its Epixenosomes?

As described above, the association between *Euplotidium* and epixenosomes appears to be obligatory in the natural environment. In a non-competitive environment, however, the ciliate survives well without its symbionts. Epixenosomes may therefore provide their host with an important ecological advantage, such as defence against predators. This hypothesis was experimentally verified by comparing the behaviour of *Litonotus lamella* when preying upon *E. itoi* with and without epixenosomes. *L. lamella*, a raptorial feeding ciliate that shares its habitat with *E. itoi*, was chosen because its feeding behaviour is well known (Ricci and Verni 1988; Ricci et al. 1996). It has been observed (Rosati et al. 1999) that, upon direct cell-to-cell contact, *L. lamella* discharges its toxicysts and thereby paralyses *E. itoi* both in the presence as well as the absence of epixenosomes. However, while *L. lamella* is able to ingest prey cells without epixenosomes, it never eats those with epibionts. It can be inferred that the presence of epixenosomes prevents the engulfment of euplotidia stricken by the predator, leading to a high probability of survival.

Indeed, about 60% of these prey cells are capable of resuming a normal behaviour.

How can epixenosomes mediate this defensive function? The possible role of the epixenosomal *EA* in defence was tested. For this purpose, the feeding behaviour of *L. lamella* against *E. itoi* with unaltered epixensomes and against *E. itoi* with epixenosomes whose ejecting capability was inhibited were compared (Rosati et al. 1997). Ejection was prevented by a treatment with the adenylate cyclase inhibitor alloxan. While, as expected, *L. lamella* did not ingest untreated euplotidia, it consumed some of those treated. Thus the ejection itself, rather than the mere presence of epixenosomes is critical for the defence. It can be hypothesised that the discharge of the predator toxicysts triggers the ejection and that this process interferes with the recognition of the prey by the predator, which is necessary for finding and engulfing the stricken prey (Rosati et al. 1999). Toxic substances are unlikely to be involved, however, since neither predators nor other ciliate species put in the same container with *Euplotidium* appear to be inhibited.

The never-ending predator–prey struggle has certainly provided a powerful stimulus for evolution also in the microbial world. Indeed several types of anti-predator defences have been reported for ciliates. It has been demonstrated that trichocyst discharge of *Paramecium* quickly dislocates the cell out of the range of the offensive organelles of an attacking predator (Miyake and Harumoto 1996). A similar function has been supposed for flagellate ejectisomes (Kugrens et al. 1994). The extrusive process of epixenosomes is involved in the host defence, although, epixenosome ejection does not dislocate *Euplotidium*.

As far as we know, the only case in ciliates in which protection against predation is provided by symbionts is that of *Paramecium bursaria* and its endosymbiotic zoochlorellae. The latter organisms in some way discourage predation by *Didinium nasutum* by releasing distasteful metabolites that repel the latter (Berger 1980). The defensive role of episymbiotic organisms described here is presently the only one reported for ciliates. The defensive function could account for the absence of *E. itoi* and *E. arenarium* cells devoid of epixenosomes in the tide pools from which all our experimental organisms have been collected, and very likely is an important factor for stabilising and maintaining such a specialised symbiotic relationship.

8
Concluding Remarks and Future Perspectives

The study of symbiotic associations between ciliates and prokaryotes is of interest from ecological and the evolutionary points of view. Especially, the ectosymbiotic relationship between ciliates of *Euplotidium* spp. and epixenosomes offers novel cues for studying both these aspects since apparently both partners benefit from living together. For epixenosomes, the association might be even vital, whilst for the ciliate it represents an ecological advantage, a powerful defensive tool in the struggle for survival.

The symbiosis is well established and specialised. The specificity of the association is also exemplified by the fact that epixenosomes are located in depressions of the host cortex which exactly match their size, but which do not form in host cells in the absence of the symbionts. Although the epixenosomes maybe functionally and genetically less integrated than intracellular symbionts, a signal exchange with the ciliate host cell is very likely to exist. The ciliate cortex typically widens to host the increased numbers of epibionts at the epixenosomal band. The higher numbers of epibionts occur prior to binary fission of the host cell. This widening and narrowing of the cortical region corresponding to the epixenosomal band does not occur if epixenosomes are not present. Subsequently, the distribution of epixenosomes between the two daughter host cells, the transformation of epixenosomes from form I to form II, and the recovery of the typical non-dividing features of the cortical band all proceed in an order that is precisely synchronised with the stages of *Euplotidium* morphogenesis (Giambelluca and Rosati 1996). Obviously, *Euplotidium* is capable of perceiving the presence of epixenosomes and epixenosomes recognise the particular stages of *Euplotidium* life cycle.

So far, the molecular mechanisms of the signal exchange between *Euplotidium* and its epixenosomes remain obscure. Membrane bound vesicles (80–100 nm in diameter) have often been observed as emerging from (or fusing with) the cell membrane of epixenosomes, or are attached to *Euplotidium* cell membrane in the epixenosomal band (Rosati 1999). It can be supposed that these vesicles are formed by evagination of the cell membrane of one organism and fuse with the cell membrane of the other. Whether the vesicles emerge from *Euplotidium* or from epixenosomes or from both organisms is not known. These vesicles may provide a direct passage of substances between the two associated organisms.

Furthermore, the molecular basis of the extremely stable cell-cell-contact between the two organisms are still unknown. A fusion between the plasma membranes of the two organisms has never been observed. In the narrow clear space that always separates the two organisms, specific substances have never been detected. Yet, we never succeeded in quantitatively separating epixenosomes from their host by means of mechanical, chemical or physical treatments. The only feature which differentiates those regions of the *Euplotidium* cortex where epixenosomes are inserted from the other portions of the cortex is the presence, roughly in the centre of each depression, of at least one coated vesicle containing granular material. These vesicles, passing through the alveolar system, subsequently fuse with the plasma membrane and open towards the exterior, i.e., toward the epixenosome located on top (they are visible in Fig. 4B) and indicated as parasomal sac-vesicle. The vesicles are permanent structures and are also present in specimens devoid of epixenosomes in which the cortical depressions are no more visible. The features of these vesicles closely recall those of parasomal sacs, which in ciliates are always associated to ciliary structures and are considered sites of exocytosis and/or pinocytosis. The fact that parasomal sac-like vesicles in *Euplotidium* (at least *E. itoi* and *E. arenarium* the only species analysed at the ultrastructural level) has been observed at the level of the epixenosomal band, not associated to ciliary structures, led us to hypothesise that these structures are relevant for symbiosis. These vesicles might also have represented a prerequisite for establishing the symbiosis. The ultrastructural analysis of other *Euplotidium* species could provide more information in this respect.

Epixenosomes by themselves are puzzling for their highly complex cell compartmentalisation and microtubular-like structures, which are both unusual features in prokaryotes. Moreover, there are indications for the presence of actin (unpubl. data). The molecular characterisation of a group of epixenosome genes coding for these eukaryotic-like features will be required to assess whether these genes are ancestral (as tubulin-like genes in *Prosthecobacter* are interpreted), true eukaryotic homologues, or have been secondarily acquired through horizontal gene transfer from the host. The results will have general implications for the understanding of microbial evolution.

Acknowledgements. The author is indebted to all those who have been involved in this study during the years. Particular thanks are due to K.H. Schleifer group, F. Verni, G. Petroni and L. Modeo who are still working on epixenosomes with me. I would also like to mention J. Staley and C. Jenkins for their cooperation and L. Margulis for her enthusiastic interest in epixenosomes.

References

Amann RI, Springer N, Ludwig W, Görtz HD, Schleifer KH (1991) Identification in situ and phylogeny of uncultured bacterial endosymbionts. Nature 351:161–164

Andersson SGE, Stothard DR, Fuerst P, Kurland C (1999) Molecular phylogeny and rearrangement of rRNA genes in *Rickettsia* species. Mol Biol Evol 16:987–995

Bauer-Nebelsick M, Bardele CF, Ott J (1996) Electron microscopic studies of *Zoothamnium niveum* (Hemprich and Erhemberg 1831) Ehremberg 1838 (Oligohymenophorea, Peritrichida), a ciliate with ectosymbiotic, chemioautotrophic bacteria. Eur J Protistol 32:202–215

Beier CL, Horn M, Michel R, Schweikert M, Görtz HD, Wagner M (2002) The genus *Caedibacter* comprises endosymbionts of *Paramecium* spp. related to the

Rickettsiales (alphaproteobacteria) and to *Francisella tularensis* (gammaproteobacteria). Appl Environ Microbiol 68:6043–6050

Berger J (1980) Feeding behavior of *Didinium nasutum* on *Paramecium bursaria* with normal or apochlorotic zooclorellae. J Gen Microbiol 118:397–404

Bermudes D, Hinkle G, Margulis L (1994) Do prokaryotes contain microtubules? Microbiol Rev 58:387–400

Bramhill D, Thompson CM (1994) GTP-dependent polymerization of *Escherichia coli* FtsZ protein to form tubules. Proc Natl Acad Sci USA 91:5813–5817

de Boer P, Crossley R, Rothfield L (1992) The essential bacterial cell-division protein FtsZ is a GTPase. Nature 359:254–256

Chin K-J, Liesack W, Janssen H (2001) *Opitutus terrae* gen nov., sp. nov., to accommodate novel strains of the division "Verrucomicrobia" isolated from rice paddy soil. Int J Syst Evol Microbiol 51:1965–1968

Daims H, Brühl A, Amann R., Schleifer K-H, Wagner M (1999) The domain-specific probe EUB338 is insufficient for detection of all Bacteria: development and evaluation of a more comprehensive probe set. Syst Appl Microbiol 23:434–444

Erickson HP (1995) FtsZ, a prokaryotic homolog of tubulin? Cell 80:367–370

Esteban G, Finlay BJ (1994). A new genus of anaerobic scuticociliate with endosymbiotic methanogens and ectobiotic bacteria. Arch Protistenkd 144:350–356

Favinger J, Stackebrandt R, Gest H (1989) *Rhodospirillum centenum* sp. nov., a thermotolerant cyst-forming anoxygenic photosynthetic bacterium. Antonie van Leeuwenhoek 55:291–296

Fenchel T, Finlay BJ (1991) The biology of free-living anaerobic ciliates. Eur J Protistol 26:200–216

Fenchel T, Finlay JB (1995) Symbiosis with prokaryotes: intracellular syntrophy. In: May RM, Harvey PH (eds) Ecology and evolution in the anoxic worlds. Oxford Univ Press, Oxford, pp 135–161

Fenchel T, Perry T, Thane A (1977) Anaerobiosis and symbiosis with bacteria in free-living ciliates. J Protozool 24:154–163

Finlay JB, Clarke KJ, Vicente E, Miracle MR (1991) Anaerobic ciliates from sulphide-rich solution lake in Spain. Eur J Protistol 27:148–159

Giambelluca MA, Rosati G (1996) Behavior of epixenosomes and epixenosomal band during divisional morphogenesis in *Euplotidium itoi* (Ciliata, Hypotrichida). Eur J Protistol 32:77–80

Hausmann K (1978) Extrusive organelles in Protists. Int Rev Cytol 52:197–276

Hedlund BP, Gosink JJ, Staley JT (1996) Phylogeny of *Prosthecobacter*, the fusiform Caulobacters: members of a recently discovered division of the bacteria. Int J Syst Bacteriol 46:960–966

Hernández-Romero D, Lucas-Elio P, López-Serrano D, Solano F, Sanchez-Amat A (2003) *Marinomonas mediterranea* is a lysogenic bacterium that synthesizes R-bodies. Microbiology UK 149:2679–2686

Heruth DP, Pond FR, Dilts JA, Quackenbush RL (1994) Characterization of genetic determinants for R body synthesis and assembly in *Caedubacter taeniospiralis* 47 and 116. J Bacteriol 176:3559–3567

Ito S (1958) Two new species of marine ciliate. *Euplotidium itoi* sp. nov. and *Gastrocirrhus* sp nov. Zool Mag Tokyo 67:184–187

Jenkins C, Samudrala R, Anderson I, Hedlund BP, Petroni G, Michailova N, Pinel N, Overbeek R, Rosati G, Staley JT (2002) Genes for the cytoskeletal protein tubulin in the bacterial genus *Prosthecobacter*. Proc Nat Acad Sci USA 99:17049–17054

Jensen TE (1984) Cyanobacterial cell inclusions of irregular occurrence: systematic and evolutionary implications. Cytobiosis 39:35–62

Kugrens P, Lee RE, Corliss JO (1994) Ultrastructure, biogenesis and functions of extrusive organelles in selected non-ciliate protists. Protoplasma 181:164–190

Lalucat J, Pares R, Schlegel HG (1982) *Pseudomonas taeniospiralis* sp. nov., an R-body-containing hydrogen bacterium. Int J Syst Bacteriol 32:232–238

Lenard J (1992) Mammalian hormones in microbial cells. Trends Biochem Sci 17:147–150

Ludwig W, Strunk O (1997) ARB: a software environment for sequence data (http://www.mikro.biologie.tu.muenchen de/pub/ARB/documentation/arb/ps)

Miyake A, Harumoto T (1996) Defensive function of trichocysts in *Paramecium* against the predatory ciliate *Monodoinium balbiani*. Eur J Protistol 32:128–123

Noland LE (1937) Observation on marine ciliates of the gulf coast of Florida. Trans Am Microsc Soc 56:160–171

Ott JA, Bright M, Schiemer F (1998) The ecology of a novel symbiosis between a marine peritrich ciliate and chemoautotrophic bacteria. Mar Ecol 19:229–243

Petroni G, Spring S, Schleifer K-H, Verni F, Rosati G (2000) Defensive extrusive ectosymbionts of *Euplotidium* (Ciliophora) that contain microtubule-like structures are bacteria related to *Verrucomicrobia*. Proc Natl Acad Sci USA 97:1813–1817

Pond FR, Gibson I, Lalucat G, Quackembush RL (1989) R-body-producing bacteria. Microbiol Re 53:25–67

Preer JR, Preer LB, Jurand A (1974) Kappa and other endosymbionts in *Paramecium aurelia*. Bacteriol Rev 38:113–163

Preer Jr JR, Preer LB (1982) Revival of names of protozooan endosymbionts and proposal of *Holospora caryophila* nom nov. Int J Syst Bacteriol 32:140–141

Quackembush RL, Burbach J (1983) Cloning and expression of DNA sequence associated with the killer trait of *Paramecium tetraurelia* sock 47. Proc Natl Acad Sci USA 80:250–254

Raikov IB (1971) Bactéries épizoiques et mode de nutrition du cilié psammophile *Kentrophorus fistulosum* Fauré-Fremiet (étude au microscope electronique). Protistologica 7:365–378

Ricci N, Verni F (1988) Motor and predatory behavior of *Litonotus lamella* (Protozoa, Ciliata). Can J Zool 66:1973–1981

Ricci N, Morelli A, Verni F (1996) The predation of *Litonotus* on *Euplotes*: a two-step cell-cell recognition process. Acta Protozool 35:201–208

Rinke C, Ott JA, Bright M (2003) Fatal picnic on a high speed elevator: nutrition in the chemoautotrophic *Zoothamnium niveum* symbiosis. IV Int Congress of ISS Halifax, p 83

Rosati G (1999) Epixenosomes: symbionts of the hypotrich ciliate *Euplotidium itoi*. Symbiosis 26:1–23

Rosati G, Verni F, Lenzi P (1993a) "Epixenosomes": peculiar epibionts of the ciliate *Euplotidium itoi*. The formation of the extrusive apparatus and the ejecting mechanism. Eur J Protistol 29:238–245

Rosati G, Lenzi P, Verni F (1993b) "Epixenosomes": peculiar epibionts of the ciliate *Euplotidium itoi*: do their cytoplasmic tubules consist of tubulin? Micron 24:465–471

Rosati G, Giambelluca MA, Taiti E (1996) Epixenosomes: peculiar epibionts of the ciliate *Euplotidium itoi*: morphological and functional cell compartmentalization. Tissue Cell 28:313–320

Rosati G, Giambelluca MA, Grossi M, Morelli A (1997) "Epixenosomes": peculiar epibionts of the ciliate *Euplotidium itoi*: involvement of membrane receptors and the adenylate-cyclase cyclic-AMP system in the ejecting process. Protoplasma 147:57–63

Rosati G, Verni F, Lenzi P, Giambelluca MA, Sironi M, Bandi C (1998) Epixenosomes, peculiar epibionts of the ciliated protozoon *Euplotidium itoi*: what kind of organisms are they? Protoplasma 201:38–44

Rosati G, Petroni G, Quochi S, Modeo L, Verni F (1999) Epixenosomes: peculiar epibionts of the hypotrich ciliate *Euplotidium itoi* defend their host against predators. J Eukaryot Microbiol 46:278–282

Tuffrau M (1985) Une nouvelle éspece du genre *Euplotidium* (Noland 1937: *Euplotidium prosaltans* n, sp (Cilié Hypotriche). Cah Biol Mar 26:53–62

Vandekerckhove TTM, Willems A, Gillis M, Coomans A (2000) Occurrence of novel verrucomicrobial species, endosymbiotic and associated with parthenogenesis in *Xiphinema americanum*-group species (nematoda, Longidoridae). Int J Syst Evol Microbiol 50:2197–2205

Verni F, Rosati G (1990) Peculiar epibionts in *Euplotidium itoi* (Ciliata, Hypotrichida). J Protozool 37:337–343

Wells B, Horne RW (1983) The ultrastructure of *Pseudomonas avenae*. II. Intracellular refractile (R-body) structure. Micron Microsc Acta 14:329–344

Hydrogenosomes and Symbiosis

Johannes H.P. Hackstein, Nigel Yarlett

1
Introduction

Hydrogenosomes are organelles approximately 1–2 µm in size that compartmentalize the terminal reactions of the anaerobic cellular energy metabolism. They were first described in the parabasilid flagellate, *Tritrichomonas foetus*, in a seminal publication by Lindmark and Müller (1973) as subcellular compartments that produce hydrogen and ATP. Since this time, hydrogenosomes or variations of them have been described in quite a number of rather different unicellular eukaryotes, which are adapted to microaerobic or anoxic environments (Roger 1999; Yarlett 2004). Several researchers considered these organelles as variations of mitochondria adapted to anaerobic environments (Biagini et al. 1997; Embley et al. 1997, 2003), but it is still a matter of debate whether or not these ancestral "mitochondria" had an aerobic metabolism using oxygen as a terminal electron acceptor – or whether these organelles functioned anaerobically, and, consequently, produced hydrogen like present-day hydrogenosomes (Tjaden et al. 2004).

The identification of rudimentary, mitochondrial-remnant organelles in organisms previously considered devoid of mitochondria (named "archaezoa" by Cavalier-Smith 1993) further complicated the considerations about the nature of the "universal" ancestor, since these mitochondrial-remnant organelles (named mitosomes or cryptons) do neither produce hydrogen nor ATP (Mai et al. 1999; Tovar et al. 1999, 2003; Yarlett 2004). Notably, some of these organisms, such as for example *Giardia* or *Entamoeba* can produce hydrogen with the aid of a cytoplasmic hydrogenase, while others, such as the Microsporidia, do not exhibit any hydrogenase activity, or do not possess hydrogenase genes, respectively (Katinka et al. 2001; Lloyd et al. 2002; Williams et al. 2002; Nixon et al. 2003). The relationship of these elusive organelles to each other,

---

J.H.P. Hackstein (e-mail: j.hackstein@science.ru.nl)
Department of Evolutionary Microbiology, Faculty of Science, Radboud University Nijmegen, Toernooiveld 1, NL 6525 ED Nijmegen, The Netherlands
Nigel Yarlett
Haskins Laboratories and Department of Chemistry and Physical Sciences, Pace University, New York, USA

regardless whether hydrogen-producing or not, and to the mitochondrion has been examined by many researchers, and several theories have been forward in an attempt to explain the origins of all of these organelles (Martin and Müller 1998; Hackstein et al. 1999, 2001; Dyall and Johnson 2000; Martin et al. 2001; Tielens et al. 2002; Embley et al. 2003; Dyall et al. 2004). However, since all mitochondrial-remnant organelles and all hydrogenosomes (except in two host taxa, i.e., *Nyctotherus/Metopus* and *Blastocystis,* see below) lack an organelle genome (Embley et al. 2003; Dyall et al. 2004), a straightforward analysis of their evolutionary history is seriously hampered. It is obvious that evidence derived from the ultrastructure of the organelles and/or from the phylogenetic analysis of a few nuclear-encoded organelle proteins is insufficient to solve all the problems concerning the complex evolutionary history of these very diverse organelles.

In this review we will discuss not only the available evidence concerning the phylogenetic analysis of hydrogenosomal/mitosomal proteins, but also the structure and function of the various hydrogenosomes, which clearly shows that not all hydrogenosomes are the same. Although the mitochondrial-remnant organelles, the mitosomes and the hydrogenosomes appeared to be evolutionarily related to mitochondria, there is persuading evidence that they evolved repeatedly and independently as adaptations to anoxic environments. In essence, hydrogenosomes are organelles that facilitate a compartmentalized energy metabolism in anoxic niches, whereas mitosomes or cryptons retained essential functions in iron-sulphur metabolism, but none related to the energy metabolism (Müller 1998; Vanacova et al. 2003; Bakatselou et al. 2003; Henze and Martin 2003; Balk and Lill 2004). These morphologically and physiologically extremely diverse subcellular compartments are the products of evolutionary tinkering, which allowed eukaryotic cells to adapt to life without oxygen. This tinkering resulted in alternative solutions for the same problem, i.e., how to maintain the homeostasis of the cellular metabolism under anaerobic conditions. Consequently, it is not surprising that the various hydrogenosomes are physiologically significantly different – with obvious consequences for symbiotic associations with organisms that rely on the metabolic products provided by the various hydrogenosomes.

2
Hydrogenosomes and Mitochondrial Remnant Organelles Evolved Repeatedly

Initially, the arguments that both mitosomes, the mitochondrial remnant organelles, and hydrogenosomes evolved several times were based on the observation that hydrogenosomes and mitosomes, respectively, were found in a broad spectrum of rather unrelated taxa of unicellular organisms, such as

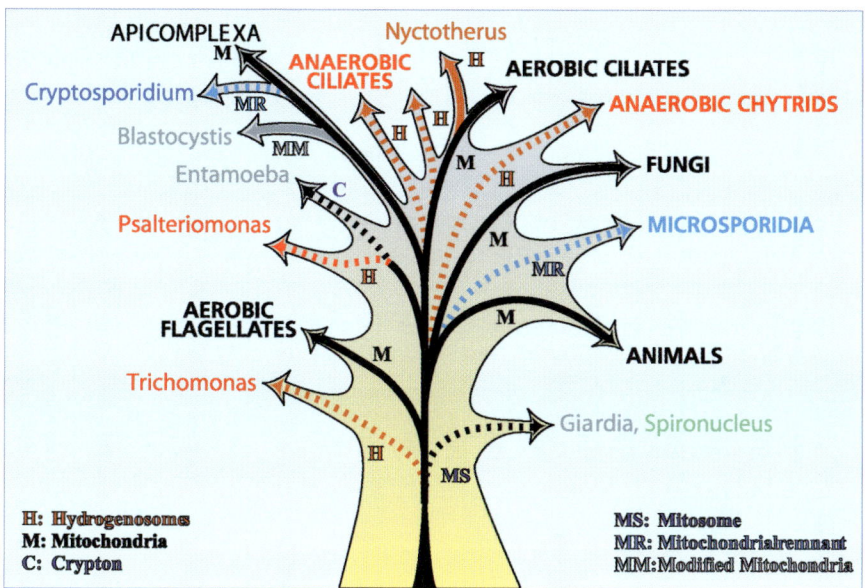

Fig. 1. Illustration displaying the phylogenetic relationships between aerobic and anaerobic protists (based on a variety of molecular data) together with a tentative evolutionary tree of mitochondria, modified mitochondria, mitochondrial remnants, and hydrogenosomes. The *solid lines* indicate phylogenetic relationships that are based on the analysis of "mitochondrial" genomes. *Dashed lines* indicate the loss of organellar genomes. The tentative phylogenetic relationships between the various types of organelles belonging to the "mitochondrial" family are therefore based on the analysis of nuclear genes encoding organellar proteins. These parts of the tree might be flawed by lateral gene transfers, evolutionary changes in the targeting to the various subcellular compartments, and biased evolutionary rates. See Table 1 and the text for details

trichomonads, diplomonads, sarcodina (entamoebids), flagellates, ciliates and chytrids (Fig. 1; Biagini et al. 1997; Embley et al. 1997; Roger 1999).

In addition, the presence of an anaerobic mitochondrion-like organelle in *Blastocystis hominis* (Straminopiles) has been proposed - mainly based upon redox sensitive dyes (Nasirudeen and Tan 2004). However, since most of these organelles did not retain a genome, the only common diagnostic characters seemed to be (i) the presence of "mitochondrial-type" chaperonines (Clark and Roger 1995; Bui et al. 1996, Germot et al. 1996), (ii) the fact that these organelles were membrane bounded, and (iii), in the case of hydrogenosomes, that these organelles produced ATP (Müller 1993). The phylogenetic analysis of HSP(cpn) 60 supported different "mitochondrial" origins of the various organelles (Voncken 2001; Voncken et al. 2002a; van der Giezen et al. 2003).

Table 1. Properties of various hydrogenosomes and mitochondrial-remnant organelles

Phylum	Order	Species	Organelle	HSP60	Hydrogenase	Pyruvate metab.	Comp. Energy metab.	Organelle genome	Methanogenic endosymbiont
Mastigophora	Trichomonadida	*Trichomonas* (Trichomonadinae)	Hydrogenosome	HSP60	[Fe]	PFO	Yes	No	No
	Diplomonadida	*Giardia* (Giardiinae)	Crypton	cytopl.HSP60	[Fe] cytopl.	PFO	No	No	No
		Spironucleus (Hexamitidae)	None identified	cytopl.HSP60	[Fe]	PFO	No	No	No
	Pelobiontida	*Mastigamoeba* (Mastigamoebidae)	None identified	ND	[Fe]	PFO	ND	No	No
Ciliophora	Armophorida	*Nyctotherus* (Clevelandellidae)	Hydrogenosome	HSP60	[Fe] 24kD+51kD	PDH	Yes	Yes	Yes
		Metopus (Armophoridae)	Hydrogenosome	ND	[Fe]	ND	Yes	Yes	Yes
	Vestibuliferida	*Dasytricha* (Isotrichidae)	Hydrogenosome	ND	[Fe]	PFO	Yes	No	No
	Plagiopylea	*Trimyema* (Plagiopylidae)	Hydrogenosome	ND	Yes	PFL	Yes	Yes	Yes
		Plagiopyla (Plagiopylidae)	Hydrogenosome	ND	Yes	ND	ND	ND	Yes
Sarcodina	Amoebida	*Entamoeba* (Lobosea)	Mitosome	HSP60	[Fe] cytopl.	PFO	No	No	No
Percolozoa	Schizopyrenida	*Psalteriomonas* (Vahlkamphiidae)	Hydrogenosome	ND	[Fe]	ND	Yes	No	Yes
Alveolata	Apicomplexa	*Cryptosporidia* (Cryptosporidiidae)	Relict-mitochondrion	Cpn60	NARF	PNOR	No	No	No
	Stramenopiles	*Blastocystis* (Stramenopiles)	Modified mitochondrion	cytopl.HSP70	No	Unknown	ND	Yes	No
Microsporidia	Pansporablastina	*Trachipleistophora* (Pleistophoridae)	Relict-mitotochondrion	No HSP70	No	Rudiment PDH	No	No	No
Chytridiomycota	Spizellomycetales	*Piromyces* (Neomasticalligales)	Hydrogenosome	HSP60	[Fe]	PFL	Yes	No	No
		Neocallimastix (Neomasticalligales)	Hydrogenosome	HSP60	[Fe]	PFL	Yes	No	No

Abbreviations:; Pyruvate ferredoxin oxidoreductase (PFO); Pyruvate dehydrogenase (PDH); Pyruvate formate lyase (PFL); Heat shock protein (HSP); Chaperonin (Cpn); Pyruvate NADH oxidoreductase (PNOR); Hydrogenosomal membrane protein (HMP); Not determined (ND).

However, a functional and phylogenetic analysis of hydrogenosomal ADP/ATP carriers (AAC) so far revealed that only the hydrogenosomes of chytrids and ciliates possess genuine mitochondrial AAC's, which cluster with the mitochondrial homologues of their aerobic, mitochondria-bearing relatives (Voncken 2001; Voncken et al. 2002a; Haferkamp et al. 2002; van der Giezen et al. 2002). *Trichomonas* uses alternative – potentially pre-mitochondrial – AACs for the transport of ATP across its hydrogenosomal membranes (Dyall et al. 2000; Tjaden et al. 2004). Furthermore, there is no evidence for the presence of true mitochondrial AACs in any of the mitosomes/cryptons or mitochondrial remnant organelles until now. Of course, the latter do not produce ATP, and therefore, might not require mitochondrial-type AACs (Tovar et al. 1999, 2003; Katinka et al. 2001; Williams et al. 2002). Also, the genome projects of *Giardia* and *Entamoeba* could not reveal the presence of true mitochondrial AACs, see http://www.NCBI.nih.gov. Thus, these observations might either argue for a deep evolutionary divergence of these organelles from a (facultatively) anaerobic, hydrogen-producing, pre-mitochondrial ancestor, or, alternatively, for a secondary loss of true mitochondrial AACs in all of these organelles (Tjaden et al. 2004). In any case, the available evidence strongly supports at least three to four (but potentially much more) independent origins of hydrogenosomes and mitochondrial-remnant organelles from organelles belonging to the "mitochondrial" family – regardless whether these ancestral organelles were strictly aerobic or facultatively anaerobic (Fig. 1; Embley et al. 2003).

Assuming an aerobic, mitochondrial ancestor, the host could rely on a splendid supply with ATP generated with the aid of the mitochondrial electron transport chain. Under anaerobic conditions, however, the mitochondrial electron transport chain cannot use oxygen as a terminal electron acceptor and the energy conservation function by the generation of ATP cannot be fulfilled (Hackstein et al. 1999, 2001). Such a cell faces a dramatic accumulation of reduction equivalents in the mitochondrion, and, consequently, it has to rely on glycolysis in the cytoplasm, which can provide only a very limited amount of ATP. Notably, certain mitochondria can cope with an anoxic environment maintaining certain mitochondrial functions even under such adverse conditions. These mitochondria are able to use alternative environmental electron acceptors (e.g., nitrate), or metabolic (Krebs cycle) intermediates such as fumarate as an endogenous electron acceptor allowing a rudimentary function of the electron transport chain even in the complete absence of oxygen (Tielens et al. 2002). In other eukaryotes, however, the adaptation to anoxic niches caused a degeneration of the mitochondrion to rather inconspicuous cellular compartments with a concomitant loss of the electron transport chain and its energy conservation capacities, as discussed earlier (Embley and Martin 1998; Rotte et al. 2000). Certain mitochondrial functions such as the production of acetyl-CoA and keto-acids for lipid and carbohydrate biosynthesis, respectively, were maintained even in the absence of a functioning electron transport chain – albeit in an alternative subcellular compartment (Akhmanova et al.

1998b; Müller 1998; Hackstein et al. 1999; Henze and Martin 2003; Dyall et al. 2004).

3
Hydrogenosomes: Organelles that Can Use Protons as Electron Acceptors

In marked contrast to the mitochondrial remnant organelles mentioned above, hydrogenosomes retained an ATP generating function (Müller 1993, 1998). They compartmentalize the terminal reactions of the cellular energy metabolism. Characteristically, hydrogenosomes import pyruvate (or malate) which is oxidatively decarboxylated to acetyl-CoA by the action of a pyruvate:ferredoxin oxidoreductase (PFO), and not, as in aerobic mitochondria, by the action of a pyruvate dehydrogenase (PDH). An acetate:succinate CoA-transferase and a succinate thiokinase are believed to mediate the formation of acetate and ATP (Müller 1993, 1998), similar to the situation in the mitochondria of the kinetoplastidae and helminths (van Hellemond et al. 1998; Tielens et al. 2002). The reduction equivalents that are formed during the decarboxylation of pyruvate are not used to fuel an electron transport chain as in mitochondria; rather they are removed from the hydrogenosome by a hydrogenase, which reduces protons resulting in the formation of molecular hydrogen (Fig. 2; Müller 1993; Embley and Martin 1998; Martin and Müller 1998; Vignais et al. 2001). Notably, anaerobic chytridiomycete fungi followed a different evolutionary strategy: they avoid the generation of reduction equivalents by using pyruvate:formate lyase (PFL) instead of pyruvate ferredoxin oxidoreductase (PFO) for the non-oxidative splitting of pyruvate into acetate and formate rendering hydrogen production a marginal metabolic route (Fig. 3; Akhmanova et al. 1999; Boxma et al. 2004). The hydrogenosomes of certain ciliates such as *Nyctotherus ovalis*, on the other hand, retained a much more mitochondrial-type of hydrogenosomal metabolism (see below). Clearly, hydrogenosomes are not the same (Coombs and Hackstein 1995; Embley et al. 1997; Hackstein et al. 1999, 2001). Therefore, it is necessary to review publications dealing with the various hydrogenosomes in more detail, and, in particular, to discuss the metabolic variants with respect to their potential to support the growth of endo- or ectosymbionts, respectively.

3.1
Hydrogenosomes of *Trichomonas vaginalis*

The hydrogenosomes of the trichomonads (Parabasalia) have been studied intensively for more than 30 years (Lindmark and Müller 1973; Müller 1993).

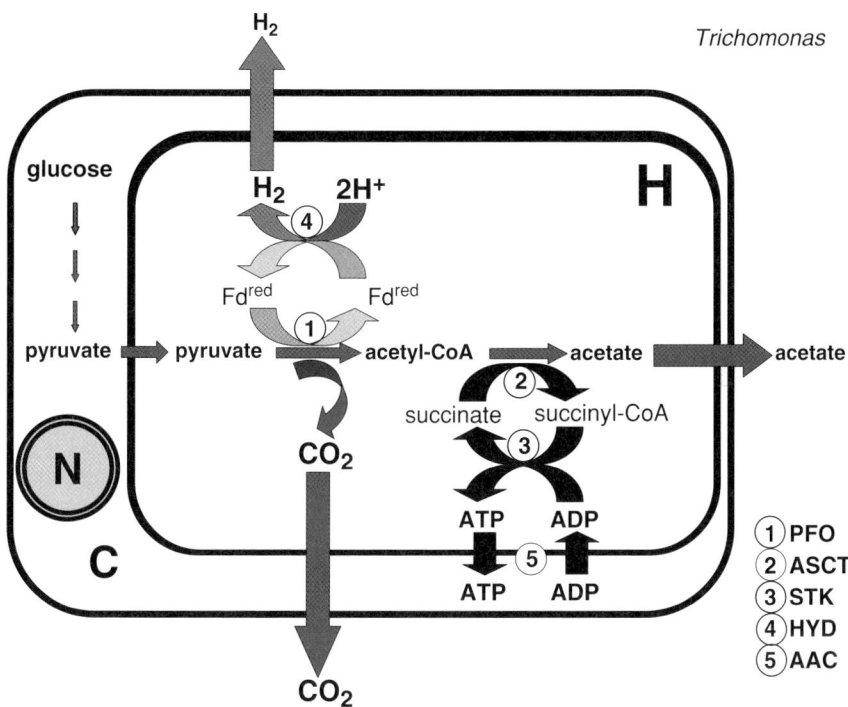

Fig. 2. Metabolic scheme of a generalised, anaerobic "textbook" protist (like *Trichomonas vaginalis*) with a hydrogenosome ("type II anaerobe"; Müller 1993, 1998). Pyruvate is formed in the cytoplasm (*C*) by glycolysis, imported into the hydrogenosome (*H*) and metabolised to acetate and CO_2 under formation of H_2. ATP is formed by substrate level phosphorylation by the enzymes acetate succinyl-CoA transferase (ASCT, *2*) and succinate thiokinase (STK, *3*). ATP is exported by an ADP-ATP carrier (AAC, *5*). The electrons resulting from the oxidative decarboxylation of pyruvate are transferred to a ferredoxin by pyruvate:ferredoxin oxidoreductase (PFO, *1*) and to protons by a Fe-hydrogenase (HYD, *4*). N: nucleus. From Hackstein et al. (2001), modified

Upon initial inspection, with the exception of a double membrane, these organelles were considered, both morphologically and biochemically, distinct from mitochondria (Fig. 4; Benchimol et al. 1996; Benchimol and Engelke 2003). Subsequent biochemical and molecular studies have changed this view, albeit sometimes in a way of wishful thinking. Trichomonad hydrogenosomes possess mitochondrial-like chaperonins, HSP 10, HSP 60 and HSP 70 (Clark and Roger 1995; Bui et al. 1996; Germot et al. 1996), proteins of the mitochondrial carrier family (HMP 31; see Dyall et al. 2000, 2004; Tjaden et al. 2004). The N-terminal extensions of many hydrogenosomal proteins favoured the assumption that a mitochondrial-type import machinery would facilitate

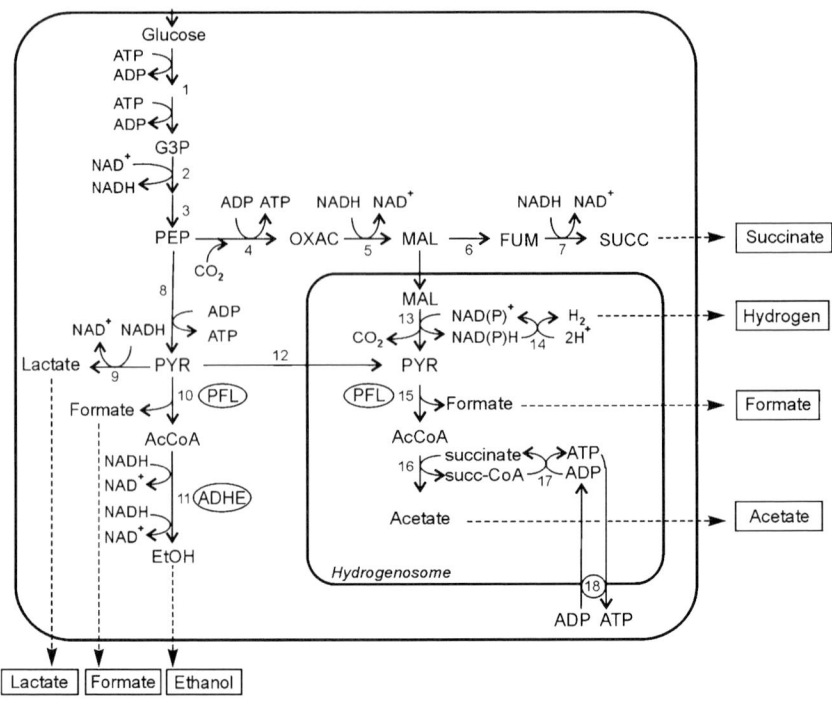

Fig. 3. Energy metabolism of the anaerobic chytridiomycete fungus *Piromyces* sp. E2. The panel shows a scheme of the metabolic pathways involved in the production of the major end products, which are representative for a bacterial-type mixed acids fermentation. The *numbers* in panel A indicate the following enzymes: (*1*) hexokinase, glucose-6-phosphate isomerase, phosphofructokinase 1, aldolase and triose phosphate isomerase; (*2*) glyceraldehyde 3-phosphate dehydrogenase; (*3*) phosphoglycerate kinase, phosphoglycerate mutase and enolase; (*4*) phosphoenolpyruvate carboxykinase; (*5*) malate dehydrogenase; (*6*) fumarase; (*7*) fumarate reductase; (*8*) pyruvate kinase; (*9*) lactate dehydrogenase; (*10*) cytosolic pyruvate:formate lyase; (*11*) alcohol dehydrogenase E; (*12*) pyruvate import into hydrogenosomes; (*13*) malic enzyme; (*14*) hydrogenase; (*15*) hydrogenosomal pyruvate:formate lyase; (*16*) acetate:succinate CoA-transferase; (*17*) succinyl-CoA synthethase; (*18*) ADP/ATP carrier. Abbreviations; *AcCoA* acetyl-CoA; *EtOH* ethanol; *FUM* fumarate; *G3P* glyceraldehyde-3-phosphate; *MAL* malate; *OXAC* oxaloacetate; *PEP* phosphoenolpyruvate; *PYR* pyruvate; *SUCC* succinate. An analysis of the metabolic fluxes revealed that the formation of hydrogen via the malate route can become marginal under certain culture conditions (Boxma et al. 2004). From Boxma et al. (2004), modified.

the import of nuclear-encoded hydrogenosomal proteins into the organelle (Dyall and Johnson 2000; Dyall et al. 2000, 2004; Embley et al. 2003). Also, the presence of acetate:succinate CoA-transferase activity (Müller 1993, 1998; van Hellemond et al. 1998), an enzyme activity that is shared by these organelles

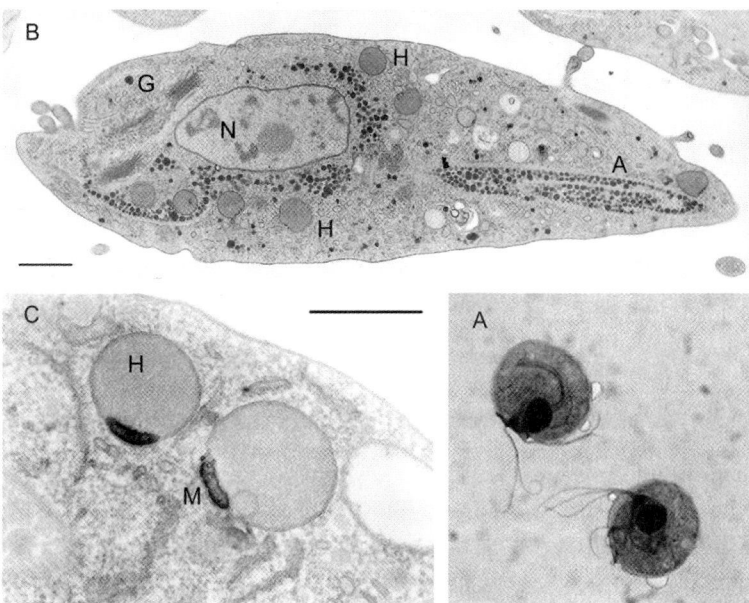

Fig. 4. A *Trichomonas vaginalis,* light microscopical picture of eosin-stained cells; natural size approximately 10 × 45 µm (courtesy of H. Aspöck, Vienna). **B** Electron micrograph of *Tritrichomonas foetus:* seven hydrogenosomes (*H*) can be identified in the cytoplasm (*N* nucleus; *G* Golgi apparatus; *A* axostyl). **C** A higher magnification reveals that a double membrane surrounds the hydrogenosomes. (*M* marginal plate). B and C were kindly provided by M. Benchimol, Rio de Janeiro *Bar*: in B and C 1 µm. From Hackstein et al. (2001), modified.

and certain mitochondria, seemed to suggest a "mitochondrial" ancestry for the hydrogenosomes of *Trichomonas* (Müller 1993, 1998; Dyall and Johnson 2000; Rotte et al. 2000). However, trichomonad hydrogenosomes are clearly different from mitochondria since they lack a genome, ribosomes, cytochromes, an electron transport chain, cardiolipin and cristae (Müller 1993, 1998; Benchimol et al. 1996; Clemens and Johnson 2000; Voncken et al. 2002a; Benchimol and Engelke 2003). Moreover, their import machinery seems to exhibit rather peculiar characteristics that are not shared with mitochondria (Dyall et al. 2003, 2004). Like mitochondria, they import pyruvate, which results from glycolysis, but trichomonad hydrogenosomes do not use a pyruvate dehydrogenase (PDH) for the catabolism of pyruvate. Rather, these hydrogenosomes metabolize pyruvate through pyruvate:ferredoxin oxidoreductase (PFO) and hydrogenase to acetate, carbon dioxide and hydrogen (Fig. 2; Müller 1993, 1998). Acetate formation from acetyl-CoA is believed to be coupled to the substrate level phosphorylation of succinate via the enzyme acetate:succinate CoA transferase (ASCT; Müller 1993); this route should yield 1 ATP per mol of pyruvate consumed. ASCT is one of the few enzymes, which are known to be shared between hydrogenosomes and certain mitochondria. However, it is still unknown whether or not the genes encoding proteins with ASCT activity are the same in trichomonads and kinetoplastids (cf. Riviere et al. 2004). Additional ATP formation seems to be feasible by the generation of a PMF as in mitochondria (Humphreys et al.

1998). The generation of a PMF has not yet been studied in detail, but the generation of a proton gradient by trichomonad hydrogenosomes is likely (Turner and Lushbaugh 1991). Notably, potential F_1F_0 ATP synthases of *Trichomonas* have not been identified so far, and the observation that trichomonad hydrogenosomes can serve as cellular Ca^{2+}-stores (Biagini et al. 1997) is rather circumstantial with respect of a "mitochondrial" ancestry. However, it might be concluded that trichomonad hydrogenosomes are capable of generating a PMF and/or a H^+ gradient, although a mitochondrial-like electron-transport chain is absent (Humphreys et al. 1998). Thus, the relationship between the hydrogenosomes of *Trichomonas* and aerobically-functioning mitochondria is much less evident than suggested in many publications (see below).

3.2
Hydrogenosomes of Anaerobic Ciliates: at Least One Appears to be a Mitochondrion that Produces Hydrogen

Ciliates belong to the "crown group" of eukaryotes (Sogin 1991), and in at least 8 of the 22 orders of ciliates as classified by Corliss (1979), anaerobic species evolved (Fenchel and Finlay 1995). There is a certain agreement that anaerobic ciliates evolved secondarily from aerobic ancestors since some higher ciliate taxa comprise both aerobic and anaerobic species (Embley et al. 1995, 1997, 2003; Fenchel and Finlay 1995; Hackstein et al. 2001, 2002). *Bona fide* hydrogenosomes are present in 7 of the 22 orders, but both the evidence that these hydrogenosomes evolved independently and repeatedly from mitochondria is rather circumstantial. Notably, Akhmanova et al. (1998a) and van Hoek et al. (2000a) have presented straightforward evidence for the presence of a mitochondrial genome in the hydrogenosomes of *Nyctotherus ovalis*, an anaerobic, heterotrichous ciliate that inhabits the intestinal tract of cockroaches (Gijzen et al. 1991; Akhmanova et al. 1998a; van Hoek et al. 1998, 1999, 2000b). This genome, identified by immunocytochemistry, hosts rRNA genes that are abundantly expressed, and phylogenetic analysis reveals a clustering among the mitochondrial rRNA genes of aerobic ciliates (van Hoek et al. 2000a; Hackstein et al. 2001). Since the phylogenies of the nuclear 18S rRNA genes of the ciliates are congruent with the SSU rRNA genes of their mitochondria and hydrogenosomes (Akhmanova et al. 1998a; van Hoek et al. 1998, 2000a,b; Hackstein et al. 2001), it is likely that the hydrogenosomes of *N. ovalis* evolved from the mitochondria of aerobic ciliates. Moreover, the hydrogenosomes of *N. ovalis* possess cristae, and thus morphologically resemble mitochondria (Fig. 5; Akhmanova et al. 1998a; Hackstein et al. 2001). In addition, there is evidence for quite a number of genes encoding mitochondrial proteins (located on the macronuclear or organelle genome, respectively (Boxma et al. 2005). Therefore, it seems reasonable to assume that the hydrogenosomes of heterotrichous ciliates evolved from mitochondria that adapted to anaerobic environments. There is evidence that the hydrogenosomes of rumen ciliates and plagiopylid ciliates, too, evolved from mitochondria (Embley et al. 1995). However, these hydrogenosomes are quite different from those of *N. ovalis* and it is likely that they evolved independently (see below).

Akhmanova et al. (1998a) have shown that the ciliate's hydrogenosomes possess an [Fe]-hydrogenase that is encoded by a macronuclear gene-sized chromosome. This hydrogenase represents a novel type of [Fe]-hydrogenase that allows H_2-formation to be coupled directly to the reoxidation of NADH. The [Fe]-hydrogenase has been linked covalently with a protein, which possesses NAD and FMN binding sites, and a ferredoxin-like module that allows transferring electrons to the catalytic site of the hydrogenase (Akhmanova et al. 1998a; Vignais et al. 2001; Voncken et al. 2002b). The origin of the hydrogenase(s) is of central importance not only for the hydrogen hypothesis, but also for an understanding of the evolution of the various types of hydrogenosomes. All eukaryotic [Fe] hydrogenases, including the hydrogenase-like proteins ("NARFs", cf. Balk et al. 2004), which neither produce nor consume hydrogen, seem to form a monophyletic cluster, with the possible exclusion of the *N. ovalis* hydrogenase (Akhmanova et al. 1998a; Horner et al. 2000, 2002; Nixon et al. 2003). However, due to the high conservation of the hydrogenases

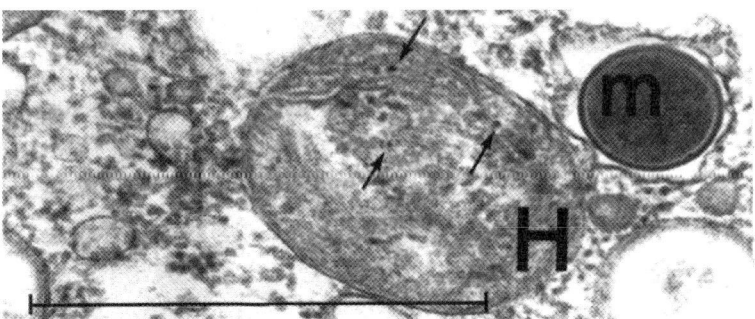

Fig. 5. The hydrogenosome (*H*) of *Nyctotherus ovalis* at higher magnification looks like a mitochondrium (glutaraldehyde/OsO_4 fixation). The inner and outer membrane, crista-like invaginations of the inner membrane (*arrowheads*), and putative 70S ribosomes can be identified (*arrows*). *m* methanogenic endosymbiont. *Bar* 1 μm. (From Akhmanova et al. 1998a, modified)

and the sensitivity for species sampling, statistical support for either of these phylogenies is poor (Voncken et al. 2002b; Embley et al. 2003). Even the identification of the first α proteobacterial hydrogenase could not solve the phylogenetic puzzle (Davidson et al. 2002; Embley et al. 2003). The recent isolation of the genes encoding hydrogenases of rumen ciliates (Boxma 2004) supports a common eukaryotic origin of all eukaryotic hydrogenases and NARFs – with the exception of the hydrogenase of *N. ovalis*, which appears to be a mosaic of proteins of δ proteobacterial and β proteobacterial origins (Akhmanova et al. 1998a; Horner et al. 2000, 2002; Voncken et al. 2002b). The [Fe]-hydrogenases of rumen ciliates and anaerobic chytrids clearly belong to the eukaryotic cluster. They are similar to the "long-type" [Fe] hydrogenases from *Trichomonas*. Interestingly, phylogenetic analysis of the hydrogenases of anaerobic chytrids clusters them with the extremely short hydrogenases of green algae, which function in the plastidic (and not in the mitochondrial) electron transport (Florin et al. 2001; Horner et al. 2002; Voncken et al. 2002b; Nixon et al. 2003). However, the origin of these eukaryotic hydrogenases from a hypothetical hydrogenase-containing universal endosymbiont remains unclear; in particular, there is no statistical support for the assumption that all eukaryotic hydrogenases (including the NARFs) evolved from an α-proteobacterial ancestor (Horner et al. 2002; Voncken et al. 2002b; Embley et al. 2003; Stejskal et al. 2003).

3.3
Hydrogenosomes of Anaerobic Chytrids: an Alternative Way to Adapt to Anaerobic Environments

Anaerobic chytrids are important symbionts in the gastro-intestinal tract of many herbivorous mammals. Their life cycle consists of an alternating flagellated zoospore stage and a vegetative phase when a multi-nucleated mycelium is formed. The hyphae of the rhizomycelial system attach to the digesta and secrete a broad spectrum of fibrolytic enzymes that is very efficient in digesting plant polymers (Teunissen et al. 1991; Orpin 1994; Yarlett 1994). These organisms are highly adapted to intestinal environments; their optimal growth temperature coincides with the body temperature of their mammalian hosts, and during almost their whole life cycle, they live and multiply under anoxic conditions (Orpin 1994). The anaerobic chytrids evolved from mitochondria-bearing ancestors, since DNA sequence analysis reveals a clustering of both aerobic and anaerobic chytrids (Bowman et al. 1992; cf. Paquin et al. 1995; Paquin and Lang 1996). Also an analysis of biochemical and morphological traits consistently establishes a close relationship between chytrids and other fungi (Ragan and Chapman 1978), and Akhmanova et al. (1998b) demonstrated that several enzymes of mitochondrial origin, which lack putative targeting signals, were retargeted to the cytoplasm (in active form) being no longer present in the hydrogenosomes. Consequently, there is little doubt that the chytrids living in the gastro-intestinal tract of herbivorous mammals have secondarily adopted an anaerobic life style (Hackstein et al. 1999).

Anaerobic chytrids such as, for example, *Neocallimastix* and *Piromyces* possess hydrogenosomes, which, however, are structurally and functionally clearly different from the hydrogenosomes of the ciliate *N. ovalis*, the amoeboflagellate *Psalteriomonas lanterna* and the parabasalid *T. vaginalis* (Fig. 6; Coombs and Hackstein 1995; Hackstein et al. 1999, 2001). Like the hydrogenosomes of the amoeboflagellate *P. lanterna* (Hackstein, unpubl.)*,* and of the parabasalid *T. vaginalis* (Clemens and Johnson 2000) the hydrogenosomes of *Neocallimastix* and *Piromyces* lack a genome (van der Giezen et al. 1997; Hackstein, unpubl.). But unlike *T. vaginalis* hydrogenosomes, the chytrid hydrogenosomes rely on malate and not pyruvate for hydrogen formation. The imported malate is oxidatively decarboxylated by a hydrogenosomal malic enzyme, and it had been postulated that the resulting pyruvate is oxidized further by pyruvate:ferredoxin oxidoreductase (PFO) to acetyl-CoA. The reduced equivalents should be transferred via ferredoxin to hydrogenase thus maintaining the redox balance (Marvin-Sikkema et al. 1992, 1993, 1994). However, Akhmanova et al. (1999) and Boxma et al. (2004) showed that the hydrogenosomes of anaerobic chytrids perform a bacterial-type mixed acid fermentation during which pyruvate is split into acetyl-CoA and formate by a pyruvate:formate lyase (PFL), and not oxidatively decarboxylated by a PFO. Consequently, the formation of reducing equivalents is avoided, and the hydrogenosome excretes formate and acetate as end products of its energy metabolism (Fig. 3). Moreover, the vast majority of the carbon flow through the hydrogenosome is mediated by pyruvate, which is imported from the cytosol and metabolized in the hydrogenosome without hydrogen formation (Boxma et al. 2004). Obviously, the hydrogenosomes of anaerobic chytrids followed a different strategy when adapting to anaerobic environments: avoiding the formation of reduced equivalents renders hydrogen production a rudimentary metabolic activity in these organelles.

Functional and phylogenetic analysis of the ADP/ATP carriers from anaerobic chytrid hydrogenosomes clearly supports a fungal mitochondrial origin for these organelles (van der Giezen et al. 2002; Voncken et al. 2002a; Tjaden et al. 2004). Given that chytrid hydrogenosomes lack a genome, the ADP/ATP carriers and HSP 60 are the "second-best" markers for tracing the evolutionary history of these organelles (Andersson and Kurland 1999). Phylogenetic analysis of both genes unequivocally reveal a fungal mitochondrial ancestry (Hackstein et al. 1999; Voncken 2001; van der Giezen et al. 2002, 2003; Voncken et al. 2002a), in agreement with the earlier finding that typical mitochondrial enzymes had been retargeted to the cytoplasm in the course of the evolution of the chytrid hydrogenosomes (Akhmanova et al. 1998b).

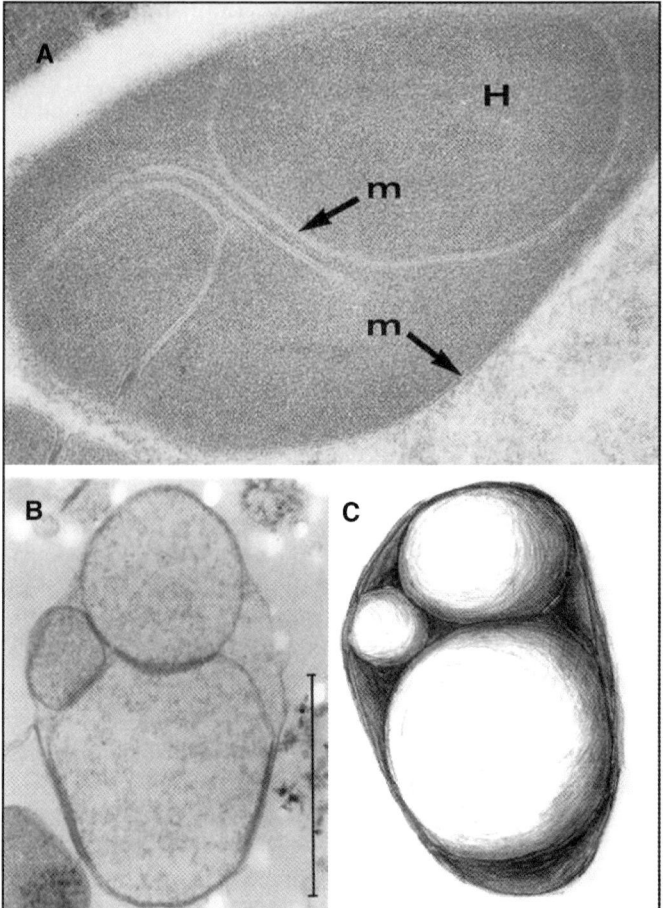

Fig. 6. A Electron micrograph of a hydrogenosome of the anaerobic chytrid *Neocallimastix* sp. L2, isolated from the faeces of a llama. This type of hydrogenosomes has a morphology very different from that of *Trichomonas* sp. or *Nyctotherus ovalis*. *Arrows* in A indicate the membranes (*m*) and vesicles. From Hackstein et al. 2001, modified. **B** Section through a hydrogenosome of *Neocallimastix* subjected to (hypotonic) osmotic treatment. The "peas in a pod" organisation of the hydrogenosome becomes visible. *Bar* 0.5 μm. **C** An artist's view of the organelle shown in B. From Voncken et al. (2002a), modified

Chytrid hydrogenosomes are therefore clearly distinct from the hydrogenosomes of *Trichomonas* that formed the basis for the Martin-Müller 'hydrogen hypothesis' for the evolution of the eukaryotic cell. Both the origin of the organelle i.e., the universal endosymbiont in the case of *Trichomonas* and a differentiated fungal mitochondrion in the case of the anaerobic chytrids, and the evolutionary strategies to adapt to anoxic environments are different. Since also the hydrogenosomes of *Trichomonas* lack a genome (Clemens and

Johnson 2000), an analysis of the hydrogenosomal ADP/ATP carriers should provide clues for or against the hydrogen hypothesis. Surprisingly, trichomonad hydrogenosomes did not host mitochondrial type ADP/ATP carriers (Tjaden et al. 2004). Rather they use a different member of the mitochondrial carrier family, the hydrogenosomal membrane protein (HMP 31) for ADP/ATP exchange that is functionally *and* phylogenetically distinct from the mitochondrial-type ADP/ATP carriers (Dyall et al. 2000; Tjaden et al. 2004). The gene encoding HMP 31 branches earlier than the mitochondrial-type ADP/ATP carriers, in agreement of the predictions derived from the hydrogen-hypothesis, and the deviating hydrogenosomal import machinery discussed by Dyall et al. (2003, 2004). Notably, also HSP 60 phylogenies cluster *Trichomonas* with HSP(cpn) 60 genes of "amitochondriate" (but mitochondrial-remnant-bearing) taxa (diplomonads, entamoebids) rather than with "true" mitochondrial chaperonines (Voncken 2001; Voncken et al. 2002a; van der Giezen et al. 2003).

4
A Common Ancestor that Produced Hydrogen

To date, it would appear that there are no present-day protozoa without mitochondria or at least rudiments of this organelle (Fig. 1; cf. Martin and Müller 1998; Embley et al. 2003). The presence of mitochondrial remnant organelles in organisms once considered as amitochondriate is an indication that the progenitor of all extant protozoa was a 'mitochondrion-containing' cell and, likewise, an indication of the evolutionary success of this acquisition. However, the experimental data and phylogenetic issues discussed above, strongly suggest that the ancestral "mitochondrion" was a facultative anaerobic symbiont, which produced hydrogen under anaerobic conditions. Since all "surviving" descendants are the products of evolutionary tinkering, it is not surprising that these organelles are of varying complexity and appearance (Table 1). For example, the mitochondrial remnant organelles of the intestinal parasites, *Entamoeba* (crypton/mitosomes) and *Giardia* (mitosomes) function in iron sulphur cluster formation, and not in redox-coupled energy conservation (Tovar et al. 2003; Vanacova et al. 2003). The identification of remnant mitochondria in *C. parvum* (Cryptosporidia) is based upon the targeting of the heat shock protein Cpn60 (Riordan et al. 2003) but no redox-coupled energy conservation has been demonstrated to date for this or any other of these rudimentary organelles (Abrahamsen et al. 2004). Moreover, *C. parvum* has been reported to lack both mitochondrial and apicoplast genomes; there are also no indications for any tRNA synthetases with mitochondrial or apicoplast targeting signals (Abrahamsen et al. 2004). Notably, *C. parvum* does not host a hydrogenase. Rather this organism retained a "NARF" gene, which seems to be characteristic for eukaryotes with "true" mitochondria (Stejskal et al. 2003). Interestingly, the HSP 60 and Fe-S machinery-hosting compartments in *Giardia* are not the same as for mitochondrial containing cells. Antibodies against either of these proteins failed to reveal a co-localization at the subcellular level. Rather the HSP 60 and "Fe-S machinery"- antibodies recognise different, clearly distinct compartments (Tovar et al. 2003). Also, a co-localization of the hydrogenase, the HSP 60-containing compartments, and the potential redox-functions with the Fe-S cluster enzymes has not been shown until now.

A mitochondrial remnant organelle has also been described in the microsporidia *Trachipleistophora hominis* and *Nosema locustae*, which host a mitochondrial type HSP 70 (Williams et al. 2002). However, a HSP 60 gene has not been found in any of the microsporidia studied so far, and the Fe-S enzymes do not possess a potential (mitochondrial-type) N-terminal targeting signal. These enzymes have not been localised to any subcellular compartment until now. Moreover, there is no evidence that the HSP 70 positive organelles posses a membrane potential, since they do not stain with Mitotracker Red (Williams et al. 2002). Thus, the available data support the assumption that all these organelles evolved from "the same" facultatively anaerobic endosymbiont, but independently and at different times. Consequently, the present day organelles are the product of evolutionary tinkering: they are rather different – with obvious consequences for the symbiotic associations found in some of these organisms.

5
Symbiotic Associations that Depend on Intracellularly Generated Hydrogen

Aerobically functioning mitochondria are unlikely to support any symbiotic association since – besides a number of functions in anabolic and catabolic cellular pathways – they produce ATP to the benefit of their hosts and, in addition, just CO_2 and H_2O. Anaerobically functioning mitochondria, on the other hand, may excrete succinate and acetate besides CO_2, and a hydrogenosome is likely to excrete substantial amounts of acetate, (formate), CO_2, and hydrogen (Müller 1993; Tielens et al. 2002). These "waste" compounds, and in particular hydrogen, could support the growth of a variety of symbionts, but only a limited number of eukaryotic host taxa seem to be able to live in symbiosis, i.e., in a long-lasting association with other organisms (de Bary 1879). For example, not a single symbiotic association (neither endo- nor ectosymbiotic) has been reported for the parasitic trichomonads of mammals or rumen-dwelling chytridiomycete fungi. However, with the growing data base of the symbiotic (mainly termite) flagellates, the phylogenetic relationships of these flagellates become more complex and subject to controversial discussions (Hampl et al. 2004; Kleina et al. 2004). Concomitantly, novel insights into the diversity of their endo- and episymbiotic relationships arise, but it remains unclear to date in how far these associations depend on intracellular hydrogen.

Until now, the physiological characteristics of their hydrogenosomes are unknown, and straightforward clues for the symbiotic associations (or their absence) are lacking. Obviously, the various flagellates from the termite gut only rarely host *endosymbiotic* methanogens, while numerous (free-living or gut wall associated), intestinal methanogens thrive in the gut (Table 1; Ohkuma et al. 1999; Tokura et al. 2000).

Associations with symbiotic eubacteria seem to prevail, but since the hydrogenosomes of these flagellates have been identified only tentatively by (electron) microscopy so far (Radek and Hausmann 1994; Radek 1997; Wier et al. 2004), the lack of metabolic information allows only speculating as to whether the endo- and episymbiotic methanogens are outcompeted by the eubacterial symbionts (Noda et al. 2003; Stingl et al. 2004; Wier et al. 2004).

Since hydrogen production is a reversible reaction (Adams 1990; Vignais et al. 2001), the metabolic activities of methanogens (or other hydrogen-consuming organisms) can pull the reaction to the "right" side. Therefore, the advantages of a symbiosis between hydrogenosomes-bearing eukaryotes and hydrogen-consuming methanogenic archaea are very suggestive. Notably, experimental data suggest that the presence of endosymbiotic methanogens in ciliates can account for a 25% increase of the growth rate and changes in the metabolic profile (Fenchel and Finlay 1992; Yamada et al. 1997). Data regarding alternative associations with non-methanogenic, hydrogen-consuming bacteria are lacking. Therefore, it is not yet possible to predict whether an organism can become engaged in a symbiotic association with a hydrogen-bearing eukaryote, and why certain unicellular eukaryotes seem to be favourite partners in symbiotic associations.

Ciliates, for example, exhibit a fascinating spectrum of symbiotic associations with prokaryotes but also with unicellular eukaryotes. Certain aerobic ciliates seem to be able to collect chloroplasts and to keep them in a functional state as *kleptochloroplasts* over extended periods of time (Stoecker et al. 1987; Schüssler and Schnepf 1992; Gustafson et al. 2000). Symbiotic associations with green, unicellular algae transform *Paramecium bursaria* and a variety of other ciliates into photosynthetic, autotrophic organisms that can thrive even in anoxic niches thanks to their endogenous oxygen production (Finlay et al. 1996; Nishihara et al. 1998). The association between various (aerobic) *Paramecium* species and eubacterial "symbionts" belonging to the genera *Holospora* and *Caedibacter* might be interpreted as a chronic infection by obligate intracellular parasites. The only benefit to their hosts seems to be the acquisition of a killer-phenotype that is provided by the bacterial "symbiont" (Kusch et al. 2000). Certain associations between ciliates and eubacteria resulted in a vital dependence from the endosymbiont; e.g., the β-proteobacterium *Polynucleobacter necessarius* appeared to be essential for the survival of its host, *Euplotes aediculatus* (Springer et al. 1996). Other ciliates, such as *Strombidium purpureum* host endosymbiotic purple-non-sulphur photosynthetic bacteria (Fenchel and Bernard 1993a,b), whereas the surface of the mouthless ciliate *Kentrophoros fasciolata* is covered by a dense layer of sulphide-oxidizing bacteria (Fenchel and Finlay 1989). Lastly, many anaerobic ciliates live in a symbiotic association with methanogenic Archaea (Vogels et al. 1980; van Bruggen et al. 1983; Embley and Finlay 1994): these consortia produce methane, one of the most important gases involved in global warming.

Hundreds to thousands of methanogens were found in close association not only with anaerobic ciliates, but also intracellularly with the hydrogen-producing organelles, sometimes even penetrating into complexes of hydrogenosomes (van Bruggen et al. 1986; Goosen et al. 1988, 1990b; Broers et al. 1990; Gijzen et al. 1991; for reviews see Fenchel and Finlay 1995; Hackstein et al. 2002). Notably, these methanogens are phylogenetically related to, but clearly distinct from their free-living relatives in the particular environments (van Hoek et al. 2000b).

The few studies that have been performed with anaerobic ciliates revealed obvious differences in metabolism, but most of the anaerobic ciliates have never been studied in detail (cf. Yarlett et al. 1985; Lloyd et al. 1989; Goosen et al. 1990a; Akhmanova et al. 1998a; Hackstein et al. 2002). The vast majority of ciliates with hydrogenosomes were only identified by their association with endosymbiotic methanogens (see below). A small number of species have been studied by electron microscopy: their hydrogenosomes clearly resembled mitochondria (van Bruggen et al. 1983, 1984, 1986; Goosen et al. 1988, 1990b; Zwart et al. 1988; Finlay and Fenchel 1989; Gijzen et al. 1991; Akhmanova et al. 1998a). In a few species of ciliates, hydrogen/methane production has been measured using gas chromatography (Fenchel and Finlay 1992; van Hoek et al. 2000b). In *Plagiopyla nasuta, Trimyema compressum* and *Trichomonas vaginalis* hydrogenase activity has been demonstrated to reside inside the hydrogenosomes with the aid of a cytochemical reaction (Zwart et al. 1988), and hydrogenase-encoding genes haven been identified in a few additional species of anaerobic ciliates (Akhmanova et al. 1998a; Voncken et al. 2002b). Notably, with respect to anaerobic ciliates, we are dealing with the unique phenomenon that the presence of a cellular organelle, the hydrogenosome, is deduced from the presence of characteristic symbionts, i.e., endosymbiotic methanogens. On the other hand, it remains unclear why so many rumen ciliates (which are monophyletic and believed to posses hydrogenosomes altogether, Moon-van der Staay et al. 2002) host only small numbers of endosymbiotic methanogens if at all (Lloyd et al. 1996; Chagan et al. 1999; Tokura et al. 1999; Kisidayova et al. 2000). Rather, ectosymbiotic associations with methanogenic archaea seem to predominate among these ciliates (Vogels et al. 1980; van Bruggen et al. 1983). The molecular identification of these associations, however, is still in its beginnings (Regensbogenova et al. 2004).

6
Conclusions

Hydrogenosomes are not the same. They evolved several times – independently – from mitochondria or the common ancestor of hydrogenosomes and mitochondria. This process, in general, involved the loss of the organellar genome together with the mitochondrial electron transport chain, and metabolic adaptations to anoxic environments such as the use of protons as terminal electron acceptors. Substantial differences in the physiological capacities of the various hydrogenosomes reflect their independent evolution through evolutionary tinkering. Notably, even the common denominator of these organelles, i.e., the production of hydrogen, can become marginal in certain hydrogenosomes.

The hydrogenosomal metabolism is crucial for the establishment of symbiotic associations, and sometimes differences in the host's metabolism seem to be able to provide the clues for an understanding of the presence or absence of symbionts in some cases. However, the physiology of the hydrogenosomes appears to be insufficient to explain the observations. Obviously, intrinsic properties of the various hosts and their symbionts play an important role, which are, at the moment, clearly not even beyond the level of a preliminary phenomenological description. Intensive efforts are required to analyse the elusive molecular basis of symbiotic associations. Many fascinating insights into the secrets of symbiotic associations await their discovery.

References

Abrahamsen MS, Templeton TJ, Enomoto S, Abrahante JE, Zhu G, Lancto CA, Deng M, Liu C, Widmer G, Tzipori S, Buck GA, Xu P, Bankier AT, Dear PH, Konfortov BA, Spriggs HF, Iyer L, Anantharaman V, Aravind L, Kapur V (2004) Complete genome sequence of the Apicomplexan, *Cryptosporidium parvum*. Science 304:441–445

Adams MWW (1990) The structure and mechanism of iron-hydrogenases. Biochim Biophys Acta 1020:115–145

Akhmanova A, Voncken FG, van Alen A, van Hoek A, Boxma B, Vogels GD, Veenhuis M, Hackstein JHP (1998a) A hydrogenosome with a genome. Nature 396:527–528

Akhmanova A, Voncken FGJ, Harhangi H, Hosea KM, Vogels GD, Hackstein JHP (1998b) Cytosolic enzymes with a mitochondrial ancestry from the anaerobic chytrid *Piromyces* sp E2. Mol Microbiol 30:1017–1027

Akhmanova A, Voncken FGJ, Hosea KM, Harhangi H, Keltjens JT, op den Camp HJ, Vogels GD, Hackstein JHP (1999) A hydrogenosome with pyruvate formate-lyase: anaerobic chytrid fungi use an alternative route for pyruvate catabolism. Mol Microbiol 32(5):1103–1114

Andersson SGE, Kurland CG (1999) Origin of mitochondria and hydrogenosomes. Curr Opin Microbiol 2:535–541

Balk J, Lill R (2004) The cell's cookbook for iron-sulfur clusters: recipes for fool's gold? Chem Biochem 5:1044–1049

Balk J, Pierik AJ, Netz DJA, Muhlenhoff U, Lill R (2004) The hydrogenase-like Nar1p is essential for maturation of cytosolic and nuclear iron-sulphur proteins. EMBO J 23(10):2105–2115

Bakatselou C, Beste D, Kadri AO, Somanath S, Clark CG (2003) Analysis of genes of mitochondrial origin in the genus *Entamoeba*. J Eukaryot Microbiol 50(3):210–214

Benchimol M, Engelke F (2003) Hydrogenosome behavior during the cell cycle in *Tritrichomonas foetus*. Biol Cell 95(5):283–293

Benchimol M, Johnson PJ, de Souza W (1996) Morphogenesis of the hydrogenosome: an ultrastructural study. Biol Cell 87:197–205

Biagini GA, Finlay BJ, Lloyd D (1997) Evolution of the hydrogenosome. FEMS Microbiol Lett 155(2):133–140

Bowman BH, Taylor JW, Brownlee AG, Lee J, Lu SD, White TJ (1992) Molecular evolution of the fungi: relationship of the Basidiomycetes, Ascomycetes, and Chytridiomycetes. Mol Biol Evol 9:285–296

Boxma B (2004) The origin of hydrogenosomes. Thesis Nijmegen. ISBN90–9018595-x, PPI (Print Partners Ipskamp), Enschede, The Netherlands

Boxma B, de Graaf RM, van der Staay GWM, van Alen TA, Ricard G, Gabaldón T, van Hoek AHAM, Moon-van der Staay SY, Koopman WJH, van Hellemond JJ, Tielens AGM, Friedrich T, Veenhuis M, Huynen MA, Hackstein JHP (2005) An anaerobic mitochondrion that produces hydrogen. Nature 434, 74–79

Boxma B, Voncken F, Jannink S, van Alen T, Akhmanova A, van Weelden SWH, van Hellemond JJ, Ricard G, Huynen M, Tielens AGM, Hackstein JHP (2004) The anaerobic chytridiomycete fungus *Piromyces* sp E2 produces ethanol via pyruvate:formate lyase and an alcohol dehydrogenase E. Mol Microbiol 51:1389–1399

Broers CAM, Stumm CK, Vogels GD, Brugerolle G (1990) *Psalteriomonas lanterna* gen. nov., sp. nov. a free-living amoeboflagellate isolated from fresh-water anaerobic sediments. Eur J Protistol 25:69–380

Bui ETN, Bradley PJ, Johnson PJ (1996) A common evolutionary origin for the mitochondria and hydrogenosomes. Proc Natl Acad Sci USA 93:9651–9656

Cavalier-Smith T (1993) Kingdom protozoa and its 18 phyla. Microbiol Rev 57:953–994

Chagan I, Tokura M, Jouany JP, Ushida K (1999) Detection of methanogenic archaea associated with rumen ciliate protozoa. J Gen Appl Microbiol 45(6):305–308

Clark CG, Roger AJ (1995) Direct evidence for secondary loss of mitochondria in *Entamoeba histolytica*. Proc Natl Acad Sci USA 92:6518–6521

Clemens DL, Johnson PJ (2000) Failure to detect DNA in hydrogenosomes of *Trichomonas vaginalis* by nick translation and immunomicroscopy. Mol Biochem Parasitol 106:307–313

Coombs GH, Hackstein JHP (1995) Anaerobic protists and anaerobic ecosystems. In: Brugerolle G, Mignot J-P (eds) Protistological actualities. Proceedings of the 2nd European Congress of Protistology, Clermont-Ferrand, France. Universite Blaise Pascal, Clermont-Ferrand, pp 90–101

Corliss JO (1979) The ciliated protozoa. Pergamon Press, Oxford

Davidson EA, van der Giezen M, Horner DS, Embley TM, Howe CJ (2002) An [Fe] hydrogenase from the anaerobic hydrogenosome-containing fungus *Neocallimastix frontalis* L2. Gene 296(1–2):45–52

De Bary A (1879) Die Erscheinung der Symbiose. Trubner, Strassburg

Dyall SD, Johnson PJ (2000) Origins of hydrogenosomes and mitochondria: evolution and organelle biogenesis. Curr Opin Microbiol 3:404–411

Dyall SD, Koehler CM, Delgadillo-Correa MG, Bradley PJ, Plümper E, Leuenberger D, Turck CW, Johnson PJ (2000) Presence of a member of the mitochondrial carrier family in hydrogenosomes: conservation of membrane-targeting pathways between hydrogenosomes and mitochondria. Mol Cell Biol 20:2488–2497

Dyall SD, Lester DC, Schneider RE, Delgadillo-Correa MG, Plümper E, Martinez A, Koehler CM, Johnson PJ (2003) *Trichomonas vaginalis* HMP35, a putative pore-forming hydrogenosomal membrane protein, can form a complex in yeast mitochondria. J Biol Chem 278:30548–30561

Dyall SD, Brown, MT, Johnson PJ (2004) Ancient invasions: from endosymbionts to organelles. Science 304:253–257

Embley TM, Finlay BJ (1994) The use of small-subunit ribosomal-RNA sequences to unravel the relationships between anaerobic ciliates and their methanogen endosymbionts. Microbiology UK 140:225–235

Embley TM, Martin W (1998) A hydrogen-producing mitochondrion. Nature 396:517–519

Embley TM, Finlay BJ, Dyal PL, Hirt RP, Wilkinson M, Williams AG (1995) Multiple origins of anaerobic ciliates with hydrogenosomes within the radiation of aerobic ciliates. Proc R Soc Lond B 262:87–93

Embley TM, Horner DA, Hirt RP (1997) Anaerobic eukaryote evolution: hydrogenosomes as biochemically modified mitochondria? Trends Ecol Evol 12:437–441

Embley TM, van der Giezen M, Horner DS, Dyal PL, Bell S, Foster PG (2003) Hydrogenosomes, mitochondria and early eukaryotic evolution. IUBMB Life 55:387–395

Fenchel T, Finlay BJ (1989) *Kentrophoros* - a mouthless ciliate with a symbiotic kitchen garden. Ophelia 30:75–93

Fenchel T, Finlay BJ (1992) Production of methane and hydrogen by anaerobic ciliates containing symbiotic methanogens. Arch Microbiol 157:475–480

Fenchel T, Bernard C (1993a) A purple protist. Nature 362:300

Fenchel T, Bernard C (1993b) Endosymbiotic purple non-sulfur bacteria in an anaerobic ciliated protozoan. FEMS Microbiol Lett 110(1):21–25

Fenchel T, Finlay BJ (1995) Ecology and evolution in anoxic worlds. Oxford Univ Press, Oxford

Finlay BJ, Fenchel T (1989) Hydrogenosomes in some anaerobic protozoa resemble mitochondria. FEMS Microbiol Lett 65:311–314

Finlay BJ, Maberly SC, Esteban GF (1996) Spectacular abundance of ciliates in anoxic pond water: contribution of symbiont photosynthesis to host respiratory oxygen requirements. FEMS Microbiol Ecol 20:229–235

Florin L, Tsokoglou A, Happe T (2001) A novel type of iron hydrogenase in the green alga *Scenedesmus obliquus* is linked to the photosynthetic electron transport chain. J Biol Chem 276(9):6125–6132

Germot A, Philippe H, LeGuyader H (1996) Presence of a mitochondrial-type 70-kDa heat shock protein in *Trichomonas vaginalis* suggests a very early mitochondrial endosymbiosis in eukaryotes. Proc Natl Acad Sci USA 93:14614–1467

Gijzen HJ, Broers CAM, Barughare M, Stumm CK (1991) Methanogenic bacteria as endosymbionts of the ciliate *Nyctotherus ovalis* in the cockroach hindgut. Appl Environ Microbiol 57:1630–1634

Goosen NK, Horemans AMC, Hillebrand SJW, Stumm CK, Vogels GD (1988) Cultivation of the sapropelic ciliate *Plagiopyla nasuta* Stein and isolation of the endosymbiont *Methanobacterium formicicum*. Arch Microbiol 150:165–170

Goosen NK, van der Drift C, Stumm CK, Vogels GD (1990a) End products of metabolism in the anaerobic ciliate *Trimyema compressum*. FEMS Microbiol Lett 69:171–175

Goosen NK, Wagener S, Stumm CK (1990b) A comparison of 2 strains of the anaerobic ciliate *Trimyema compressum*. Arch Microbiol 153:187–192

Gustafson DE, Stoecker DK, Johnson MD, van Heukelem WF, Sneider K (2000) Cryptophyte algae are robbed of their organelles by the marine ciliate *Mesodinium rubrum*. Nature 405:1049–1052

Hackstein JHP, Akhmanova A, Boxma B, Harhangi HR, Voncken FG (1999) Hydrogenosomes: eukaryotic adaptations to anaerobic environments. Trends Microbiol 7:441–447

Hackstein JHP, Akhmanova A, Voncken F, van Hoek A, van Alen T, Boxma B, Moon-van der Staay SY, van der Staay G, Leunissen J, Huynen M, Rosenberg J, Veenhuis M (2001) Hydrogenosomes: convergent adaptations of mitochondria to anaerobic environments. Zoology 104:290–302

Hackstein JHP, van Hoek AHAM, Leunissen JAM, Huynen M (2002) Anaerobic ciliates and their methanogenic endosymbionts. In: Seckbach J (ed) Symbiosis: mechanisms and model systems. Kluwer Academic Publ, Dordrecht, pp 451–464

Haferkamp I, Hackstein JHP, Voncken FGJ, Schmit G, Tjaden J (2002) Functional integration of mitochondrial and hydrogenosomal ADP/ATP carriers in the *Escherichia coli* membrane reveals different biochemical characteristics for plants, mammals and anaerobic chytrids. Eur J Biochem 269(13):3172–3181

Hampl V, Cepicka I, Flegr J, Tachezy J, Kulda J (2004) Critical analysis of the topology and rooting of the parabasalian 16S rRNA tree. Mol Phylogenet Evol 32(3):711–723

Henze K, Martin W (2003) Evolutionary biology: essence of mitochondria. Nature 426:127–128

Horner DS, Foster PG, Embley TM (2000) Iron hydrogenases and the evolution of anaerobic eukaryotes. Mol Biol Evol 17:1695–1705

Horner DS, Heil B, Happe T, Embley TM (2002) Iron hydrogenases – ancient enzymes in modern eukaryotes. Trends Biochem Sci 27(3):148–153

Humphreys M, Ralphs J, Durrant L, Lloyd D (1998) Confocal laser scanning microscopy of trichomonads: Hydrogenosomes store calcium and show a membrane potential. Eur J Protistol 34(4):356–362

Katinka MD, Duprat S, Cornillot E, Metenier G, Thomarat F, Prensier G, Barbe V, Peyretaillade E, Brottier P, Wincker P, Delbac F, El Alaoui H, Peyret P, Saurin W, Gouy M, Weissenbach J, Vivares CP (2001) Genome sequence and gene compaction of the eukaryote parasite *Encephalitozoon cuniculi*. Nature 414:450–453

Kisidayova S, Varadyova Z, Zelenak I, Siroka P (2000) Methanogenesis in rumen ciliate cultures of *Entodinium caudatum* and *Epidinium ecaudatum* after long-term cultivation in a chemically defined medium. Folia Microbiologica 45(3):269–274

Kleina P, Bettim-Bandinelli J, Bonatto SL, Benchimol M, Bogo MR (2004) Molecular phylogeny of Trichomonadidae family inferred from ITS-1, 5.8S rRNA and ITS-2 sequences. Int J Parasitol 34(8):963–970

Kusch J, Stremmel M, Breiner HW, Adams V, Schweikert M, Schmidt HJ (2000) The toxic symbiont *Caedibacter caryophila* in the cytoplasm of *Paramecium novaurelia*. Microb Ecol 40:330–335

Lindmark DG, Müller M (1973) Hydrogenosome, a cytoplasmic organelle of the anaerobic flagellate *Tritrichomonas foetus*, and its role in pyruvate metabolism. J Biol Chem 248:7724–7728

Lloyd D, Hillman K, Yarlett N, Williams AG (1989) Hydrogen-production by rumen holotrich protozoa – effects of oxygen and implications for metabolic control by in situ conditions. J Protozool 36:205–213

Lloyd D, Williams AG, Amann R, Hayes AJ, Durrant L, Ralphs JR (1996) Intracellular prokaryotes in rumen ciliate protozoa: detection by confocal laser scanning microscopy after in situ hybridization with fluorescent 16S rRNA probes. Eur J Protistol 32(4):523–531

Lloyd D, Ralphs JR, Harris JC (2002) Hydrogen production in *Giardia intestinalis*, a eukaryote with no hydrogenosomes. Trends Parasitol 18(4):155–156

Mai ZM, Ghosh S, Frisardi M, Rosenthal B, Rogers R, Samuelson J (1999) Hsp60 is targeted to a cryptic mitochondrion-derived organelle ("crypton") in the microaerophilic protozoan parasite *Entamoeba histolytica*. Mol Cell Biol 19(3):2198–2205

Martin W, Müller M (1998) The hydrogen hypothesis for the first eukaryote. Nature 392:37–41

Martin W, Hoffmeister M, Rotte C, Henze K (2001) An overview of endosymbiotic models for the origins of eukaryotes, their ATP producing organelles (mitochondria and hydrogenosomes), and their heterotrophic lifestyle. Biol Chem 382:1521–1539

Marvin-Sikkema FD, Lahpor GA, Kraak MN, Gottschal JC, Prins RA (1992) Characterization of an anaerobic fungus from llama faeces. J Gen Microbiol 138:2235–2241

Marvin-Sikkema FD, Kraak MN, Veenhuis M, Gottschal JC, Prins RA (1993) The hydrogenosomal enzyme hydrogenase from the anaerobic fungus *Neocallimastix* sp. L2 is recognized by antibodies directed against the C-terminal microbody targeting signal SKL. Eur J Cell Biol 61:86–91

Marvin-Sikkema FD, Driessen AJM, Gottschal JC, Prins RA (1994) Metabolic energy generation in hydrogenosomes of the anaerobic fungus *Neocallimastix*: evidence for a functional relationship with mitochondria. Mycol Res 98:205–212

Moon-van der Staay SY, van der Staay GWM, Javorsky P, Jouany J-P, Michalowski T, Nsabimana E, Macheboeuf D, Kisidayová S, Varadyova Z, McEwan NR, Newbold CJ, Hackstein JHP (2002) Diversity of rumen ciliates revealed by 18S ribosomal DNA analysis. Reprod Nutr Dev 42 [Suppl 1]:S76

Müller M (1993) The hydrogenosome. J Gen Microbiol 139:2879–2889

Müller M (1998) Enzymes and compartmentalization of core energy metabolism of anaerobic protists – a special case in eukaryotic evolution? In: Coombs GH, Vickerman K, Sleigh MA, Warren A (eds) Evolutionary relationships among protozoa. The Systematics Association, special volume series 56. Kluwer Academic Publ, Dordrecht, pp 109–132

Nasirudeen AMA, Tan KSW (2004) Isolation and characterization of the mitochondrion-like organelle from *Blastocystis hominis*. J Microbiol Methods 58(1):101–109

Nishihara N, Horiike S, Takahashi T, Kosaka T, Shigenaka Y, Hosoya H (1998) Cloning and characterization of endosymbiotic algae isolated from *Paramecium bursaria*. Protoplasma 203:91–99

Nixon JE, Field J, McArthur AG, Sogin ML, Yarlett N, Loftus BJ, Samuelson J (2003) Iron-dependent hydrogenases of *Entamoeba histolytica* and *Giardia lambia*: activity of the recombinant entamoebic enzyme and evidence for lateral gene transfer. Biol Bull 204:1–9

Noda S, Ohkuma M, Yamada A, Hongoh Y, Kudo T (2003) Phylogenetic position and in situ identification of ectosymbiotic spirochetes on protists in the termite gut. Appl Environ Microbiol 69(1):625–633

Ohkuma M, Noda S, Kudo T (1999) Phylogenetic relationships of symbiotic methanogens in diverse termites. FEMS Microbiol Lett 171(2):147–153

Orpin CG (1994) Anaerobic fungi: taxonomy, biology and distribution in nature. In: Mountfort DO, Orpin CG (eds) Anaerobic fungi. Dekker, New York, pp 1–45

Paquin B, Lang BF (1996) The mitochondrial DNA of *Allomyces macrogynus*: the complete genomic sequence from an ancestral fungus. J Mol Biol 255:688–701

Paquin B, Forget L, Roewer I, Lang BF (1995) Molecular phylogeny of *Allomyces macrogynus*: congruency between nuclear ribosomal RNA- and mitochondrial protein-based trees. J Mol Evol 41:657–665

Radek R (1997) *Spirotrichonympha minor* n. sp., a new hypermastigote termite flagellate. Eur J Protistol 33(4):360–374

Radek R, Hausmann K (1994) *Placojoenia sinaica* n g, n sp, a symbiotic flagellate from the termite *Kalotermes sinaicus*. Eur J Protistol 30(1):25–37

Ragan MA, Chapman DJ (1978) A biochemical phylogeny of the protists. Academic Press, New York

Regensbogenova M, McEwan N, Javorsky P, Kisidayova S, Michalowski T, Newbold CJ, Hackstein JHP, Pristas P (2004b) A re-appraisal of the diversity of the methanogens associated with the rumen ciliates. FEMS Microbiol Lett 238:307–313

Riordan CE, Ault JG, Langreth SG, Keithley JG (2003) *Cryptosporidium parvum* Cpn60 targets a relict organelle. Curr Genet 44:138–147

Rivière L, van Weelden SWH, Glass P, Vegh P, Coustou V, Biran M, van Hellemond JJ, Bringaud F, Tielens AGM, Boshart M (2004) Acetate:succinate CoA-transferase in procyclic *Trypanosoma brucei*: gene identification and role in carbohydrate metabolism. J Biol Chem 279(44):45337–45346

Roger AJ (1999) Reconstructing early events in eukaryotic evolution. Am Nat 154:S146–S163

Rotte C, Henze K, Müller M, Martin W (2000) Origins of hydrogenosomes and mitochondria – commentary. Curr Opin Microbiol 3:481–486

Schüssler A, Schnepf E (1992) Photosynthesis dependent acidification of perialgal vacuoles in the *Paramecium bursaria* - *Chlorella* symbiosis – visualization by monensin. Protoplasma 166(3–4):218–222

Sogin ML (1991) Early evolution and the origin of eukaryotes. Curr Opin Genet Dev 1:457–463

Springer N, Amann R, Ludwig W, Schleifer KH, Schmidt H (1996) *Polynucleobacter necessarius*, an obligate bacterial endosymbiont of the hypotrichous ciliate *Euplotes aediculatus*, is a member of the beta-subclass of Proteobacteria. FEMS Microbiol Lett 135:333–336

Stejskal F, Slapeta J, Ctrnacta V, Keithly JS (2003) A Narf-like gene from *Cryptosporidium parvum* resembles homologues observed in aerobic protists and higher eukaryotes. FEMS Microbiol Lett 229(1):91–96

Stingl U, Maass A, Radek R, Brune A (2004) Symbionts of the gut flagellate *Stauroojoenina* sp from *Neotermes cubanus* represent a novel, termite-associated lineage of Bacteroidales: description of '*Candidatus Vestibaculum illigatum*'. Microbiology SGM 150:2229–2235

Stoecker DK, Michaels AE, Davis LH (1987) Large proportion of marine planktonic ciliates found to contain functional chloroplasts. Nature 326:790–792

Teunissen MJ, Smits AA, Op den Camp HJ, Huis in't Veld JH, Vogels GD (1991) Fermentation of cellulose and production of cellulolytic and xylanolytic enzymes

by anaerobic fungi from ruminant and non-ruminant herbivores. Arch Microbiol 156:290–296
Tielens AG, Rotte C, van Hellemond JJ, Martin W (2002) Mitochondria as we don't know them. Trends Biochem Sci 27:564–572
Tjaden J, Haferkamp I, Boxma B, Tielens AGM, Huynen M, Hackstein JHP (2004) A divergent ADP/ATP carrier in the hydrogenosomes of *Trichomonas gallinae* argues for an independent origin of these organelles. Mol Microbiol 51:1439–1446
Tokura M, Chagan I, Ushida K, Kojima Y (1999) Phylogenetic study of methanogens associated with rumen ciliates. Curr Microbiol 39(3):123–128
Tokura M, Ohkuma M, Kudo T (2000) Molecular phylogeny of methanogens associated with flagellated protists in the gut and with the gut epithelium of termites. FEMS Microbiol Ecol 33(3):233–240
Tovar J, Fischer A, Clark CG (1999) The mitosome, a novel organelle related to mitochondria in the amitochondrial parasite *Entamoeba histolytica*. Mol Microbiol 32:1013–1021
Tovar J, Leon-Avila G, Sanchez LB, Sutak R, Tachezy J, van der Giezen M, Hernandez Müller M, Lucocq JM (2003) Mitochondrial remnant organelles of *Giardia* function in iron sulfur protein maturation. Nature 426:172–176
Turner AC, Lushbaugh WB (1991) Three aspecific ATPases in *Trichomonas vaginalis*. Comp Biochem Physiol B 100:691–696
Vanacova S, Liston DR, Tachezy J, Johnson PJ (2003) Molecular biology of the amitochondriate parasites, *Giardia intestinalis, Entamoeba histolytica* and *Trichomonas vaginalis*. Int J Parasitol 33:235–255
Van Bruggen JJA, Stumm CK, Vogels GD (1983) Symbiosis of methanogenic bacteria and sapropelic protozoa. Arch Microbiol 136:89–95
Van Bruggen JJA, Zwart KB, van Assema RM, Stumm CK, Vogels GD (1984) *Methanobacterium formicicum*, an endosymbiont of the anaerobic ciliate *Metopus striatus* McMurrich. Arch Microbiol 139:1–7
Van Bruggen JJA, Zwart KB, Herman JGF, van Hove EM, Assema RM, Stumm CK, Vogels GD (1986) Isolation and characterization of *Methanoplanus endosymbiosus* sp.nov., an endosymbiont of the marine sapropelic ciliate *Metopus contortus* Quennerstedt. Arch Microbiol 144:367–374
Van der Giezen M, Sjollema KA, Artz RRE, Alkema W, Prins RA (1997) Hydrogenosomes in the anaerobic fungus *Neocallimastix frontalis* have a double membrane but lack an associated organelle genome. FEBS Lett 408(2):147–150
Van der Giezen M, Slotboom DJ, Horner DS, Dyal PL, Harding M, Xue GP, Embley TM, Kunji ERS (2002) Conserved properties of hydrogenosomal and mitochondrial ADP/ATP carriers: a common origin for both organelles. EMBO J 21(4):572–579
Van der Giezen M, Birdsey GM, Horner DS, Lucocq J, Dyal PL, Benchimol M, Danpure CJ, Embley TM (2003) Fungal hydrogenosomes contain mitochondrial heat-shock proteins. Mol Biol Evol 20(7):1051–1061
Van Hellemond JJ, Opperdoes FR, Tielens AGM (1998) Trypanosomatidae produce acetate via a mitochondrial acetate: succinate CoA transferase. Proc Natl Acad Sci USA 95(6):3036–3041
Van Hoek AHAM, van Alen TA, Sprakel VSI, Hackstein JHP, Vogels GD (1998) Evolution of anaerobic ciliates from the gastrointestinal tract: phylogenetic analysis of the ribosomal repeat from *Nyctotherus ovalis* and its relatives. Mol Biol Evol 15:1195–1206

Van Hoek AHAM, Sprakel VSI, van Alen TA, Theuvenet APR, Vogels GD, Hackstein JHP (1999) Voltage dependent reversal of anodic galvanotaxis in *Nyctotherus ovalis*. J Euk Microbiol 46:427–433

Van Hoek AHAM, Akhmanova AS, Huynen M, Hackstein JHP (2000a) A mitochondrial ancestry of the hydrogenosomes of *Nyctotherus ovalis*. Mol Biol Evol 17:202–206

Van Hoek AHAM, van Alen TA, Sprakel VSI, Leunissen JAM, Brigge T, Vogels GD, Hackstein JHP (2000b) Multiple acquisition of methanogenic archaeal symbionts by anaerobic ciliates. Mol Biol Evol 17:251–258

Vignais PM, Billoud B, Meyer J (2001) Classification and phylogeny of hydrogenases. FEMS Microbiol Rev 25(4):455–501

Vogels GD, Hoppe WF, Stumm CK (1980) Association of methanogenic bacteria with rumen ciliates. Appl Environ Microbiol 40:608–612

Voncken FGJ (2001) Hydrogenosomes: eukaryotic adaptations to anaerobic environments. Thesis Nijmegen. Ponsen and Looien BV, Wageningen, The Netherlands

Voncken FGJ, Boxma B, Tjaden J, Akhmanova AS, Huynen M, Verbeek F, Tielens AGM, Haferkamp I, Neuhaus HE, Vogels G, Veenhuis M, Hackstein JHP (2002a) Multiple origins of hydrogenosomes: functional and phylogenetic evidence from the ADP/ATP carrier of the anaerobic chytrid *Neocallimastix* sp. Mol Microbiol 44:1441–1454

Voncken FGJ, Boxma B, van Hoek AHAM, Akhmanova AS, Vogels GD, Huynen M, Veenhuis M, Hackstein JHP (2002b) A hydrogenosomal [Fe]-hydrogenase from the anaerobic chytrid *Neocallimastix* sp L2. Gene 284(1–2):103–112

Wier AM, Dolan MF, Margulis L (2004) Cortical symbionts and hydrogenosomes of the amitochondriate protist *Staurojoenina assimilis*. Symbiosis 36(2):153–168

Williams BAP, Hirt RP, Lucocq JM, Embley TM (2002) A mitochondrial remnant in the microsporidian *Trachipleistophora hominis*. Nature 418:865–869

Yamada K, Kamagata Y, Nakamura K (1997) The effect of endosymbiotic methanogens on the growth and metabolic profile of the anaerobic free-living ciliate *Trimyema compressum*. FEMS Microbiol Lett 149(1):129–132

Yarlett N (1994) Fermentation product generation in rumen chytridiomycetes. In: Mountfort DO, Orpin CG (eds) Anaerobic fungi. Dekker, New York, pp 129–146

Yarlett N (2004) Anaerobic protists and hidden mitochondria. Microbiology SGM 150:1127–1129

Yarlett N, Lloyd D, Williams AG (1985) Butyrate formation from glucose by the rumen protozoan *Dasytricha ruminantium*. Biochem J 228(1):187–192

Yarlett N, Orpin CG, Munn EA, Yarlett NC, Greenwood CA (1986) Hydrogenosomes in the rumen fungus *Neocallimastix patriciarum*. Biochem J 236:729–739

Zwart KB, Goosen NK, van Schijndel MW, Broers CAM, Stumm CK, Vogels GD (1988) Cytochemical-localization of hydrogenase activity in the anaerobic protozoa *Trichomonas vaginalis*, *Plagiopyla nasuta* and *Trimyema compressum*. J Gen Microbiol 134:2165–2170

Molecular Interactions between *Rhizobium* and Legumes

Peter Skorpil, William J. Broughton

1
Introduction

Nitrogen availability limits the biosynthesis of proteins, amino acids, nucleotides and vitamins in plants more than any other element. Plants assimilate nitrogen as nitrate, and modern agriculture is highly dependent on nitrogenous fertiliser at a global annual cost exceeding 300 million $ US. Production of nitrogenous fertilisers requires huge energy inputs, and leaching of nitrate causes environmental problems such as contamination of ground water and surface streams (Gresshoff 2003). More than a century ago, Hellriegel and Wilfarth (1888) identified rhizobia as a source of nitrogen fixation. Bacterial nitrogen fixation contributes approximately the same amount of nitrate to agriculture as the application of chemical fertilisers, yet it is free of charge and without adverse environmental effects (Gage 2004). Gram-negative soil bacteria, collectively called rhizobia, induce the formation of nodules on many, but not all, leguminous plants [e.g., soybeans (*Glycine max*), common beans (*Phaseolus vulgaris*), peas (*Pisum sativum*), alfalfa (*Medicago sativa*), etc] and one non-legume, *Parasponia andersonii* of the elm family (*Ulmaceae*). Another micro-symbiont, *Frankia* spp. (Actinomycetales), nodulates and fixes nitrogen in eight different families of flowering plants including the *Betulaceae, Casuarinaceae, Coriariaceae, Dastiscaceae, Elaeagnaceae, Myricaceae, Rhamnaceae,* and *Rosaceae* (Hirsch et al. 2001). Nevertheless, due to the importance of legumes in modern agriculture, and since rhizobia are much easier to cultivate than frankia, much of our understanding of biological nitrogen fixation (BNF) in plants derives from research on legume-*Rhizobium* associations. Accordingly, this review will be concentrated on a few selected aspects of the *Rhizobium*-legume symbioses such as bacterial symbiotic signals (Nod-factors, surface polysaccharides and secreted proteins) and the plants' perception of Nod-factors.

P. Skorpil, W.J. Broughton (e-mail: william.broughton@bioveg.unige.ch)
Laboratoire de Biologie Moléculaire des Plantes Supérieures (LBMPS), Sciences III, Université de Genève, 1212 Genève 4, Switzerland

Progress in Molecular and Subcellular Biology
Jörg Overmann (Ed.)
Molecular Basis of Symbiosis
© Springer-Verlag Berlin Heidelberg 2005

2
Nod-Factors

Plants initiate the molecular dialogue with rhizobia by releasing flavonoids into their rhizospheres (Schultze and Kondorosi 1998; Broughton et al. 2003). These flavonoids are then taken up by the bacteria where they bind NodD proteins of the LysR family of transcriptional regulators (Broughton et al. 2000). The promoters of genes relevant for Nod-factor synthesis (*nol*, *noe* and *nod* genes) contain conserved 49 bp motifs called *nod*-boxes (Feng et al. 2003). NodD proteins bind *nod*-boxes as tetramers even in the absence of flavonoids, but activate *nod*-box controlled genes only in the presence of flavonoids (Fisher and Long 1993). This way, the timing and levels of production of Nod-factors are carefully controlled (Kobayashi et al. 2004) since overproduction of Nod-factors may provoke plant defence reactions and lead to the abortion of the infection process (Savouré et al. 1997; Hogg et al. 2002; Ramu et al. 2002). Regulation of Nod-factor production has been well studied in *Rhizobium* sp. NGR234 (from now on referred to as NGR234) (Fellay et al. 1995; Perret et al. 1999; Kobayashi et al. 2004). The symbiotic plasmid of NGR234 (pNGR234*a*) contains 19 *nod*-boxes, 18 of which are functional. Upon binding to flavonoids, NodD1 activates transcription of the genes required for Nod-factor synthesis (*nod*-genes). NodD1 also activates transcription of the *ttsI* and *syrM2* genes. TtsI is a transcriptional regulator that activates genes controlled by *tts*-boxes, leading to the induction of the type three secretion system (T3SS) (Marie et al. 2004) (Section 8.5 and Fig. 1). In turn SyrM2, another transcriptional activator, modulates transcription of the *nodD2* gene. In concert with TtsI, NodD2 also up-regulates transcription of genes required for the synthesis of rhamnose-rich lipo-poly-saccharides (LPS) (Marie et al. 2004) (section 8.4) and ultimately represses NodD1. In this way, the Nod-factor regulatory circuit is eventually closed (Kobayashi et al. 2004). Thus, flavonoids induce the production of Nod-factors, which are the first bacterial signals perceived by the plant and are crucial to nodulation (Relić et al. 1994a). Nod-factors are acylated lipo-chito-oligosaccharides made of 2 to 6 β-1,4-linked *N*-acetyl-D-glucosamine units carrying a fatty acid chain at the non-reducing terminus (Perret et al. 2000). Some rare Nod-factors have a slightly different oligo-saccharide backbone (Demont et al. 1994; Bec-Ferté et al. 1996; Olsthoorn et al. 1998; Pacios-Bras et al. 2002). Chain length of the oligo-saccharide, the type of the acyl moiety, the nature of the fatty acid and various decorations resulting from fucosylation, sulphation, acetlyation, *N*-methylation, 3-, 4-, or 6-, *O*-carbamoylation, 6-*O*-glycosylation, *D*-arabinosylation or 2-*O* methylation all contribute to the great variety of Nod-factors known (Perret et al. 2000). Nod-factor structures and Nod-factor biosynthesis have been extensively reviewed (Dénarié et al. 1996; Broughton et al. 2000; Perret et al. 2000). Unfortunately, there are few obvious correlations between the diversity of Nod-factors produced by one bacterial strain and its host-range. NGR234 as one example produces a wide range of different Nod-factors (Price et al. 1992) and has an exceptionally broad host-range that includes more than 112 genera of legumes. Some NGR234 hosts have determinate nodules, others indeterminate structures with a permanent meristem. In addition, NGR234 nodulates the non-legume *Parasponia andersonii* (Pueppke and Broughton 1999). At the other extreme, stem nodules on *Sesbania rostrata* can only be induced by *Azorhizobium caulinodans* (Mergaert et al. 1993). *R. fredii* strain USDA257 nodulates an exact subset of the NGR234 hosts and excretes a subset of the Nod-factors produced by NGR234 (Pueppke and Broughton 1999). In spite of this, Nod-factor requirements cannot be assigned to specific legumes and Nod-factor structures cannot be used as a tool to predict host-ranges. For example, *R. etli* and *R. tropici* produce different Nod-factors but both efficiently nodulate *Phaseolus vulgaris* (Poupot et al. 1993, 1995). The quantity of Nod-factors produced also helps determine the spectrum of hosts. USDA257 produces one-fortieth the amount of Nod-factors secreted by NGR234. Variation in the amount of Nod-factors produced may depend on NodD1. If *nodD*1 of NGR234 is transferred into *R. meliloti*, it produces two times more Nod-factors than usual and gains the ability to nodulate *Vigna unguiculata*, normally a non-host of *R. meliloti* (Relić et al. 1994b). A strain of *Mesorhizobium loti*, NZP2213, produces a much greater variety of Nod-factors than any other *M. loti* strain described (Olsthoorn et al. 1998). This does not, however, result in an enlarged spectrum of hosts. Furthermore, even an artificial change in the spectrum of Nod-factors does not necessarily lead to an altered host-range. NodL of *R. leguminosarum* bv. *viciae* is an enzyme that adds an *O*-acetyl group to C-6 of the non-reducing *N*-acetylglucosamine residue of Nod-factors. Introduction of the *nodL* gene into *M. loti* E1R did not affect the host spectrum either, although the transformed *M. loti* strain also produced *O*-acetylated Nod-factors (López-Lara et al. 2001). That said, some Nod-factor substituents have been reported to play roles in successful nodulation: *R. meliloti*, which produces sulphated Nod-factors, nodulates *Medicago sativa*. *nodH* mutants produce non-sulphated Nod-factors, and lose the ability to nodulate *M. sativa*, but gain the capacity to nodulate *Vicia sativa* subsp. *nigra*, which the wild-type *R. meliloti* strain cannot nodulate (Roche et al. 1991). For efficient stem and root nodulation on *Sesbania rostrata*, *Azorhizobium caulinodans* must produce Nod-factors which carry a *D*-arabinosyl or a *L*-fucosyl group, or both, at the reducing terminal residue. Bacterial mutants that produce Nod-factors lacking either the *D*-arabinosyl or the *L*-fucosyl group nodulate *S. rostrata* less efficiently, indicating that none of these decorations is strictly required for normal nodule formation, but that they act synergistically (D'Haeze et al. 2000).

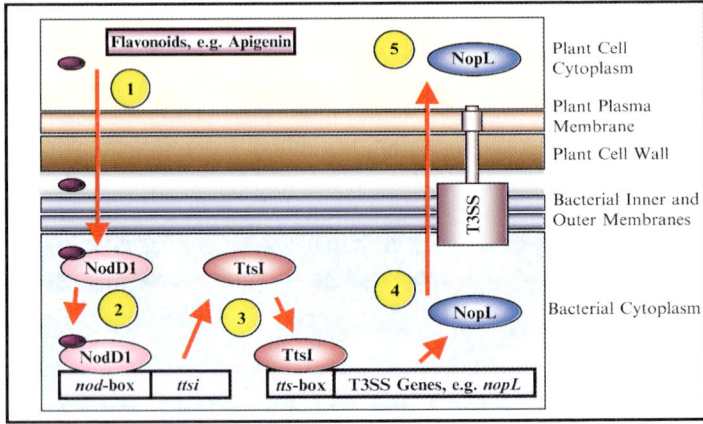

Fig. 1. Model of induction of the T3SS of *Rhizobium* sp. NGR234 by flavonoids. The host plant secretes flavonoids (*1*), e.g., apigenin, that are captured by the bacterial NodD1 transcriptional activator. NodD1 binds to *nod*-boxes and hence *nod-box* dependent genes are transcribed (*2*). This starts the production of Nod-factors and, amongst others, of the transcriptional activator TtsI. TtsI then binds to *tts*-boxes and so T3SS related genes are transcribed (*3*). This includes the T3SS machinery itself (*4*), but also T3SS effector proteins such as NopL (*5*)

3
Nod-Factor Perception

In some plants, homologous Nod-factors (i.e., Nod-factors produced by the native symbiont of the plant) can induce a variety of effects, such as root-hair curling, induction of pre-infection threads (Catoira et al. 2001), division of cortical cells and nodule morphogenesis (Dénarié et al. 1996; Bladergroen and Spaink 1998; Schultze and Kondorosi 1998; Downie and Walker 1999). Root-hair curling is provoked by depolarisation of the plasma-membrane of the root-hair, which allows an influx and accumulation of calcium at the tip (Felle et al. 1996, 1998; Gehring et al. 1997). Then, possibly mediated by G-proteins, calcium spiking is induced (Ehrhardt et al. 1996; Pringet et al. 1998). Nod-factors of NGR234 have been shown to stimulate the activity of a phospholipase C and of both heterotrimeric and monomeric G-proteins in *Vigna unguiculata* (Kelly and Irving 2001, 2003). This leads to rearrangements of the cytoskeleton (Miller et al. 1999; Cárdenas et al. 2003), provoking the typical shepherd's crook-like root-hair curling. Many transcripts are induced early in nodulation and the genes that encode them are commonly referred to as early nodulin genes (*ENOD*s) (Schultze and Kondorosi 1998; Downie and Walker 1999). Expression of early nodulin genes and division of the cortical cells follows the extreme curling of the root hairs and eventually leads to nodule formation (Yang et al. 1994).

Many *ENODs* have been identified, but the exact function of none of them is known. *ENOD40* for example, is expressed before cell division in the pericycle opposite the protoxylem poles occurs, but it is also expressed in dividing cells (Fang and Hirsch 1998; Kouchi et al. 1999). In transgenic legumes that over-express *ENDO40*, cortical-cell division is initiated at many places throughout the root (Charon et al. 1997). *ENOD40* encodes a very short peptide, but it is more likely that the main function of ENOD40 resides in a long, non-coding RNA that appears to have a regulatory role. *ENOD40* RNA causes *Mt*RBP1 (*Medicago truncatula* RNA Binding Protein 1) to be relocated from the nucleus into the cytoplasm (Campalans et al. 2004). Other plant responses to Nod-factors include a plant-dependent increase in flavonoid production (Recourt et al. 1991; Schmidt et al. 1994). Nod-factors also appear to trigger their own hydrolysis by the plant (Staehelin et al. 1995; Ovtsyna et al. 2000).

How the plant perceives Nod-factors is the subject of intense research. A *nodF-nodL* double mutant of *R. meliloti*, which produces Nod-factors lacking the acetyl group at the non-reducing terminus and which contain an alternative lipid, loses the ability to nodulate *M. sativa*. This mutant is still able to elicit root-hair deformation and inner cortical cell division however (Ardourel et al. 1994). An analogous finding has been made with a *R. leguminosarum* bv. *viciae nodFEMNTLO* mutant inoculated on *Vicia hirsuta* (Walker and Downie 2000). Interestingly, nodulation by the *nodFEMNTLO* mutant can be partially rescued by over-expression of *nodO*. NodO is secreted from bacteria and makes pores in membranes (Vlassak et al. 1998). NodO re-establishes the ion-influx that the non-substituted Nod-factors of the *nodFEMNTLO* mutant can no longer provoke. A two-step model for Nod-factor perception has thus been proposed: first, a less stringent Nod-factor receptor prepares the plant for bacterial invasion; second, a more stringent Nod-factor receptor actually allows bacterial entry and infection thread growth (Ardourel et al. 1994). Alternatively, one Nod-factor receptor could perform both tasks by being activated first in a less stringent manner, but would subsequently need strong activation by tight binding to a Nod-factor to sustain bacterial entry and infection-thread growth. The latter model is supported by recent findings on Nod-factor receptors containing LysM domains. In 2003, the group of Jens Stougaard (University of Aarhus, Denmark) discovered what are most likely the high affinity Nod-factor receptors that have been sought for so long (Madsen et al. 2003; Radutoiu et al. 2003). Using a genetic approach in *L. japonicus*, they identified two genes, *nfr1* and *nfr5*. Mutants of *nfr1* or *nfr5* fail to respond to Nod-factors. NFR1 and NFR5 code for two receptor kinases with a LysM motif in the extra-cellular domain. NFR5, however, lacks a kinase-activation loop in its kinase domain and it is no longer an active kinase. Upon binding to Nod-factors, NFR1 and NFR5 are predicted to form hetero-dimers via their LysM motifs. Formation of hetero-dimers may then activate the intra-cellular kinase domain of NFR1 (Parniske and Downie 2003). *P. sativum* has a homologue of NFR5, called SYM10 (Madsen et al. 2003). Another NFR1 homologue, LYK3 was also identified in *M. trunculata* (Limpens et al. 2003). Although NFR1 and NFR5 are early high-affinity receptors for Nod-factors, they also appear to be required throughout the infection process (Parniske and Downie 2003). So far, direct biochemical evidence supporting neither NFR1/NFR5-hetero-dimer formation, nor Nod-factor binding to the hetero-dimer or to either NFR1 or NFR5 has been found. LysM domains have been identified in an *Escherichia coli* protein, MltD, that binds peptidoglycan (Steen et al. 2003). Additionally, LysM domains have also been found in at least two chitin-binding proteins (Ponting et al. 1999). LysM domains are thus good candidates for Nod-factor binding sites. Furthermore, chitin is able to induce calcium-spiking in legumes (Pringet et al. 1998; Walker et al. 2000). Together, the very good genetic evidence and the circumstantial data inferred from structural similarities strongly suggest that NFR1 and NFR5 are Nod-factor receptors (Cullimore and Dénarié 2003; Oldroyd and Downie 2004), but hard biochemical evidence for this is eagerly awaited.

Several other approaches have been used to identify Nod-factor receptors (Geurts and Bisseling 2002). A Nod-factor binding site (called NFBS1 for Nod-Factor Binding Site 1) has been characterised in root extracts of *M. trunculata* (Bono et al. 1995). NFBS1 has also been identified in tomato (*Lycopersicon esculentum*) however, which suggests that NFBS1 plays a more general role. NFBS1 would fit into the model of Ardourel et al. (1994) as the less-stringent Nod-factor receptor. Another Nod-factor binding site, NFBS2, has been identified in the microsomal fraction of *Medicago varia* cell suspension cultures (Niebel et al. 1997), although its affinity for Nod-factors is not higher than that of NFBS1. Another possible Nod-factor binding-protein is a lectin-nucleotide-phosphohydrolase (LNP), which was isolated from the roots of the legume *Dolichos biflorus* (Etzler et al. 1999). LNP is an ATPase and it's enzymatic activity increases upon binding of Nod-factors to the lectin domain of LNP. Furthermore, LNP is present on the surface of root hairs, but in presence of Nod-factors it accumulates at their tips (Day et al. 2000; Kalsi and Etzler 2000).

Several genes controlling the very early stages of symbiosis have also been cloned. *M. trunculata* mutants in three genes called *dmi1* (does not make infections), *dmi2*, and *dmi3* have pleiotropic effects on Nod-factor responses and are unable to establish a symbiotic association with endomycorrhizal fungi (Catoira et al. 2000; Mitra et al. 2004). This suggests that many of the early steps in symbioses with rhizobia and endomycorrhizal fungi are the same (Albrecht et al. 1999). Grafting experiments have shown that the genetic control of *dmi1*, *dmi2* and *dmi3* is determined at the root level (Ané et al. 2002). DMI1 has slight similarities to a ligand-gated cation channel of archaea

(Ané et al. 2004). DMI3 shows high similarity to calmodulin-dependent protein kinases (Lévy et al. 2004). Various orthologues of DMI2 have been found: a nodulation receptor kinase (NORK) has been identified in *M. sativa* and, by analogy, in *M. trunculata* and *P. sativum* (Endre et al. 2002). NORK extra-cellular sequence-like (*NSL*) genes are widely distributed throughout the plant kingdom and it thus appears that an ancient system of *NSL* genes has been modified to serve as symbiotic Nod-factor receptors (Endre et al. 2002). In another study, symbiosis receptor-like kinase (*SYMRK*) genes have been cloned from *L. japonicus* and *P. sativum* (Stracke et al. 2002). *SYMRK* mutants of *L. japonicus* are unable to establish symbiosis with either rhizobia or endomycorrhizal fungi. In addition, an orthologue of the *DMI2* gene of *M. trunculata* has been cloned from *P. sativum* (*Sym19*) (Schneider et al. 1999). *SYMRK* of *L. japonicus*, *Sym19* of *P. sativum*, *NORK* of *M. sativa* and *DMI2* of *M. trunculata* all belong to the family of leucine-rich repeat (LLR) receptor-kinases (Endre et al. 2002; Stracke et al. 2002). This suggests that these receptor-like kinases may form homo- or hetero-dimers or interact with other proteins containing LLR domains.

4
Surface Polysaccharides

Nod-factors are essential signal molecules that permit rhizobia to enter legume roots, but they are not the only bacterial determinants required for successful symbiosis. Other determinants include secreted proteins and surface polysaccharides (SPS). SPS include exo-polysaccharides (EPS), lipo-polysaccharides (LPS), capsular polysaccharides (CPS and K-Antigens) and cyclic glucans (Fraysse et al. 2003). SPS are abundant extra-cellular components that are exported to the cell surface and into the cell's surroundings. SPS facilitate bacterial attachment to surfaces and protect the cells from environmental stresses (Fraysse et al. 2003; Spaink 2000). EPS also play essential roles in the establishment of nitrogen-fixing symbiosis on plants that form indeterminate nodules (Leigh et al. 1985; Djordjevic et al. 1987; Breedveld et al. 1993). Interestingly, with plants that form determinate nodules, EPS seem not to be essential for the formation of functional nodules (Glazebrook et al. 1990; Cheng and Walker 1998; van Workum et al. 1998; Pellock et al. 2000). It is possible in these cases that LPS may take on the role of EPS. Another possibility is that the principal role of EPS may be in penetration of cortical cells. During formation of indeterminate nodules, bacteria continuously infect new cells, while in determinate nodules they propagate via cell division of infected cortical cells (Reuber et al. 1991; Leigh and Coplin 1992).

Rhizobium meliloti produces two main classes of EPS: EPSI and EPSII. EPSI consist of high molecular weight polysaccharides of 10^6 to 10^7 Da (Leigh and Lee 1988) which are called succinoglycans. Succinoglycans consist of repeats of octasaccharide subunits (Reuber and Walker 1993; Reinhold et al. 1994). Succinoglycans also occur as monomers, dimers and trimers, called the low molecular weight fraction of EPSI (Wang et al. 1999). EPSI mutants of *R. meliloti* lead to the formation of empty nodules on *M. sativa* (Finan et al. 1985; Leigh et al. 1985). Nodule formation can be rescued by addition of the low molecular fraction of EPSI (Battisti et al. 1992; Wang et al. 1999), which suggests that the low molecular-weight form of EPSI is the symbiotically active form. EPSI succinoglycan can be decorated with succinyl, pyruvyl or acetyl substituents. An *exoH* mutant of *R. meliloti* (which lacks the succinyl decoration on its EPSI) is still able to induce infection threads on *M. sativa*, but the elongation of the infection thread is greatly impaired (Cheng and Walker 1998). A possible role for EPSI might be the suppression of plant defences or the avoidance of recognition by the plant as a pathogen since EPSI mutants of *R. meliloti* trigger high levels of plant defence reactions on *M. sativa* (Niehaus et al. 1993).

Some *R. meliloti* strains produce EPSII in addition to EPSI. EPSII are low molecular weight polymers consisting of repeated galactose-glucose subunits with acetyl and pyruvyl substituents (Her et al. 1990; Levery et al. 1991). *R. meliloti* EPSI mutants in the background of EPSII producing strains, still lead to the formation of nitrogen-fixing nodules on *M. sativa*. Hence EPSII are able to substitute for the absence of EPSI (González et al. 1996; Pellock et al. 2000).

Although EPSI of *R. meliloti* are constitutively produced, it cannot be excluded that the abundance and composition of many SPS changes upon interaction with the host plant. For example, in NGR234 synthesis of a rhamnose-rich LPS is under transcriptional control of NodD1 and TtsI. It is thus only produced in the presence of inducing flavonoids (Fraysse et al. 2002; Marie et al. 2004). Bacteroids of *B. japonicum* within nodules of *Glycine max* produce EPS that are structurally different from those found in free-living bacteria (Streeter et al. 1992; An et al. 1995). In numerous symbiotic bacteria-plant interactions, LPS structures change greatly during establishment of symbiosis. Under low oxygen conditions, *R. leguminosarum* strain 3841 produces more hydrophobic LPS. This is true for bacteroids as well as for free-living bacteria, when grown under low oxygen pressures (Kannenberg and Carlson 2001). It is thus possible that low oxygen is one of the clues that bacteria use to adapt themselves to the new environment within nodules.

LPS make up a large fraction of the outer leaflet of the lipid bilayer of bacterial outer-membranes (see Whitfield and Valvano 1993). Smooth LPS (S-LPS) molecules are made of three main moieties: the lipid A membrane anchor, a core oligo-saccharide (which is linked to the lipid A via 3-deoxy-D-*manno*-2-octulosonic acid), and an antigenic polysaccharide, the

O-Antigen. The rough LPS (R-LPS) molecules are identical to the S-LPS but lack the O-Antigen.

LPSs play significant roles in the interaction of bacteria and plants. A *R. meliloti lpsB* mutant is still able to invade *M. sativa* nodules, but is unable to establish the normal intracellular infection required for bacteroid development (Campbell et al. 2002). *lpsB* encodes a glycosyltransferase that, when mutated, significantly alters the carbohydrate core of the LPS. One way to explain the *lpsB* phenotype is that this mutant is less resistant to the innate plant immune system, especially to cationic peptides. Cationic peptides have a high affinity for the negatively charged bacterial outer envelope and act as antimicrobial compounds. Thus, it is not surprising that free-living *lpsB* mutant bacteria are much more sensitive than the wild-type *R. meliloti* to cationic peptides added to the culture medium (Campbell et al. 2002). LPS may also play roles in plant-bacteria recognition processes (as components of signalling pathways), they may camouflage the bacteria in order to avoid eliciting plant defences, or they might be implicated in the resistance of bacteria to plant defences.

The bacterial surface is often covered by a layer of capsular polysaccharides (CPS) and K-antigens. K-antigens are polysaccharides containing a high proportion of 3-deoxy-D-*manno*-2-octulosonic acid (Kdo) and show structural analogy to a subgroup of K-antigens in *E. coli* (Reuhs et al. 1993). K-antigens may play roles in the recognition of bacteria by plants (Forsberg and Reuhs 1997; Forsberg and Carlson 1998) since they adhere to the outer surface of the bacteria and are not secreted into the medium. In general, growth conditions affect K-antigen production, which in *R. fredii* USDA205 at least, is also controlled by flavonoid-induction (Reuhs et al. 1993, 1995). Infiltration of *M. sativa* leaves with *R. meliloti* and measurement of induction of the isoflavonoid pathway has shown that the K-antigen is necessary for rapid recognition by the plant (Becquart-de Kozak et al. 1997). This finding cannot be automatically extrapolated to root cells however, and further experiments are required to determine the role of K-antigens in rhizobia-plant recognition.

A *R. meliloti* EPS deficient strain is known which is still able to establish nitrogen-fixing nodules with *M. sativa*. It is likely that, in this peculiar case, CPS at least partially substitute for EPS (Putnoky et al. 1990; Kiss et al. 2001). And this points to the complications in the analysis of SPS function – often, different SPS act synergistically. CPS and EPSI (Putnoky et al. 1990; Reuhs et al. 1995) or EPSI and EPSII (Glazebrook and Walker 1989; González, JE. et al. 1996) seem to be able to substitute for one another in *R. meliloti*. Undoubtedly, this synergy has limits since K-Antigens (in contrast to EPS) appear to be strain-specific rather than species-specific (Reuhs et al. 1993, 1995). Another difficulty in analysing SPS function is that their respective biosynthetic pathways are interconnected in manifold ways (Kereszt et al. 1998; Reuhs et al. 1998).

5
Secreted Proteins

Bacteria have evolved several different ways of secreting proteins into the environment including the type I, II, III, IV and V secretion systems. Not all bacteria dispose of every one of the five secretion systems and so far, only type I, III and IV secretion systems have been shown to be relevant for the establishment of symbioses. Type IV secretion systems (T4SSs) are being unveiled in ever more bacteria (Nagai and Roy 2003) and a T4SS was found to be encoded by the symbiotic plasmid of *R. etli* strain CNF42 (González V. et al. 2003). *M. loti* strain R7A possesses a T4SS, which is detrimental to nodulation of *Leucaena leucocephala,* but favours nodulation of *Lotus corniculatus* (Hubber et al. 2004).

To date, a symbiotic type I secretion system (T1SS) has only been found in *R. leguminosarum* bv. *viciae*. Here, a glycanase and at least one other protein necessary for symbiotic nitrogen fixation are secreted in a T1SS-dependent manner (Finnie et al. 1997; Vlassak et al. 1998; Walker and Downie 2000). More information is available on symbiotically relevant T3SSs. T3SSs are present in animal and plant pathogens, where their function is to subvert the host's immune system (Kempf et al. 2002; Büttner and Bonas 2003). Bacterial T3SSs consist of a core secretion machine, which spans both bacterial membranes and a pilus that establishes a link to the host plasma membrane on which a pore is formed (Fig. 1).

The core T3SS secretion machine is composed of a minimal set of ten genes that were originally identified in bacteria pathogenic to animals as *ysc* genes and in plants as *hrc* genes (Hueck 1998). *hrc* homologues of rhizobia are known as *rhc* (for *Rhizobium* conserved) in order to distinguish them from their pathogenic counterparts (Bogdanove et al. 1996; Viprey et al. 1998). In NGR234, the pilus is thought to be made of two secreted proteins, Nodulation Outer Protein A (NopA) (Deakin et al. 2005) and NopB (Saad et al. 2005). NopA is probably the main component of the pilus, while NopB could have a stabilizing role either as a part of the pilus itself or in association with it. The pilus is thought to connect to a pore in the plant plasma membrane, formed by NopX. NopX is a protein secreted by the T3SS of NGR234 (Marie et al. 2003) and by the T3SS of *R. fredii* USDA257 (Krishnan 2002). NopX is homologous to HrpF proteins of *Xanthomonas* and these are also thought to form a pore in the host's plasma-membrane (Rossier et al. 2000). This system, which resembles a syringe needle, allows the bacteria to secrete effector proteins directly into the cytoplasm of the host-cell (Hueck 1998).

Before the complete nucleotide sequence of the symbiotic plasmid pNGR234*a* was published (Freiberg et al. 1997), T3SSs were only known from pathogens. Since then, T3SSs have also been identified in a number of other rhizobia including *M. loti* MAFF 303099 (Kaneko et al. 2000a,b), *B. japonicum* USDA110 (Göttfert et al. 2001; Kaneko et al. 2002a,b; Krause et al. 2002), *R. etli* CNF42 (Genebank U80928 - the complete genome sequence of *R. etli* will be published shortly by the group of Guillermo Dávila, UNAM, Cuernavaca, Mexico) and *R. fredii* strains USDA257 and HH103 (Krishnan et al. 2003; Vinardell et al. 2004). Interestingly, the complete nucleotide sequence of *R. meliloti* strain 1021, did not reveal the presence of a T3SS (Galibert et al. 2001). As more and more bacterial genomes are being sequenced, our understanding of the evolution and phylogenetic distribution of T3SSs is bound to increase (Pühler et al. 2004). In pathogenic bacteria, T3SSs are not constitutively produced, but are induced by contact with host cells (He 1998). In NGR234, the T3SS is under the control of the transcriptional regulator TtsI, which itself is inducible in a flavonoid-NodD1-dependent manner (Marie et al. 2004) (Fig. 1).

What kinds of proteins are secreted by T3SSs? They can be grouped into three classes: external components of the T3SS as well as translocator and effector proteins (He 1998; Hueck 1998). External components of the T3SS include the pilus. Translocator proteins are those that facilitate the translocation of effector proteins into the host cytoplasm like NopX. And finally, the effector proteins, as their name implies, are active in the cytoplasm of the host cell. In pathogenic bacteria, effectors subvert the host cell's metabolism and immune system.

Perhaps surprisingly, the symbiotic relevance of effector proteins varies with the host-plant. Mutation of the *rhcN* gene of NGR234 renders the T3SS non-functional, but vastly augments nodulation of the tropical legume *Pachyrhizus tuberosus*. *Tephrosia vogelii* on the other hand, requires a functional T3SS for optimal nodulation, while *L. japonicus* and *Vigna unguiculata* are indifferent to the presence or absence of T3SS proteins (Viprey et al. 1998). Absence of the effector protein NopL of NGR234 leads to fewer nodules on *Flemingia congesta* (Marie et al. 2003). T3SS secreted proteins do not only affect the number of nodules formed. Inoculation of *Crotalaria juncea* with either NGR234 or the *rhcN* mutant leads to similar numbers of nodules but only those formed by the *rhcN* mutant are Fix$^+$ (Marie et al. 2003). NopL, a protein secreted by the T3SS of NGR234, thwarts plant defence reactions (Bartsev et al. 2004). NopL probably exercises this function by interfering with the plant's MAP-kinase signalling pathway (Bartsev et al. 2003). NopP, a second effector protein of NGR234 has been identified (Ausmees et al. 2004; Skorpil et al. 2005), and it is likely that more effector proteins will be characterised (Marie et al. 2003).

6
Conclusions

Tremendous advances in our understanding of the *Rhizobium*-legume symbioses have been made. The probable identification of NFR1 and NFR5 as

Nod-factor receptors and the discovery of symbiotically relevant protein secretion systems in rhizobia, are but two of the milestones in this field of research. Light is being shed on the fundamental role of SPS in symbiosis. Probable commonalities between the infection of legumes by arbuscular mycorrhiza and rhizobia suggest similar evolutionary mechanisms. In the years that lay ahead, more parts of the symbiotic puzzle will be put together at an ever-increasing pace.

Acknowledgements. We would like to thank Drs. Christian Staehelin and Maged Saad (University of Geneva) for helpful comments and advice. We are grateful to Daw Yin-Yin Aung and Florencia Ares-Orpel for their help with plant tests and to Dora Gerber for her general support. Financial assistance was provided by the Université de Genève and the Fonds National Suisse de la Recherche Scientifique (projects 31–63893.00 and 3100AO-104097/1).

References

Albrecht C, Geurts R, Bisseling T (1999) Legume nodulation and mycorrhizae formation; two extremes in host specificity meet. EMBO J 18:281–288

An J, Carlson RW, Glushka J, Streeter JG (1995) The structure of a novel polysaccharide produced by *Bradyrhizobium* species within soybean nodules. Carbohydr Res 269:303–317

Ané J-M, Lévy J, Thoquet P, Kulikova O, de Billy F, Penmetsa V, Kim D-J, Debellé F, Rosenberg C, Cook DR, Bisseling T, Huguet T, Dénarié J (2002) Genetic and cytogenetic mapping of *DMI1*, *DMI2* and *DMI3* genes of *Medicago trunculata* involved in Nod factor transduction, nodulation and mycorrhization. Mol. Plant Microbe Interactions 15:1108–1118

Ané J-M, Kiss GB, Riely BK, Penmetsa RV, Oldroyd GED, Ayax C, Lévy J, Debellé F, Baek J-M, Kalo P, Rosenberg C, Roe BA, Long SR, Dénarié J, Cook DR (2004) *Medicago truncatula DMI1* required for bacterial and fungal symbioses in legumes. Science 303:1364–1367

Ardourel M, Demont N, Debellé F, Maillet F, de Billy F, Promé J-C, Dénarié J, Truchet G (1994) *Rhizobium meliloti* lipooligosaccharide nodulation factors: different structural requirements for bacterial entry into target root hair cells and induction of plant symbiotic developmental responses. Plant Cell 6:1357–1374

Ausmees N, Kobayashi H, Deakin WJ, Marie C, Krishnan HB, Broughton WJ, Perret X (2004) Characterization of NopP, a type III secreted effector of *Rhizobium* sp. strain NGR234. J Bacteriol 186:4774–4780

Bartsev AV, Boukli NM, Deakin WJ, Staehelin C, Broughton WJ (2003) Purification and phosphorylation of the effector protein NopL from *Rhizobium* sp. NGR234. FEBS Lett 554:271–274

Bartsev AV, Deakin WJ, Boukli NM, McAlvin CB, Stacey G, Malnoë P, Broughton WJ, Staehelin C (2004) NopL, an effector protein of *Rhizobium* sp. NGR234, thwarts activation of plant defense reactions. Plant Physiol 134:871–879

Battisti L, Lara JC, Leigh JA (1992) Specific oligosaccharide form of the *Rhizobium meliloti* exopolysaccharide promotes nodule invasion in alfalfa. Proc Natl Acad Sci USA 89:5625–5629

Bec-Ferté M-P, Krishnan HB, Savagnac A, Pueppke SG, Promé J-C (1996) *Rhizobium fredii* synthesizes an array of lipooligosaccharides, including a novel compound with glucose inserted into the backbone of the molecule. FEBS Lett 393:273–279

Becquart-de Kozak I, Reuhs BL, Buffard D, Breda C, Kim JS, Esnault R, Kondorosi A (1997) Role of the K-antigen subgroup of capsular polysaccharides in the early recognition process between *Rhizobium meliloti* and alfalfa leaves. Mol Plant Microbe Interact 10:114–123

Bladergroen MR, Spaink HP (1998) Genes and signal molecules involved in the rhizobia-leguminoseae symbiosis. Curr Opin Plant Biol 1:353–359

Bogdanove AJ, Beer SV, Bonas U, Boucher CA, Collmer A, Coplin DL, Cornelis GR, Huang H-C, Hutcheson SW, Panopoulos NJ, van Gijsegem F (1996) Unified nomenclature for broadly conserved *hrp* genes of phytopathogenic bacteria. Mol Microbiol 20:681–683

Bono J-J, Riond J, Nicolaou KC, Bockovich NJ, Estevez VA, Cullimore JV, Ranjeva R (1995) Characterization of a binding site for chemically synthesized lipo-oligosaccharidic NodRm factors in particulate fractions prepared from roots. Plant J 7:253–260

Breedveld MW, Cremers HCJC, Batley M, Posthumus MA, Zevenhuizen LPTM, Wijffelman CA, Zehnder AJB (1993) Polysaccharide synthesis in relation to nodulation behavior of *Rhizobium leguminosarum*. J Bacteriol 175:750–757

Broughton WJ, Jabbouri S, Perret X (2000) Keys to symbiotic harmony. J Bacteriol 182:5641–5652

Broughton WJ, Zhang F, Perret X, Staehelin C (2003) Signals exchanged between legumes and *Rhizobium*: agricultural uses and perspectives. Plant Soil 252:129–137

Büttner D, Bonas U (2003) Common infection strategies of plant and animal pathogenic bacteria. Curr Opin Plant Biol 6:312–319

Campalans A, Kondorosi A, Crespi MD (2004) *Enod40*, a short open reading frame-containing mRNA, induces cytoplasmic localization of a nuclear RNA binding protein in *Medicago trunculata*. Plant Cell 16:1047–1059

Campbell GRO, Reuhs BL, Walker GC (2002) Chronic intracellular infection of alfalfa nodules by *Sinorhizobium meliloti* requires correct lipopolysaccharide core. Proc Natl Acad Sci USA 99:3938–3943

Cárdenas L, Thomas-Oates JE, Nava N, López IM, Hepler PK, Quinto C (2003) The role of Nod factor substituents in actin cytoskeleton rearrangements in *Phaseolus vulgaris*. Mol Plant Microbe Interact 16:326–334

Catoira R, Galera C, de Billy F, Penmetsa RV, Journet E-P, Maillet F, Rosenberg C, Cook DR, Gough C, Dénarié J (2000) Four genes of *Medicago truncatula* controlling components of a Nod factor transduction pathway. Plant Cell 12:1647–1666

Catoira R, Timmers ACJ, Maillet F, Galera C, Penmetsa RV, Cook DR, Dénarié J, Gough C (2001) The *HCL* gene of *Medicago truncatula* controls *Rhizobium*-induced root hair curling. Development 128:1507–1518

Charon C, Johansson C, Kondorosi E, Kondorosi A, Crespi MD (1997) *enod40* induces dedifferentiation and division of root cortical cells in legumes. Proc Natl Acad Sci USA 94:8901–8906

Cheng H-P, Walker GC (1998) Succinoglycan is required for initiation and elongation of infection threads during nodulation of alfalfa by *Rhizobium meliloti*. J Bacteriol 180:5183–5191

Cullimore JV, Dénarié J (2003) Plant sciences. How legumes select their sweet talking symbionts. Science 302:575–578

Day RB, McAlvin CB, Loh JT, Denny RL, Wood TC, Young ND, Stacey G (2000) Differential expression of two soybean apyrases, one of which is an early nodulin. Mol Plant Microbe Interact 13:1053–1070

Deakin, WJ, Marie, C, Saad, MM, Krishnan, HB, and Broughton, WJ (2005) NopA is associated with cell surface appendages produced by the Type III Secretion System of *Rhizobium* sp. strain NGR234. *Mol Plant-Microbe Interact* 18: 499–507

Demont N, Ardourel M, Maillet F, Promé D, Ferro M, Promé J-C, Dénarié J (1994) The *Rhizobium meliloti* regulatory *nodD3* and *syrM* genes control the synthesis of a particular class of nodulation factors *N*-acylated by (*omega*-1)-hydroxylated fatty acids. EMBO J 13:2139–2149

Dénarié J, Debellé J, Promé D (1996) *Rhizobium* lipo-chitooligosaccharide nodulation factors: signaling molecules mediating recognition and morphogenesis. Annu Rev Biochem 65:503–535

D'Haeze W, Mergaert P, Promé J-C, Holsters M (2000) Nod factor requirements for efficient stem and root nodulation of the tropical legume *Sesbania rostrata*. J Biol Chem 275:15676–15684

Djordjevic SP, Chen H, Batley M, Redmond JW, Rolfe BG (1987) Nitrogen fixation ability of exopolysaccharide synthesis mutants of *Rhizobium* sp. strain NGR234 and *Rhizobium trifolii* is restored by the addition of homologous exopolysaccharides. J Bacteriol 169:53–60

Downie JA, Walker SA (1999) Plant responses to nodulation factors. Curr Opin Plant Biol 2:483–489

Ehrhardt DW, Wais R, Long SR (1996) Calcium spiking in plant root hairs responding to *Rhizobium* nodulation signals. Cell 85:673–681

Endre G, Kereszt A, Kevei Z, Mihacea S, Kaló P, Kiss GB (2002) A receptor kinase gene regulating symbiotic nodule development. Nature 417:962–966

Etzler ME, Kalsi G, Ewing NN, Roberts NJ, Day RB, Murphy JB (1999) A Nod factor binding lectin with apyrase activity from legume roots. Proc Natl Acad Sci USA 96:5856–5861

Fang Y, Hirsch AM (1998) Studying early nodulin gene *ENOD40* expression and induction by nodulation factor and cytokinin in transgenic alfalfa. Plant Physiol 116:53–68

Fellay R, Perret X, Viprey V, Broughton WJ, Brenner S (1995) Organization of host-inducible transcripts on the symbiotic plasmid of *Rhizobium* sp. NGR234. Mol Microbiol 16:657–667

Felle HH, Kondorosi É, Kondorosi Á, Schultze M (1996) Rapid alkalinization in alfalfa root hairs in response to rhizobial lipochitoologosaccharide signals. Plant J 10:295–301

Felle HH, Kondorosi É, Kondorosi Á, Schultze M (1998) The role of ion fluxes in Nod factor signalling in *Medicago sativa*. Plant J 13:455–463

Feng J, Li Q, Hu H-L, Chen X-C, Hong G-F (2003) Inactivation of the *nod* box distal half-site allows tetrameric NodD to activate *nodA* transcription in an inducer-independent manner. Nucleic Acids Res 31:3143–3156

Finan TM, Hirsch AM, Leigh JA, Johansen E, Kuldau GA, Deegan S, Walker GC, Signer ER (1985) Symbiotic mutants of *Rhizobium meliloti* that uncouple plant from bacterial differentiation. Cell 40:869–877

Finnie C, Hartley NM, Findlay KC, Downie JA (1997) The *Rhizobium leguminosarum prsDE* genes are required for secretion of several proteins, some of which influence nodulation, symbiotic nitrogen fixation and exopolysaccharide modification. Mol Microbiol 25:135–146

Fisher RF, Long SR (1993) Interactions of NodD at the *nod* Box: NodD binds to two distinct sites on the same face of the helix and induces a bend in the DNA. J Mol Biol 233:336–348

Forsberg LS, Carlson RW (1998) The structures of the lipopolysaccharides from *Rhizobium etli* strains CE358 and CE359. The complete structure of the core region of *R. etli* lipopolysaccharides. J Biol Chem 273:2747–2757

Forsberg LS, Reuhs BL (1997) Structural characterization of the K antigens from *Rhizobium fredii* USDA257: evidence for a common structural motif, with strain-specific variation, in the capsular polysaccharides of *Rhizobium* spp. J Bacteriol 179:5366–5371

Fraysse N, Jabbouri S, Treilhou M, Couderc F, Poinsot V (2002) Symbiotic conditions induce structural modifications of *Sinorhizobium* sp. NGR234 surface polysaccharides. Glycobiology 12:741–748

Fraysse N, Couderc F, Poinsot V (2003) Surface polysaccharide involvement in establishing the *Rhizobium*-legume symbiosis. Eur J Biochem 270:1365–1380

Freiberg C, Fellay R, Bairoch A, Broughton WJ, Rosenthal A, Perret X (1997) Molecular basis of symbiosis between *Rhizobium* and legumes. Nature 387:394–401

Gage DJ (2004) Infection and invasion of roots by symbiotic, nitrogen-fixing rhizobia during nodulation of temperate legumes. Microbiol Mol Biol Rev 68:280–300

Galibert F, Finan TM, Long SR, Pühler A, Abola P, Ampe F, Barloy-Hubler F, Barnett MJ, Becker A, Boistard P, Bothe G, Boutry M, Bowser L, Buhrmester J, Cadieu E, Capela D, Chain P, Cowie A, Davis RW, Dréano S, Federspiel NA, Fisher RF, Gloux S, Godric T, Goffeau A, Golding B, Gouzy J, Gurjal M, Hernandez-Lucas I, Hong A, Huizar L, Hyman RW, Jones T, Kahn D, Kahn ML, Kalman S, Keating DH, Kiss E, Komp C, Lelaure V, Masuy D, Palm C, Peck MC, Pohl TM, Portetelle D, Purnelle B, Ramsperger U, Surzycki R, Thébault P, Vandenbol M, Vorhölter F-J, Weidner S, Wells DH, Wong K, Yeh K-C, Batut J (2001) The composite genome of the legume symbiont *Sinorhizobium meliloti*. Science 293:668–672

Gehring CA, Irving HR, Kabbara AA, Parish RW, Boukli NM, Broughton WJ (1997) Rapid, plateau-like increases in intracelluar free calcium are associated with Nod-factor-induced root-hair-deformation. Mol Plant Microbe Interact 10:791–802

Geurts R, Bisseling T (2002) *Rhizobium* Nod factor perception and signalling. Plant Cell 14:S239-S249

Glazebrook J, Walker GC (1989) A novel exopolysaccharide can function in place of the calcofluor-binding exopolysaccharide in nodulation of alfalfa by Rhizobium meliloti. Cell 56:661–672

Glazebrook J, Reed JW, Reuber TL, Walker GC (1990) Genetic analyses of *Rhizobium meliloti* exopolysaccharides. Int J Biol Macromol 12:67–70

González JE, Reuhs BL, Walker GC (1996) Low molecular weight EPS II of *Rhizobium meliloti* allows nodule invasion in *Medicago sativa*. Proc Natl Acad Sci USA 93:8636–8641

González V, Bustos P, Ramírez-Romero MA, Medrano-Soto A, Salgado H, Hernández-González I, Hernández-Celis JC, Quintero V, Moreno-Hagelsieb G, Girard L, Rodríguez O, Flores M, Cevallos MA, Collado-Vides J, Romero D, Dávila G (2003) The mosaic structure of the symbiotic plasmid of *Rhizobium etli* CFN42 and its relation to other symbiotic genome compartments. Genome Biol 4:R36

Göttfert M, Röthlisberger S, Kündig C, Beck C, Marty R, Hennecke H (2001) Potential symbiosis-specific genes uncovered by sequencing a 410-kilobase DNA region of the *Bradyrhizobium japonicum* chromosome. J Bacteriol 183:1405–1412

Gresshoff PM (2003) Post-genomic insights into plant nodulation symbioses. Genome Biol 4:201

He SY (1998) Type III protein secretion systems in plant and animal phathogenic bacteria. Annu Rev Phytopathol 36:363–392

Hellriegel H, Wilfarth H (1888) Untersuchungen über die Stickstoffnahrung der Gramineen und Leguminosen. Beilageheft zu der Zeitschrift des Vereines für die Rübenzucker-Industrie des Deutschen Reiches K. C. Buchdruckerei der "Post", Berlin, Germany

Her G-R, Glazebrook J, Walker GC, Reinhold VN (1990) Structural studies of a novel exopolysaccharide produced by a mutant of *Rhizobium meliloti* strain Rm1021. Carbohydr Res 198:305–312

Hirsch AM, Lum MR, Downie JA (2001) What makes the rhizobia-legume symbiosis so special? Plant Physiol 127:1484–1492

Hogg B, Davies AE, Wilson KE, Bisseling T, Downie JA (2002) Competitive nodulation blocking of cv. Afghanistan pea is related to high levels of nodulation factors made by some strains of *Rhizobium leguminosarum* bv. *viciae*. Mol. Plant Microbe Interact 15:60–68

Hubber A, Vergunst AC, Sullivan JT, Hooykaas PJ and Ronson CW (2004) Symbiotic phenotypes and translocated effector proteins of the *Mesorhizobium loti* strain R7A VirB/D4 type IV secretion system. Mol Microbiol 54:561–574

Hueck CJ (1998) Type III protein secretion systems in bacterial pathogens of animals and plants. Microbiol Mol Biol Rev 62:379–433

Kalsi G, Etzler ME (2000) Localization of a Nod factor-binding protein in legume roots and factors influencing its distribution and expression. Plant Physiol 124:1039–1048

Kaneko T, Nakamura Y, Sato S, Asamizu E, Kato T, Sasamoto S, Watanabe A, Idesawa K, Ishikawa A, Kawashima K, Kimura T, Kishida Y, Kiyokawa C, Kohara M, Matsumoto M, Matsuno A, Mochizuki Y, Nakayama S, Nakazaki N, Shimpo S, Sugimoto M, Takeuchi C, Yamada M, Tabata S (2000a) Complete genome structure of the nitrogen-fixing symbiotic bacterium *Mesorhizobium loti*. DNA Res 7:331–338

Kaneko T, Nakamura Y, Sato S, Asamizu E, Kato T, Sasamoto S, Watanabe A, Idesawa K, Ishikawa A, Kawashima K, Kimura T, Kishida Y, Kiyokawa C, Kohara M, Matsumoto M, Matsuno A, Mochizuki Y, Nakayama S, Nakazaki N, Shimpo S,

Sugimoto M, Takeuchi C, Yamada M, Tabata S (2000b) Complete genome structure of the nitrogen-fixing symbiotic bacterium *Mesorhizobium loti* (supplement). DNA Res 7:381–406

Kaneko T, Nakamura Y, Sato S, Minamisawa K, Uchiumi T, Sasamoto S, Watanabe A, Idesawa K, Iriguchi M, Kawashima K, Kohara M, Matsumoto M, Shimpo S, Tsuruoka H, Wada T, Yamada M, Tabata S (2002a) Complete genomic sequence of nitrogen-fixing symbiotic bacterium *Bradyrhizobium japonicum* USDA110. DNA Res 9:189–197

Kaneko T, Nakamura Y, Sato S, Minamisawa K, Uchiumi T, Sasamoto S, Watanabe A, Idesawa K, Iriguchi M, Kawashima K, Kohara M, Matsumoto M, Shimpo S, Tsuruoka H, Wada T, Yamada M, Tabata S (2002b) Complete genomic sequence of nitrogen-fixing symbiotic bacterium *Bradyrhizobium japonicum* USDA110 (supplement). DNA Res 9:225–256

Kannenberg EL, Carlson RW (2001) Lipid A and O-chain modifications cause *Rhizobium* lipopolysaccharides to become hydrophobic during bacteroid development. Mol Microbiol 39:379–391

Kelly MN, Irving HR (2001) Nod factors stimulate plasma membrane delimited phospholipase C activity in vitro. Physiol Plant 113:461–468

Kelly MN, Irving HR (2003) Nod factors activate both heterotrimeric and monomeric G-proteins in *Vigna unguiculata* (L.) Walp. Planta 216:674–685

Kempf VAJ, Hitziger N, Riess T, Autenrieth IB (2002) Do plant and human pathogens have a common pathogenicity strategy? Trends Microbiol 10:269–275

Kereszt A, Kiss E, Reuhs BL, Carlson RW, Kondorosi Á, Putnoky P (1998) Novel *rkp* gene clusters of *Sinorhizobium meliloti* involved in capsular polysaccharide production and invasion of the symbiotic nodule: the *rkpK* gene encodes a UDP-glucose dehydrogenase. J Bacteriol 180:5426–5431

Kiss E, Kereszt A, Barta F, Stephens S, Reuhs BL, Kondorosi Á, Putnoky P (2001) The *rpk-3* gene region of *Sinorhizobium meliloti* Rm 41 contains strain-specific genes that determine K Antigen structure. Mol Plant Microbe Interact 14: 1395–1403

Kobayashi H, Naciri-Graven Y, Broughton WJ, Perret X (2004) Flavonoids induce temporal shifts in gene-expression of *nod*-box controlled loci in *Rhizobium* sp. NGR234. Mol Microbiol 51:335–347

Kouchi H, Takane K-I, So RB, Ladha JK, Reddy PM (1999) Rice *ENOD40*: isolation and expression analysis in rice and transgenic soybean root nodules. Plant J 18:121–129

Krause A, Doerfel A, Göttfert M (2002) Mutational and transcriptional analysis of the type III secretion system of *Bradyrhizobium japonicum*. Mol Plant Microbe Interact 15:1228–1235

Krishnan HB (2002) NolX of *Sinorhizobium fredii* USDA257, a type III-secreted protein involved in host range determination, is localized in the infection threads of cowpea (*Vigna unguiculata* [L.] Walp) and soybean (*Glycine max* [L.] Merr.) nodules. J Bacteriol 184:831–839

Krishnan HB, Lorio J, Kim WS, Jiang G, Kim KY, DeBoer M, Pueppke SG (2003) Extracellular proteins involved in soybean cultivar-specific nodulation are associated with pilus-like surface appendages and exported by a type III protein secretion system in *Sinorhizobium fredii* USDA257. Mol Plant Microbe Interact 16:617–625

Leigh JA, Coplin DL (1992) Exopolysaccharides in plant-bacterial interactions. Annu Rev Microbiol 46:307–346

Leigh JA, Lee CC (1988) Characterization of polysaccharides of *Rhizobium meliloti exo* mutants that form ineffective nodules. J Bacteriol 170:3327–3332

Leigh JA, Singer ER, Walker GC (1985) Exopolysaccharide-deficient mutants of *Rhizobium meliloti* that form ineffective nodules. Proc Natl Acad Sci USA 82:6231–6235

Levery SB, Zhan H, Lee CC, Leigh JA, Hakomori S-I (1991) Structural analysis of a second acidic exopolysaccharide of *Rhizobium meliloti* that can function in alfalfa root nodule invasion. Carbohydr Res 210:339–347

Lévy J, Bres C, Geurts R, Chalhoub B, Kulikova O, Duc G, Journet E-P, Ané J-M, Lauber E, Bisseling T, Dénarié J, Rosenberg C, Debellé F (2004) A putative Ca2+ and calmodulin-dependent protein kinase required for bacterial and fungal symbioses. Science 303:1361–1364

Limpens E, Franken C, Smit P, Willemse J, Bisseling T, Geurts R (2003) LysM domain receptor kinases regulating rhizobial Nod factor-induced infection. Science 302:630–633

López-Lara IM, Kafetzopoulos D, Spaink HP, Thomas-Oates JE (2001) Rhizobial NodL *O*-acetyl transferase and NodS *N*-methyl transferase functionally interfere in production of modified Nod factors. J Bacteriol 183:3408–3416

Madsen EB, Madsen LH, Radutoiu S, Olbryt M, Rakwalska M, Szczyglowski K, Sato S, Kaneko T, Tabata S, Sandal N, Stougaard J (2003) A receptor kinase gene of the LysM type is involved in legume perception of rhizobial signals. Nature 425: 637–640

Marie C, Deakin WJ, Viprey V, Kopcinska J, Golinowski W, Krishnan HB, Perret X, Broughton WJ (2003) Characterization of Nops, nodulation outer proteins, secreted via the type III secretion system of NGR234. Mol Plant Microbe Interactions 16:743–751

Marie C, Deakin WJ, Ojanen-Reuhs T, Diallo E, Reuhs BL, Broughton WJ, Perret X (2004) TtsI, a key regulator of *Rhizobium* species NGR234, is required for type III-dependent protein secretion and synthesis of rhamnose-rich polysaccharides. Mol Plant Microbe Interact 17:958–966

Mergaert P, van Montagu M, Promé J-C, Holsters M (1993) Three unusual modifications, a D-arabinosyl, an N-methyl, and a carbamoyl group, are present on the Nod factors of *Azorhizobium caulinodans* strain ORS571. Proc Natl Acad Sci USA 90:1551–1555

Miller DD, de Ruijter NCA, Bisseling T, Emons AMC (1999) The role of actin in root hair morphogenesis: studies with lipochioto-oligosaccharide as a growth stimulator and cytochalasin as an actin perturbing drug. Plant J 17:141–154

Mitra RM, Shaw SL, Long SR (2004) Six nonnodulating plant mutants defective for Nod factor-induced transcriptional changes associated with the legume-rhizobia symbiosis. Proc Natl Acad Sci USA 101:10217–10222

Nagai H, Roy CR (2003) Show me the substrates: modulation of host cell function by type IV secretion systems. Cell Microbiol 5:373–383

Niebel A, Bono J-J, Ranjeva R, Cullimore JV (1997) Identification of a high affinity binding site for lipo-oligosaccharidic NodRm factors in the microsomal fraction of *Medicago* cell suspension cultures. Mol Plant Microbe Interact 10:132–134

Niehaus K, Kapp D, Pühler A (1993) Plant defence and delayed infection of alfalfa pseudonodules induced by an exopolysaccharide (EPSI)-deficient *Rhizobium meliloti* mutant. Planta 190:415–425

Oldroyd GED, Downie JA (2004) Calcium, kinases and nodulation signalling in legumes. Nat Rev Mol Cell Biol 5:566–576

Olsthoorn MM, López-Lara IM, Petersen BO, Bock K, Haverkamp J, Spaink HP, Thomas-Oates JE (1998) Novel branched Nod factor structure results from alpha-(1–>3) fucosyl transferase activity: the major lipo-chitin oligosaccharides from *Mesorhizobium loti* strain NZP2213 bear an alpha-(1–>3) fucosyl substituent on a nonterminal backbone residue. Biochemistry 37:9024–9032

Ovtsyna AO, Schultze M, Tikhonovich IA, Spaink HP, Kondorosi É, Kondorosi Á, Staehelin C (2000) Nod Factors of *Rhizobium leguminosarum* bv. *viciae* and their fucosylated derivatives stimulate a Nod factor cleaving activity in pea roots and are hydrolyzed in vitro by plant chitinases at different rates. Mol Plant Microbe Interact 13:799–807

Pacios-Bras C, van der Burgt YEM, Deelder AM, Vinuesa P, Werner D, Spaink HP (2002) Novel lipochitin oligosaccharide structures produced by *Rhizobium etli* KIM5 s. Carbohydr Res 337:1193–1202

Parniske M, Downie JA (2003) Plant biology: locks, keys and symbioses. Nature 425:569–570

Pellock BJ, Cheng H-P, Walker GC (2000) Alfalfa root nodule invasion efficiency is dependent on *Sinorhizobium meliloti* polysaccharides. J Bacteriol 182:4310–4318

Perret X, Freiberg C, Rosenthal A, Broughton WJ, Fellay R (1999) High-resolution transcriptional analysis of the symbiotic plasmid of *Rhizobium* sp. NGR234. Mol Microbiol 32:415–425

Perret X, Staehelin C, Broughton WJ (2000) Molecular basis of symbiotic promiscuity. Microbiol Mol Biol Rev 64:180–201

Ponting CP, Aravind L, Schultz J, Bork P, Koonin EV (1999) Eukaryotic signalling domain homologues in archaea and bacteria. Ancient ancestry and horizontal gene transfer. J Mol Biol 289:729–745

Poupot R, Martinez-Romero E, Promé J-C (1993) Nodulation factors from *Rhizobium tropici* are sulfated or nonsulfated chitopentasaccharides containing an *N*-methyl-*N*-acylglucosaminyl terminus. Biochemistry 32:10430–10435

Poupot R, Martinez-Romero E, Gautier N, Promé J-C (1995) Wild type *Rhizobium etli*, a bean symbiont, produces acetyl-fucosylated, *N*-methylated, and carbamoylated nodulation factors. J Biol Chem 270:6050–6055

Price NP, Relić B, Talmont F, Lewin A, Promé D, Pueppke SG, Maillet F, Dénarié J, Promé J-C, Broughton WJ (1992) Broad-host-range *Rhizobium* species strain NGR234 secretes a family of carbamoylated, and fucosylated, nodulation signals that are *O*-acetylated or sulphated. Mol Microbiol 6:3575–3584

Pringet J-L, Journet E-P, Barker DG (1998) *Rhizobium* Nod factor signaling: evidence for a G Protein-mediated transduction mechanism. Plant Cell 10:659–671

Pueppke SG, Broughton WJ (1999) *Rhizobium* sp. strain NGR234 and *R. fredii* USDA257 share exceptionally broad, nested host ranges. Mol Plant Microbe Interact 12:293–318

Pühler A, Arlat M, Becker A, Göttfert M, Morrissey JP, O'Gara F (2004) What can bacterial genome research teach us about bacteria-plant interactions? Curr Opin Plant Biol 7:137–147

Putnoky P, Petrovics G, Kereszt A, Grosskopf E, Ha DTC, Banfalvi Z, Kondorosi Á (1990) *Rhizobium meliloti* lipopolysaccharide and exopolysaccharide can have the same function in the plant-bacterium interaction. J Bacteriol 172:5450–5458

Radutoiu S, Madsen LH, Madsen EB, Felle HH, Umehara Y, Gronlund M, Sato S, Nakamura Y, Tabata S, Sandal N, Stougaard J (2003) Plant recognition of symbiotic bacteria requires two LysM receptor-like kinases. Nature 425:585–592

Ramu SK, Peng H-M, Cook DR (2002) Nod factor induction of reactive oxygen species production is correlated with expression of the early Nodulin gene *rip1* in *Medicago trunculata*. Mol Plant Microbe Interact 15:522–528

Recourt K, Schripsema J, Kijne JW, van Brussel AAN, Lugtenberg BJJ (1991) Inoculation of *Vicia sativa* subsp. *nigra* roots with *Rhizobium leguminosarum* biovar *viciae* results in release of *nod* gene activating flavanones and chalcones. Plant Mol Biol 16:841–852

Reinhold BB, Chan SY, Reuber TL, Marra A, Walker GC, Reinhold VN (1994) Detailed structural characterization of succinoglycan, the major exopolysaccharide of *Rhizobium meliloti* Rm1021. J Bacteriol 176:1997–2002

Relić B, Perret X, Estrada-García MT, Kopcinska J, Golinowski W, Krishnan HB, Pueppke SG, Broughton WJ (1994a) Nod factors of *Rhizobium* are a key to the legume door. Mol Microbiol 13:171–178

Relić B, Staehelin C, Fellay R, Jabbouri S, Boller T, Broughton WJ (1994b) Do Nod-factor levels play a role in host-specificity? In: Kiss GB, Endre G (eds) Proceedings of the 1st European nitrogen fixation conference. Officina Press, Sezged, Hungary, pp 69–75

Reuber TL, Walker GC (1993) Biosynthesis of succinoglycan, a symbiotically important exopolysaccharide of *Rhizobium meliloti*. Cell 74:269–280

Reuber TL, Long SR, Walker GC (1991) Regulation of *Rhizobium meliloti exo* genes in free-living cells and in planta examined by using Tn*phoA* fusions. J Bacteriol 173:426–434

Reuhs BL, Carlson RW, Kim JS (1993) *Rhizobium fredii* and *Rhizobium meliloti* produce 3-deoxy-D-*manno*-2-octulosonic acid-containing polysaccharides that are structurally analogous to group II K antigens (capsular polysaccharides) found in *Escherichia coli*. J Bacteriol 175:3570–3580

Reuhs BL, Williams MNV, Kim JS, Carlson RW, Côté F (1995) Suppression of the Fix- phenotype of *Rhizobium meliloti exoB* mutants by *lpsZ* is correlated to a modified expression of the K polysaccharide. J Bacteriol 177:4289–4296

Reuhs BL, Geller DP, Kim JS, Fox JE, Kolli VSK, Pueppke SG (1998) *Sinorhizobium fredii* and *Sinorhizobium meliloti* produce structurally conserved lipopolysaccharides and strain-specific K antigens. Appl Environ Microbiol 64:4930–4938

Roche P, Debellé F, Maillet F, Lerouge P, Faucher C, Truchet G, Dénarié J, Promé J-C (1991) Molecular basis of symbiotic host specificity in Rhizobium meliloti: *nodH* and *nodPQ* genes encode the sulfation of lipo-oligosaccharide signals. Cell 67:1131–1143

Rossier O, van den Ackerveken G, Bonas U (2000) HrpB2 and HrpF from *Xanthomonas* are type III-secreted proteins and essential for pathogenicity and recognition by the host plant. Mol Microbiol 38:828–838

Saad, MM, Kobayashi, H, Marie, C, Brown, IR, Mansfield, JW, Broughton, WJ, and Deakin, WJ (2005) NopB, a type III secreted protein of *Rhizobium* sp.

strain NGR234, is associated with pilus-like surface appendages. J Bacteriol 187: 1173–1181.

Savouré A, Sallaud C, El-Turk J, Zuanazzi J, Ratet P, Schultze M, Kondorosi Á, Esnault R and Kondorosi É (1997) Distinct responses of *Medicago* suspension cultures and roots to Nod factors and chitin oligomers in the elicitation of defense-related responses. Plant J 11:277–287

Schmidt PE, Broughton WJ, Werner D (1994) Nod factors of *Bradyrhizobium japonicum* and *Rhizobium* sp. NGR234 induce flavonoid accumulation in soybean root exudate. Mol Plant Microbe Interact 7:384–390

Schneider A, Walker SA, Poyser S, Sagan M, Ellis TH, Downie JA (1999) Genetic mapping and functional analysis of a nodulation-defective mutant (*sym19*) of pea (*Pisum sativum* L.). Mol Gen Genet 262:1–11

Schultze M, Kondorosi A (1998) Regulation of symbiotic root nodule development. Annu Rev Genet 32:33–57

Skorpil P, Saad MM, Boukli NM, Kobayashi H, Ares-Orpel F, Broughton WJ, Deakin WJ (2005) NopP, a phosphorylated effector of *Rhizobium* sp. strain NGR234, is a major determinant of nodulation of the tropical legumes *Flemingia congesta* and *Tephrosia vogelii*. Mol Microbiol 57:1304-1317

Spaink HP (2000) Root nodulation and infection factors produced by rhizobial bacteria. Annu Rev Microbiol 54:257–288

Staehelin C, Schultze M, Kondorosi Á, Kondorosi É (1995) Lipo-chitooligosaccharide nodulation signals from *Rhizobium meliloti* induce their rapid degradation by the host plant Alfalfa. Plant Physiol 108:1607–1614

Steen A, Buist G, Leenhouts KJ, El Khattabi M, Grijpstra F, Zomer AL, Venema G, Kuipers OP, Kok J (2003) Cell wall attachment of a widely distributed peptidoglycan binding domain is hindered by cell wall constituents. J Biol Chem 278: 23874–23881

Stracke S, Kistner C, Yoshida S, Mulder L, Sato S, Kaneko T, Tabata S, Sandal N, Stougaard J, Szczyglowski K, Parniske M (2002) A plant receptor-like kinase required for both bacterial and fungal symbiosis. Nature 417:959–962

Streeter JG, Salminen SO, Whitmoyer RE, Carlson RW (1992) Formation of novel polysaccharides by *Bradyrhizobium japonicum* bacteroids in soybean nodules. Appl Environ Microbiol 58:607–613

Van Workum WAT, van Slageren S, van Brussel AAN, Kijne JW (1998) Role of exopolysaccharides of *Rhizobium leguminosarum* bv. *viciae* as host plant-specific molecules required for infection thread formation during nodulation of *Vicia sativa*. Mol Plant Microbe Interact 11:1233–1241

Vinardell JM, Ollero FJ, Hidalgo Á, López-Baena FJ, Medina C, Ivanov-Vangelov K, Parada M, Madinabeitia N, del Rosario Espuny M, Bellogín RA, Camacho M, Rodríguez-Navarro D-N, Soria-Díaz ME, Gil-Serrano AM, Ruiz-Sainz JE (2004) NolR regulates diverse symbiotic signals of *Sinorhizobium fredii* HH103. Mol Plant Microbe Interact 17:676–685

Viprey V, Del Greco A, Golinowski W, Broughton WJ, Perret X (1998) Symbiotic implications of type III protein secretion machinery in *Rhizobium*. Mol Microbiol 28:1381–1389

Vlassak KM, Luyten E, Verreth C, van Rhijn P, Bisseling T, Vanderleyden J (1998) The *Rhizobium* sp. BR816 *nodO* gene can function as a determinant for nodulation of *Leucaena leucocephala*, *Phaseolus vulgaris*, and *Trifolium repens* by a diversity of *Rhizobium* spp. Mol Plant Microbe Interact 11:383–392

Walker SA, Downie JA (2000) Entry of *Rhizobium leguminosarum* bv *viciae* into root hairs requires minimal Nod factor specificity, but subsequent infection thread growth requires *nodO* or *nodE*. Mol Plant Microbe Interact 13:754–762

Walker SA, Viprey V, Downie JA (2000) Dissection of nodulation signaling using pea mutants defective for calcium spiking induced by nod factors and chitin oligomers. Proc Natl Acad Sci USA 97:13413–13418

Wang L-X, Wang Y, Pellock BJ, Walker GC (1999) Structural characterization of the symbiotically important low-molecular-weight succinoglycan of *Sinorhizobium meliloti*. J Bacteriol 181:6788–6796

Whitfield C, Valvano MA (1993) Biosynthesis and expression of cell-surface polysaccharides in gram-negative bacteria. Adv Microb Physiol 35:135–246

Yang W-C, de Blank C, Meskiene I, Hirt H, Bakker J, van Kammen A, Franssen H, Bisseling T (1994) *Rhizobium* Nod factors reactivate the cell-cycle during infection and nodule primordium formation, but the cycle is only completed in primordium formation. Plant Cell 6:1415–1426

Molecular mechanisms in the nitrogen-fixing *Nostoc*-Bryophyte symbiosis

John C. Meeks

1
Introduction

The metabolic products of cyanobacteria are highly coveted by organisms in their immediate environment. A basis for this directional relationship is that cyanobacteria are amongst the most nutritionally independent organisms in the biosphere. Their unifying characteristic is an oxygen-evolving photosynthetic mechanism, leading to a CO_2 fixing autotrophic metabolic mode. Thus, they require only light, water, CO_2 and a few inorganic molecules or elements for growth; consequently, cyanobacteria are ubiquitous in illuminated habitats (Whitton and Potts 2000). Cyanobacteria and their evolutionary progeny, the chloroplasts of algae and plants (Douglas 1998; Raven and Allen 2003), are the overwhelmingly dominant producers of the reduced carbon that sustains growth of heterotrophic organisms in the biosphere via detrital and grazing food chains. A subset, distributed amongst all five taxonomic orders (subdivisions) of the cyanobacteria (Castenholz 2001), are also capable of nitrogen fixation (Rippka and Herdman 1992), thereby enhancing both their nutritional independence and the desirability of their metabolites.

However, oxygenic photosynthesis and nitrogen fixation are biochemically incompatible processes, due to the oxygen liability of the nitrogenase enzyme complex. Cyanobacteria have evolved temporal behavioral and spatial morphological solutions to the oxygen incompatibility dilemma. The common behavioral solution is to photosynthesize during the day, accumulating reserves of photosynthate, and fix nitrogen at night when the cellular oxygen tension can be reduced by a lack of production, coupled with respiratory consumption, fueled by the stored photosynthate (Fay 1992; Gallon 1992). A spatial separation from concurrent oxygen production is provided by highly differentiated cells called heterocysts, which are specialized for nitrogen

J.C. Meeks (e-mail: jcmeeks@ucdavis.edu)
Section of Microbiology, University of California, Davis, CA 95616, USA

fixation. Heterocysts differ physiologically from vegetative cells by a loss of the oxygen-evolving photosynthetic mechanism, reflecting a transition to a heterotrophic metabolic mode, an increased rate of aerobic respiration, and the deposition of a solute and gas impermeable extra wall layer (this wall layer is interrupted at the junctions with adjacent vegetative cells), all of which contribute to a microoxic cytoplasm. Heterocyst formation occurs only in filamentous cyanobacteria in which they establish a reciprocal source-sink relationship. Vegetative cells provide heterocysts with reductant in the form of photosynthate, most likely as sucrose, to support nitrogen fixation and respiratory activities, and receive fixed nitrogen as glutamine in return (Wolk et al. 1994).

Similar to other bacteria, even superficial observations indicate that cyanobacteria rarely lead a solitary life in nature. They are found in nonspecific association with a wide range of organisms, especially in marine ecosystems (Carpenter and Foster 2002). A close association with an autotroph is highly beneficial to a heterotroph, be it a unicellular prokaryote or a multicellular, multitissue eukaryote, because it minimizes competition for the reduced carbon. Such a breadth of organismal associations with cyanobacteria has been documented (Adams 2000; Paerl 1992; Rai et al. 2002) and many more transient associations are predicted to occur. Nitrogen-fixing, heterocyst-forming, *Nostoc* species, and their close relatives, are also recruited into specific associations with representative nonphotosynthetic and photosynthetic eukaryotes that span the phylogenetic spectrum. The nonphotosynthetic eukaryotes include extracellular associations with lichenized fungi (Rikkinen 2002), as well as an intracellular association with the unique mycorrhizal fungus *Geosiphon pyriforme* (Mollenhauer et al. 1996). The photosynthetic eukaryotes range from aquatic unicellular forms, such as the marine diatoms *Rhizosolenium* sp. (Mague et al. 1974) and *Hemiaulus* spp. (Villareal 1991), to representatives of the major divisions of terrestrial plants. The plant representatives include the spore-producing non-vascular bryophytes, especially the hornworts (Ridgeway 1967), the spore-producing vascularized ferns, specifically the aquatic fern *Azolla* (Peters and Mayne 1974), seed-producing gymnosperms represented by cycads (Bergersen et al. 1965), and one family of angiosperms, the Gunneraceae (Silvester and McNamara 1976). The nitrogen-fixing cyanobacterial-plant associations have been more highly studied that any other cyanobacterial association and comprehensive reviews have recently been published (Adams 2000; Bergman et al. 1996; Meeks 1998, 2003; Rai et al. 2000, 2002).

2
The *Nostoc*-Bryophyte Symbiotic Experimental System

This chapter will review the results of studies primarily utilizing an association between the nitrogen-fixing cyanobacterium *Nostoc punctiforme* (strain PCC 73102, synonym ATCC 29133) and the bryophyte hornwort *Anthoceros punctatus* that is routinely reconstituted in the laboratory with the separately pure-cultured partners. The focus will be more on mechanisms of the interaction than on diversity of symbionts; the latter is an expanding topic which has recently been examined using molecular genetic techniques (West and Adams 1997; Costa et al. 2001; Rasmussen and Nilsson 2002). The complete genome sequence of *N. punctiforme* strain ATCC 29133 is now known (http://genome.jgi-psf.org/microbial/). Therefore, results and concepts will be discussed and hypotheses offered in the context of the *N. punctiforme* genome sequence, as well as genetic, physiological and biochemical traits of the two partners. When appropriate, comparisons will also be made to other cyanobacterial-plant systems.

Nostoc punctiforme has an extraordinarily wide range of physiological properties, vegetative cell developmental alternatives and ecological niches (Meeks et al. 2002; Meeks 2003). It is amenable to genetic manipulation, including random transposon mutagenesis (Cohen et al. 1994) and targeted gene replacement (Hagen and Meeks 1999). These collective characteristics enhance the scientific value of its genome sequence (Meeks et al. 2001). *N. punctiforme* represents a subgroup of cyanobacteria that can grow in complete darkness as a respiratory heterotroph, a metabolic mode that may be required in symbiotic association. Two of the developmental alternatives are essential to its symbiotic interactions; the differentiation of motile filaments called hormogonia, which serve as the infective units of plant associations, and of heterocysts, the sole sites of nitrogenase expression in nearly all *Nostoc* strains (Fig. 1). The third vegetative cell developmental alternative is the formation of resting cells termed akinetes. Akinetes can be detected in the senescing regions of plant associations, but their role, if any, in symbiotic interactions has not been addressed. Inclusive of the vegetative cell cycle, the four developmental directions of a *N. punctiforme* vegetative cell are numerically unparalleled in the bacterial world. They provide multiple stages of a cell cycle for developmental regulation, including by environmental signals from symbiotic partners. The *N. punctiforme* genome contains a remarkably high number of sensory transduction systems with approximately 156 sensor histidine kinases, 103 response regulator proteins, and 51 serine/threonine protein kinases, plus 7 putative adenylate/guanylate cyclases (Meeks et al. 2001; Meeks 2005). The environmental signals that modulate the activity of these sensory systems are largely unknown; nevertheless, there are more than ample regulatory proteins to serve as potential targets for signals from symbiotic partners.

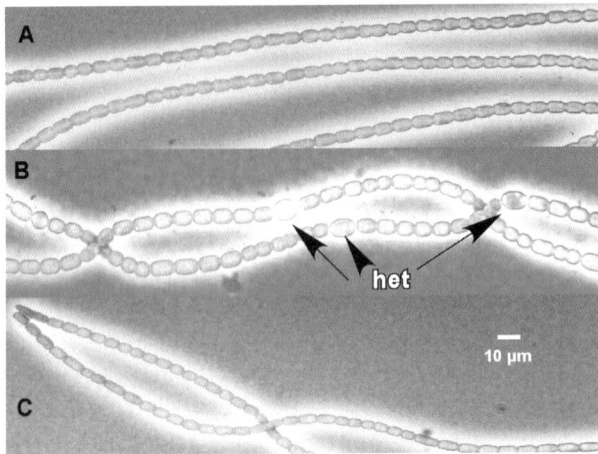

Fig. 1. Phase contrast photomicrographs of three free-living developmental states of *N. punctiforme*. **A** Ammonium grown filaments lacking any differentiated cells. **B** Dinitrogen grown filaments containing nitrogen-fixing heterocysts (*het*) present in a non-random spacing pattern. **C** Motile hormogonium filaments showing the smaller cell size relative to N_2 or NH_4^+ grown cultures. Panels A and C reproduced from Meeks et al. (2002)

Nostoc punctiforme shows broad symbiotic competence within the phylogenetic spectrum of plants, including the bryophyte hornworts (Enderlin and Meeks 1983) and liverworts (Joseph and Adams 2000), gymnosperm cycads, from whence it was isolated (Rippka et al. 1979), and the angiosperm *Gunnera* spp. (Johansson and Bergman 1994), as well as the non-lichen fungus, *G. pyriforme* (Kluge et al. 2002). Photographs of the plant associations, illustrating the different compartments in which the cyanobacterium is localized, are presented in Fig. 2. It is difficult to imagine that *N. punctiforme* successively developed specific adaptive processes for each plant group as members of the group emerged over evolutionary time. Rather, it is logical to hypothesize that the plants must have independently evolved strategies to control key regulatory and metabolic pathways of the cyanobacterium that are normally expressed in free-living growth. For example, a shift to a heterotrophic mode of carbon and energy metabolism, the differentiation of hormogonia and heterocysts, and N_2 fixation are basic processes that *Nostoc* expresses apart from a plant partner, but the responses are enhanced in the presence of the plant partner (Meeks 1998). Thus, the transition from free-living to symbiotic growth of *Nostoc* can best be characterized as changes in the *degree* of a response. In contrast, when rhizobia associate with leguminous plants, the vegetative cells change into a bacteroid state and induce nitrogenase synthesis and activity, neither of which do they express apart from the plant partner (van Rhijn and Vanderleyden

1995). These morphological and physiological transitions to the rhizobial symbiotic growth state are best characterized as a new *kind* of a response. Moreover, the *Nostoc* colonize structures or areas (called symbiotic cavities) of the plants that are present at all times and change very little in the symbiotic state. Conversely, rhizobia induce the formation of a root nodule, which can be considered as a new plant organ, structured to protect nitrogenase from oxygen, while supplying sufficient oxygen for rhizobial respiration (van Rhijn and Vanderleyden 1995). The fact that *Nostoc*, and related heterocyst-forming species, carry their own oxygen protective mechanism may account for the relative simplicity of their symbiotic associations. These differences in degree, versus kind of response, and simple structures, versus a new organ, contribute to our conclusion that the interactions in *Nostoc*-based symbioses are primarily (but not exclusively) unidirectional from the plant to the cyanobacterium (Meeks 1998), in contrast to the extensive signal exchange between partners that characterizes the rhizobia-legume associations (Perret et al. 2000).

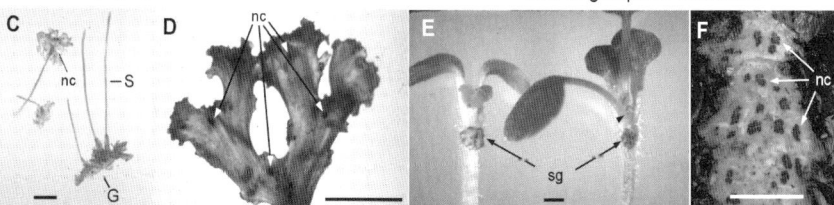

Fig. 2. Photographs of representatives of the three terrestrial groups of plant partners with which *N. punctiforme* will form a symbiotic association. The gymnosperm *Cycas* sp. showing the vegetative fronds (**A**) and (**B**) an excised coralloid root with the tip of one lobe sectioned to show the ring-shaped cyanobacterial cavity (*nc*). The *bars* in panels A and B are 0.5 m and 0.5 cm, respectively. The bryophyte hornwort *Anthoceros punctatus* showing the gametophyte (*G*) and attached sporophyte (*S*) generations in panel **C** and **D** pure cultured gametophyte tissue reconstituted in the laboratory with *N. punctiforme* localized in slime cavities (nc). The *bars* in panels C and D are both 1.0 cm. The Angiosperm *Gunnera* spp. depicting a seedling with distinctive stem glands (**E**) the sites of *Nostoc* entry into the stem and (**F**) a tangential section of a stem of a giant *Gunnera chilensis* plant showing numerous symbiotic cavities occupied by *Nostoc* (nc). The *bars* in panels E and F are both 1.0 cm. Reproduced from Meeks and Elhai (2002)

As originally defined by deBary (1879), the term symbiosis simply described the living together of differently named organisms. It is now most often applied to *mutualistic* interactions in which all of the organisms involved clearly benefit from the intimate physical association, as opposed to *parasitic* interactions in which one partner benefits and the other is slowly harmed. The mutualistic terminology may appropriately reflect the rhizobia-legume associations where the heterotrophic rhizobia benefit from a physical association with a photoautotrophic plant, while the plant benefits from the nitrogen fixation activity of the bacterium. The benefit to a photoautotrophic *Nostoc* in association with a photoautotrophic plant is not obvious, except, perhaps, as a shelter to avoid grazing predation. As an alternative to mutualistic and parasitic, we have described nitrogen-fixing *Nostoc* associations as a form of symbiosis more precisely conceptualized as plant *domestication* of a cyanobacterial ammonium-producing factory (commensal) (Meeks and Elhai 2002). The cyanobacterium neither benefits nor is harmed by the association. Consequently, the working model we are testing does not draw regulatory or structural gene analogies to rhizobial nodulation (*nod*) genes that are active only during association of the bacteria with legumes (van Rhijn and Vanderleyden 1995; Perret et al. 2000), although general strategies and chemical signals may be similar.

The above conceptualization of unidirectional flow of information in a commensal interaction leads to the hypothesis that the regulatory circuits and structural gene targets in these nitrogen-fixing associations uniquely evolved in the cyanobacterial lineage. We have suggested a similar evolutionary scenario in the context of vegetative developmental alternatives (Meeks et al. 2002; Meeks 2005). Therefore, the genes encoding the processes cannot be identified in the genome sequence by purely comparative bioinformatics approaches. Rather, functional assays will be necessary to fish the respective genes from the genome.

As guides in devising functional assays, we have developed working models. The interactions between *N. punctiforme* and *A. punctatus*, while most likely operating as a continuum, can be experimentally separated into two sequential stages, each with substages (Fig. 3). The initial stage is establishment of the association and involves the differentiation and behavior of hormogonium filaments, the infective units. The second stage is development of a functional association involving growth and metabolic alteration of vegetative cells, and the differentiation and behavior of heterocysts. This model is supported by mutants we have isolated of *N. punctiforme* that are separately defective in establishment or functional development of an association (Table 1). The physiological roles of the products of the mutated genes will be discussed in subsequent sections.

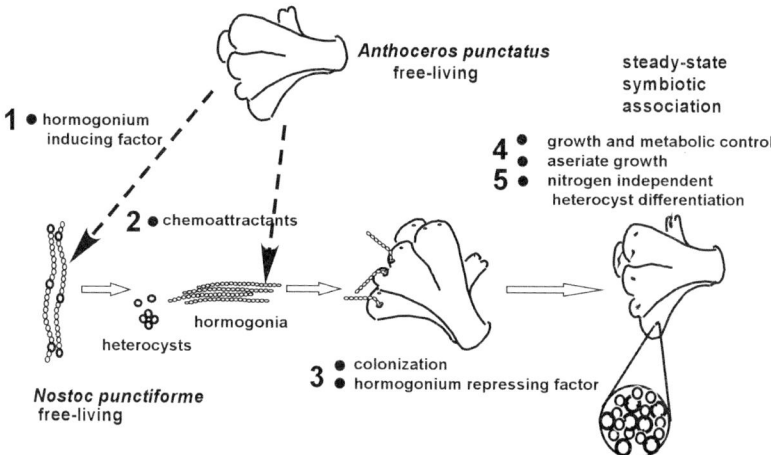

Fig. 3. Schematic of the continuum of interactions between *A. punctatus* and *N. punctiforme* leading to a N_2-fixing symbiotic association. The interactions are depicted as unidirectional from plant to cyanobacterium. The numbers refer to experimentally distinct substages that are described in the text. Reproduced with modifications from Meeks (2003)

Table 1. Phenotypes of selected *N. punctiforme* mutants altered in symbiotic infection or function. Details of the mutated genes and their functional roles are in the text

Strain	Symbiotic colonies[a]	Acetylene reduction[b]		Gene induced by[c]	Stage affected
		per g FW	colony		
ATCC 29133[d]	0.21±0.04	6.3±1.2	12.4±3.3	--	--
UCD 398 (*sigH*)[d]	1.2±0.2	8.0±3.9	10.1±4.1	HIF	infection
UCD 328 (*hrmA*)[e]	1.6±0.1	6.1±1.1	8.6±1.3	HRF	infection
UCD 444 (*ntcA*)[f]	0	0	0	--	infection
UCD 416 (*hetF*)[f]	0.26±0.06	0		--	function
UCD 464 (*hetR*)[f]	0.36±0.04	0	0	--	function

[a] Number of symbiotic colonies visible in a dissecting microscope 2 weeks after co-culture, normalized to mg dry wt of *A. punctatus* tissue counted and µg Chl *a* of the *Nostoc* inoculant
[b] In situ acetylene reduction activity given as nmol ethylene formed per g fresh weight of gametophyte tissue or as pmol ethylene formed per symbiotic *Nostoc* colony
[c] HIF is hormogonium inducing factor and refers to exudate of *A. punctatus* containing a factor(s) that induces hormogonium differentiation; HRF is hormogonium repressing factor and refers to an aqueous extract of *A. punctatus* that contains a factor(s) that represses hormogonium differentiation
[d] Campbell et al. (1998)
[e] Cohen and Meeks (1996)
[f] Wong and Meeks (2002)

3
Establishment of the Association–Differentiation and Behavior of Hormogonia

This general stage can be viewed as three substages: (1) induction of hormogonium differentiation; (2) control of the direction of hormogonium gliding; and (3) colonization and repression of hormogonium differentiation.

Physiologically, hormogonia are analogous to non-growing swarmer cells of certain prosthecate bacteria, such as *Caulobacter crescentus* (Roberts et al. 1996; Shimkets and Brun 2000). In response to a variety of environmental signals, including light quality, nutrient stress and the presence of a symbiotic plant partner (Tandeau de Marsac 1994; Meeks and Elhai 2002), the vegetative cells cease net macromolecular synthesis, initiate one or more rounds of cell division, and the filaments fragment at the vegetative cell-heterocyst junction (Herdman and Rippka 1988; Campbell and Meeks 1989). Hormogonium filaments do not fix N_2. The released filaments initiate gliding and remain motile for 48 to 72 h (the infection window), after which they cease to glide, begin to differentiate heterocysts, and then reinitiate net macromolecular synthesis and cell enlargement. This series of events defines the hormogonium developmental cycle (Tandeau de Marsac 1994; Campbell and Meeks 1989; Meeks and Elhai 2002; Meeks et al. 2002). *N. punctiforme* displays an "immunity" period after exit from the hormogonium cycle before it can reenter the cycle (Campbell and Meeks 1989). There is currently insufficient information from which to develop a mechanistic working model of the hormogonium cycle.

3.1
Substage 1. Induction of Hormogonium Differentiation

There is evidence for the production of an extracellular hormogonium-inducing factor (HIF) by *A. punctatus* (Campbell and Meeks 1989), the cycad *Zamia* (Ow et al. 1999) and the angiosperm *Gunnera* (Rasmussen et al. 1994). Increasing the frequency of hormogonia in the immediate environment would increase the probability of an infection event. HIF excreted into the culture medium appears to be a small molecule, but its identity is unknown and it could have plant-specific identities. HIF from *A. punctatus* is highly unstable, which hinders its chemical characterization and verification (Campbell and Meeks 1989). Nevertheless, the data in Table 2 support the conclusion that the *Nostoc* response to *A. punctatus* HIF in the growth-conditioned medium is biologically relevant and not an artifact. The most compelling experiments are those in which ammonium was added before or after conditioning of the medium. HIF-dependent hormogonium differentiation is insensitive to the presence of ammonium, provided the ammonium is added after the medium is conditioned, but its presence inhibits the production of HIF by *A. punctatus*. In addition, whereas hormogonia tend not to differentiate in dark incubated cultures (Tandeau de Marsac 1994), the presence of HIF stimulates differentiation in the light or dark (Campbell and Meeks 1989).

Table 2. Evidence for a hormogonium inducing activity produced into conditioned medium by *A. punctatus* (derived from Campbell and Meeks 1989)

Incubation medium[a]	Addition at time 0	% hormogonia at time 24 h[b]
Fresh	none	ca 10
Conditioned –N	none	ca 90
Conditioned –N	2.5 mM NH_4^+	ca 90
Conditioned +N	none	ca 10
Conditioned symbiotic	none	ca 90

[a]Fresh, *A. punctatus* growth medium plus or minus combined nitrogen; Conditioned, incubated with *A. punctatus* for 2 d in the absence (-N) or presence (+N) of ammonium, or with reconstituted *Nostoc-A. punctatus* in the absence of ammonium
[b]The % of hormogonium filaments in the population was determined microscopically 24 h after exposure to the incubation medium

Potential targets of HIF are largely unidentified. The only genes currently documented to be induced in *N. punctiforme* during differentiation include those encoding a sigma subunit of RNA polymerase (*sigH*) and a carboxyl terminal protease (Campbell et al. 1998). In related *Calothrix* sp. strain PCC 7601, gas vesicle proteins (Damerval et al. 1991), cell division proteins and pili are differentially synthesized in hormogonia (Doherty and Adams 1999). Conversely, there is a corresponding repression of transcription of nearly all other genes and degradation of some of the gene products. An example is the photosynthetic accessory pigment phycoerythrin, encoded by *cpeAB*. Transcription of *cpeA* is suppressed during hormogonium induction in *N. punctiforme* (E. Campbell and J. Meeks, unpubl.) and the weak fluorescence of cells in hormogonium filaments is evidence of phycobiliprotein degradation (Meeks 1998). The mechanisms of differential transcription of the target genes have yet to be defined.

3.2
Substage 2. Control of Hormogonium Behavior

Cyanobacteria are motile through a gliding mechanism which requires contact with the substratum; none has a flagellar apparatus, although one unicell is capable of swimming by an unknown mechanism (Brahamsha 1996). The mechanism of gliding motility is also unclear (Adams 2001); propulsion by polysaccharide excretion (Hoiczyk and Baumeister 1998) and type IV pili retraction have been proposed (Bhaya et al. 2000), similar to other gliding bacteria (Spormann 1999; McBride 2001). Whether both mechanisms, or only one, and which, are operative in hormogonia is unknown. The *N. punctiforme* genome does contain at least one gene encoding a putative Type IV pilin protein, plus two copies encoding PilT involved in pilus retraction.

While the differentiation of motile hormogonia is essential for an infection event in the *A. punctatus* association, it is not singularly sufficient. This conclusion is supported by a natural isolate, *Nostoc* sp. strain ATCC 27896, which differentiates hormogonia in the presence of HIF, but the hormogonia only rarely infect *A. punctatus*. When the strain from a rare infection was isolated from the association and again cocultured with *A. punctatus*, it retained the low frequency of infection phenotype; thus, there was no selection for a spontaneous symbiotically competent mutant (Meeks 1998). These observations imply the presence of a receptor that senses a plant signal, such as a chemoattractant, in *N. punctiforme*, and other competent strains, that may be absent, or has been altered in *Nostoc* sp. strain ATCC 27896. Chemotactic control would seem essential for the plant to attract *Nostoc* species that are present in relatively low abundance in the adjacent soil. Knight and Adams (1996) unequivocally established that bryophytes do produce chemoattractants that influence the direction of hormogonium gliding. The identity of the chemoattractant(s) is yet unknown (Watts et al. 1999).

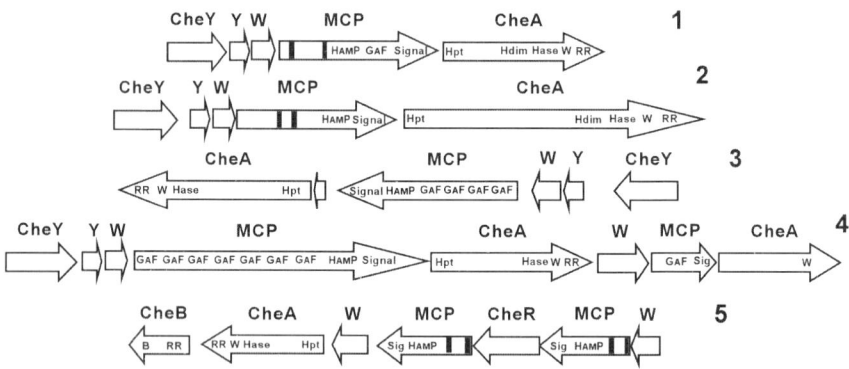

Fig. 4. Cartoon of the five taxis loci computationally predicted to be present in the genome of *N. punctiforme*. The genes in each loci are provisionally designated based on their respective sequence homology to genes in *Escherichia coli* and are discussed in the text. The *bars* in the MCPs of loci 1, 2, and 5 indicate putative transmembrane domains. The domains cited within selected genes are as follows: *B* methylesterase; *GAF* sensor – cGMP phosphodiesterase, Adenylyl cyclase, bacterial transcription factors, FhlA; *HamP* linker protein - Histidine kinase, Adenylyl cyclases, Methyl binding proteins, Phosphatates; *Hase* Histidine kinase ATPase; *Hdim* Histidine kinase dimerization; *Hpt* Histidine phosphotransfer; *PAS* sensor - Period circadian protein, Ah receptor nuclear translocator protein, Single minded protein; *Signal* output with CheW binding; *RR* response regulator receiver; *W* binding site for CheW

Computational analysis of the *N. punctiforme* genome revealed five separate clusters (physical loci) of collocated genes encoding products with significant homology to the chemo- and photo-taxis proteins from other bacteria (Fig. 4). Taxis proteins are exemplary of signal transduction pathways and need not be restricted to interactions with motility motors. Because hormogonia are the only motile state of *N. punctiforme*, those gene products essential for tactic responses are likely to be present and active only in hormogonia.

The basic bacterial chemotaxis system, defined in *Escherichia coli* (Parkinson and Kofoid 1992; Armitage 1999), consists of a methyl-accepting chemotaxis protein (MCP), which is methylated from S-adenosylmethionine by a methyl transferase (CheR). Activity of the MCP is in response to signals detected by a periplasmic sensing domain. There are three homologous *cheR* genes in the *N. punctiforme* genome, only one of which is collocated with other taxis genes. The genome contains six genes encoding putative MCPs, plus a fragment (locus 4), two of which are collocated in one locus. The signal is transferred from the MCP, through the intercession of a small protein called CheW, to a unique histidine kinase called CheA. There are seven homologous *cheW* and five *cheA* genes in *N. punctiforme* and all are collocated in the distinct loci, with *cheW* duplicated in two of the loci. CheA is the central signal transducer of chemotaxis; it autophosphorylates in response to the (lack of a) MCP signal and then transfers the phosphate to either CheY, a receiver domain only response regulator, or to CheB, a receiver domain protein with an additional methylesterase domain. The function of CheY-P is to modulate the flagellar motor, while CheB-P demethylates the MCP. There are four genes encoding putative CheB methylesterase proteins in the genome; only one is found in a cluster. There are 34 genes encoding CheY-like proteins in the genome, most of which have alternative roles in the physiology of *N. punctiforme*. However, eight genes encoding CheY-like proteins are present as duplicates in four of the five genetic loci.

Gliding motility mediated by Type IV pili and phototactic responses are being genetically analyzed in unicellular *Synechocystis* sp. strain PCC 6803 (Bhaya et al. 2000, 2001). The *Synechocystis* 6803 genome contains three taxis loci; the CheA homolog associated with locus Tax3 is located elsewhere in the chromosome (Bhaya et al. 2001). All of the MCPs have transmembrane motifs and the CheA homologs are hybrid histidine kinases with kinase dimerization domains. Genetic analyses imply that locus Tax3 controls pilus biogenesis, while locus Tax1 influences phototactic responses. The CheA homolog involved in pilus biogenesis is unique in having two response regulator receiver domains in the C-terminus.

None of the taxis loci of *N. punctiforme* are identical to those of *Synechocystis* 6803, but loci 1 & 2 (Fig. 4) have the greatest gene and protein organizational similarities. The MCP of Tax1 in *Synechocystis* 6803 contains two GAF domains; whereas the MCP in locus 1 contains but a single GAF domain, it is computationally identified as a light sensing phytochrome. However, the MCP in locus 1 also has an extensive putative periplasmic domain implying that locus 1 may be more involved in chemical than light sensory transduction. Based on their significant similarities to phytochrome, the apparently soluble MCPs in loci 3 & 4 are more likely to be involved in light sensing and phototactic responses of hormogonia (Lazaroff 1973). In the case of photosensing, the MCPs would not, a priori, require a periplasmic domain as light can readily be sensed in the cytoplasm. Homologs of the genes in locus 5 are absent in the genomes of the two other heterocyst-forming cyanobacteria that have been sequenced, *Anabaena* sp. strain PCC 7120 and *A. variabilis*. This implies that this sensory system is unlikely to have a role in cellular differentiation common to those strains, such as hormogonia, heterocysts or akinetes. The gene products encoded in this locus may be involved in a phenotypic trait unique to *N. punctiforme*, possibly chemotactic or other interactions with plants. The gene organization in locus 5 is most similar to Cluster 3 (*wsp* homologs) of *Pseudomonas* species (Parales et al. 2004) and the *che3* locus of *Myxococcus xanthus* (Kirby and Zusman 2003). The gene products encoded from these loci function in sensory transduction pathways not directly related to tactic responses. Resolution of all of these sensory transduction possibilities awaits genetic analyses in *N. punctiforme*.

The behavior of hormogonia during infection may involve more than just taxis or regulation of the combined taxis and motility system may occur at multiple levels. Mutation of an alternative sigma factor of RNA polymerase, SigH, in *N. punctiforme* resulted in a phenotype of an increased frequency of infection of *A. punctatus* relative to the wild-type (Table 1). The mutant did not make more hormogonia, nor did the hormogonia remain motile for a longer period of time; rather the mutant was simply more highly infective, suggesting a behavioral change (Campbell et al. 1998), such as a more intense response to the plant signal or more rapid gliding. Conversely, mutation of the global transcriptional regulator, NtcA, that is nominally involved in nitrogen control (Herrero et al. 2004), resulted in a strain that was less responsive to HIF in hormogonium differentiation and the hormogonia that were formed failed to infect *A. punctatus* (Table 1) (Wong and Meeks 2002). The genes that may be transcriptionally activated or repressed by SigH and NtcA are unknown, but these results are consistent with genetic control over the behavior of hormogonia in response to plant signals.

3.3
Substage 3. Colonization and the Repression of Hormogonium Differentiation

Little is known about the actual colonization of the preexisting cavity in the gametophyte tissue of *A. punctatus*. However, motile hormogonia entering the symbiotic cavities (auricules) of the liverwort *Blasia pustilla* has been documented (Kimura and Nakano 1990; Adams 2000).

Once colonization has taken place, further hormogonium differentiation appears to be suppressed. The isolation of a second *N. punctiforme* mutant, with the phenotype of a high frequency of infection (Table 1), lead to the identification of the *hrm* locus (Cohen and Meeks 1996), which is hypothetically involved in suppression of hormogonium differentiation. The physiological consequence of the suppression is to prevent continued differentiation of hormogonia once these filaments have colonized the plant tissue, thereby allowing the differentiation of heterocysts and expression of nitrogenase activity, which is instrumental in the formation of a functional association and beneficial to the controlling plant partner. Transcription of genes in the locus is induced by an aqueous extract of *A. punctatus* gametophyte tissue that blocks hormogonium formation in the wild-type strain. The activity in the extract was called a hormogonium-repressing factor (HRF) because it appears to prevent entry of vegetative filaments into the hormogonium cycle, even in the presence of HIF, by transcriptional upregulation of genes in the *hrm* locus (Cohen and Meeks 1996).

The *hrm* locus was initially defined as consisting of eight genes, with the organization as depicted in Fig. 5, extending from *hrmE* to *hrmA* (Cohen and Meeks 1996, Campbell et al. 2003). HrmR has sequence similarity to a sugar-binding transcriptional repressor in the LacI/GalR family. The structural genes *hrmK*, *hrmI*, *hrmU*, *hrmA* and *hrmE* have sequence similarity to genes encoding a kinase, an isomerase, a dehydrogenase, perhaps a dehydratase, and an aldehyde reductase, respectively, involved in hexuronic and hexonic acid metabolism in *Escherichia coli* (Lin 1996) and *Bacillus stearothermophilus* (Shulami et al. 1999). Transcription of the four genes in the downstream transcriptional frame of *hrmE*, putatively encoding a sugar ABC transporter, is also induced by HRF (E. Sano, T. Ronne, E. Campbell and J. Meeks, unpubl.); thus, these genes are tentatively included as part of the *hrm* locus. The transport genes are provisionally identified as *hrmB1*, *hrmB2*, encoding two periplasmic sugar binding proteins, *hrmT*, encoding an ATP binding protein, and *hrmP*, encoding a permease. Genes in this locus appear to be unique to *N. punctiforme* and related symbiotically competent strains (Cohen and Meeks 1996). Whereas a homolog of HrmR is present in the *A. variabilis* genome, the protein has greatest phylogenetic similarity to *Pseudomonas* and *Listeria* species (E. Campbell and J. Meeks, unpubl.). HrmI has no significant homologies in cyanobacteria and its phylogenetic relatives are in *Corynebacterium* and *Propionibacterium* species. Mutation of *N. punctiforme* genes in the *hrm* locus yields phenotypes of either high infection frequency (*hrmUA*; Cohen and Meeks 1996) or no infection (*hrmR*; Campbell et al. 2003) of *A. punctatus*.

A. hrm locus

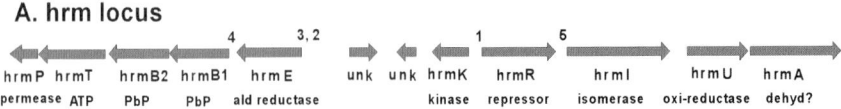

B. Known and putative regulatory protein binding sites

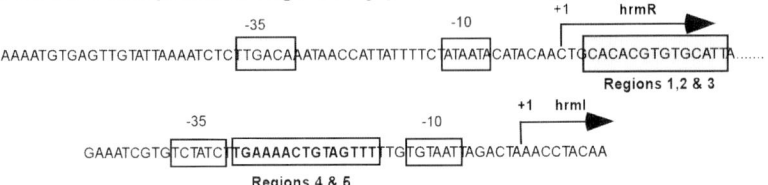

Fig. 5. The *hrm* locus of *N. punctiforme*, showing the gene organization and sites of conserved nucleotide sequences (**A**). The genes and gene products are described in the text. Transcript 5' ends (putative transcriptional start points) are known for *hrmR*, *hrmE* and *hrmI* (**B**). The binding of HrmR to regions *1* (*hrmR*), *2* and *3* (*hrmE*) (panel A) has been experimentally determined and thus is shaded in panel B. The conserved sequence in regions 4 and 5 was identified by computational screening and has not been experimentally verified as a protein binding site

An initial hypothesis was that HrmR regulated the transcription of a *hrmR-hrmI-hrmU-hrmA* operon in analogy to *Escherichia* and *Bacillus*. However, Northern analysis indicated monocistronic transcription of *hrmR* and polycistronic transcription of *hrmIUA* and *hrmUA* (Campbell et al. 2003). Gel electrophoretic mobility shift assays, established that recombinant His-HrmR, purified from *E. coli*, binds to specific 14 bp palindromic sequences in the promoter regions of itself (Fig. 5B) and *hrmE*; the 14 bp sequence is lacking in the 5' regions of *hrmI*, *hrmU* and elsewhere in the genome. HrmR binding is suppressed by cell extract from a *hrmR* mutant exposed to HRF and by galacturonate, but not by glucuronate, which indicates sugar specificity. Extracts of uninduced cells, or cells induced by naringin (Campbell et al. 2003), which activates *hrmIUA* transcription (Cohen and Yamasaki 2000), did not interfere with HrmR binding. Alignment of the 5' region upstream of *hrmB1* with that of the 5' region of *hrmI*, including the mapped 5' end of its transcript (E. Campbell and J. Meeks, unpubl.), reveals a highly conserved 15 bp nucleotide sequence (the sites are depicted as 4 and 5 in Fig. 5A, and the sequence is in Fig. 5B). The current working hypothesis is that this is the binding sequence of a transcriptional factor other than HrmR and that naringin interacts with this transcriptional factor to modulate expression of *hrmB1B2TP* and *hrmIUA*.

These observations lead to two interesting conclusions regarding the *hrm* locus. Transcription in the locus clearly involves more than one regulatory factor and expression is organized into more than one regulon. The sugar binding domains of the two periplasmic binding proteins show significant similarity to the sugar binding domain of HrmR, implying this transporter may be involved in uptake of the sugar ligand of HrmR.

An autogenic hormogonium repressing activity was previously observed in cultures of *Nostoc* species (Herdman and Rippka 1988) and such an activity could provide for an "immunity" period for recovery from an initial round of hormogonium differentiation. In the absence of an "immunity" period and in the presence of an inducting signal, a hormogonium filament could immediately reenter the hormogonium cycle. Since a period of growth is required for vegetative cells to regain their original biomass (Campbell and Meeks 1989) and chromosome copy number (Herdman and Rippka 1988) upon exit from the hormogonium cycle, continual reentry is lethal by extinction. This is the phenotype of a *hrmA* mutant in the continual presence of HIF (Cohen and Meeks 1996). Whether the *hrm* locus plays a role in the immunity period of hormogonium differentiation apart from the plant partner is unknown.

4
Development of a Functional Nitrogen-Fixing Association–Differentiation and Behavior of Heterocysts

Relative to free-living growth, the functional symbiotic state of *N. punctiforme* in association with *A. punctatus* is characterized by a decrease in the rates of growth, photosynthesis and NH_4^+ assimilation, coupled with an increase in heterocyst frequency, N_2 fixation and release of N_2-derived NH_4^+ (Meeks 1998, 2003). The stage can be divided into the two substages of growth and metabolic control (4 in Fig. 3), and heterocyst differentiation and behavior (5 in Fig. 3).

4.1
Substage 4. Growth and Metabolic Control

The physiological characteristics of *Nostoc* in symbiotic and free-living growth states are summarized in Table 3. In all plant associations, the *Nostoc* partners grow obviously slower than do free-living cultures. The mechanistic basis of this growth control is unknown, but it clearly influences the most fundamental developmental direction of *N. punctiforme*, the vegetative cell cycle. Although the rates of CO_2 and NH_4^+ assimilation are depressed (see below), they are so in proportion to the slower growth rate. One would predict just such behavior in any bacterium as it balances growth demands with metabolic flux. However, there is no obvious causal relationship between slow

Table 3. Physiological characteristics of *Nostoc* in free-living and *A. punctatus* associated growth states (derived from Meeks 1990)

Whole cell metabolic system/ in vitro enzyme[a]	Free-living state		Symbiotic state	
	Activity	Protein	Activity	Protein
Growth	24–48 h		~240 h	
Light-dependent cellular CO_2 fixation	128.0 units		15.0 units	
Rubisco	215.0 units	52 µg	25.0 units	43 µg
Cellular ammonium assimilation	13.9 units		2.9 units	
GS	130.0 units	7.1 µg	50.0 units	6.1 µg
Nitrogenase	6.3 units		23.5 units	

[a]Activities and units. Growth rate is doubling time in hours. Light-dependent cellular CO_2 fixation is the assimilation of $^{14}CO_2$ in whole cells, or isolated symbiotic colonies, corrected for dark incorporation, and the units are nmol/min/mg protein. Rubisco is in vitro $^{14}CO_2$ incorporation into acid stable material and the units are nmol/min/mg protein. Cellular ammonium assimilation is the incorporation of $^{13}NH_4^+$ into all non-volatile cellular metabolites and macromolecules by whole cells, or isolated symbiotic colonies, and the units are ^{13}N cpm incorporated/$^{13}NH_4^+$ cpm added × 100 (equals %)/5 min/mg protein. GS is in vitro activity of the biosynthetic assay reaction and the units are nmol/min/mg protein. Nitrogenase is free-living or in situ symbiotic reduction of acetylene to ethylene and the units are nmol/min/mg protein. Protein values are reported as µg/mg total cell protein, determined by enzyme linked immunosorbant assay

growth and metabolic alterations, although very few metabolic systems have been examined by direct biochemical analyses.

Two morphological changes accompany the symbiotic growth state. First, there is a three-fold, or more, greater frequency of heterocysts in the cell population. This phenomenon will be elaborated below (Section IV.b). Second, vegetative cells are markedly larger than in free-living culture and the appearance of filamentation is muted, as seen in electron micrographs of colony sections, as well as in samples examined in the light microscope (see Meeks 1990, 1998). This morphology is highly reminiscent of the aseriate growth stage in the life cycle of many *Nostoc* species (Lazaroff 1973; Potts 2000). In a rudimentary sense, the aseriate stage results from the initial formation of a contiguous sheath around the vegetative cells in each interval between heterocysts. As the interval continues to expand by vegetative cell divisions within the sheath, the chain of cells becomes contorted with a loss of the linear filamentous appearance. Overall, the previously linear filament becomes a series of distinct globular cell masses. This stage is thought to lead to the "punctiforme" macrocolony structure (Lazaroff 1973). Eventually the sheath disintegrates and filaments containing intercalary heterocysts are released. In many *Nostoc* strains, cultures may be held in the aseriate stage by green light, while exit is enhanced by red light leading to hormogonia formation (Lazaroff 1973). *Nostoc* in the *A. punctatus* gametophyte thallus is continuously exposed to the predominantly green light not absorbed by the plant. Unfortunately, because the aseriate stage is a hindrance to experimental manipulations, it tends to be selected against in liquid suspension cultures of most *Nostoc* species and remains little studied. Nevertheless, an implication from these observations is that the symbiotic growth morphology may reflect another example of *Nostoc* overexpressing a phenotypic trait that it normally expresses in the soil apart from the plant partner.

Perhaps the most dramatic physiological change in symbiotic *Nostoc* is the nearly complete loss of its independent photosynthetic potential and the assumption of a dependent heterotrophic metabolism. *Nostoc* freshly separated from *A. punctatus* has 12% of the free-living in vivo rate of light-dependent CO_2 fixation and 12% of the in vitro ribulose bisphosphate carboxylase/oxygenase (Rubisco, the primary CO_2 assimilating enzyme) specific activity (Steinberg and Meeks 1989). However, the amount of Rubisco protein is equivalent in both growth states. The photochemical reactions have not been examined, but the amounts of photosynthetic pigments (phycobiliproteins and chlorophyll) are similar in both growth states. The heterotrophic potential of symbiotic *Nostoc* was established by examining light and exogenous carbohydrate-dependent nitrogenase activity in associations with wild type and herbicide-resistant *Nostoc* mutants (Steinberg and Meeks 1991). The results of these experiments establish that *Nostoc* can directly provide less than 30% of the photosynthetically generated reductant required for maximal nitrogenase activity on a short term basis. However, the maximal steady state nitrogenase activity is dependent on exogenous carbohydrate supplied by *A. punctatus*. Since the symbiotic *Nostoc* can carry out light-dependent CO_2 fixation in situ, the shift to heterotrophic metabolism is most likely not due to light limitation, but rather that the plant exerts a regulatory role. The rates of light and dark carbohydrate-dependent symbiotic nitrogenase activity are about five-fold higher than the free-living light-dependent rate. The enhanced photoheterotrophic rate parallels the enhanced heterocyst frequency. However, dark carbohydrate-dependent nitrogenase activity is typically only 20–30% of the light-dependent rate in the free-living state (Gibson and Smith 1982). Heterotrophic metabolism in symbiotic *Nostoc* appears to be substantially enhanced.

The assimilation of NH_4^+ is similarly depressed in the *Nostoc* symbiotic growth state to about 15% of the free-living rate (Joseph and Meeks 1987). Glutamine synthetase (GS), the primary NH_4^+ assimilating enzyme, has only about 15% of the in vitro specific activity as free-living cultures, although the total amount of GS protein is statistically the same in the two growth states. Based on immunoelectronmicroscopic examination, Rai et al. (1989) concluded that symbiotic heterocysts contained about half of the GS antigen as free-living heterocysts. Conversely, there is an enhanced rate of nitrogen fixation in symbiosis (Table 3), which indicates a dramatic uncoupling between fixation and assimilation. Dinitrogen-derived ammonium accumulates in the *A. punctatus* (Meeks et al. 1985), *Azolla* (Meeks et al. 1987) and *Gunnera* (Silvester et al. 1996) associations, in contrast to its immediate incorporation into amino acids in free-living culture (Wolk et al. 1976). The pool of NH_4^+ is commonly assumed to result from a high rate of N_2 fixation coupled with a low rate of NH_4^+ assimilation due to limitations in GS activity. However, the stoichiometry between the rates of nitrogenase and GS does not support such causal relationship in the *A. punctatus* association where 80% of the fixed nitrogen accumulates as NH_4^+ in the short term (Meeks et al. 1985). The calculated symbiotic rate of in situ nitrogen fixation is 12.5 nmol of NH_4^+ produced per min per mg of protein, and GS has an in vitro NH_4^+ consumption rate of 19.8 nmol per min per mg protein when averaged across vegetative cells and heterocysts (Meeks 2003). The conclusion of a lack of a causal relationship assumes translocation of the N_2-derived NH_4^+ from the heterocysts to the adjacent vegetative cells, which would have an excess of GS, and not through the unique heterocyst envelope into the symbiotic cavity; such an assumption is consistent with the permeability properties of the heterocyst envelope (A.E. Walsby, personal communication). Thus, the mechanistic basis of the release of N_2-derived NH_4^+ is more complicated than it first appears.

It is of interest that some form of irreversible posttranslational regulation appears to operate on both Rubisco and GS in the *A. punctatus* defined symbiotic growth state of *Nostoc*. The two enzymes and their metabolic pathways appear to be modulated differently in other plant associations (summarized in Meeks 1998). The symbionts immediately separated from Cycads and *Gunnera* have 70 to 100% of the in vitro GS activity of free-living cultures and 100% of the in vitro Rubisco activity, but neither show light-dependent in vivo CO_2 fixation. The GS activity is consistent with evidence that the symbiont in Cycads releases citrulline and other amino acids to the plant partner (Pate et al. 1988). However, it does not account for the release of NH_4^+ in *Gunnera* associations (Silvester et al. 1996). This apparent diversity of regulatory targets, and possibly mechanisms, contributes to the idea that the plant partners evolved different ways of controlling growth and metabolism in the symbiotic *Nostoc*. The extent of differential gene expression in the *Nostoc* partner, as opposed to posttranslational modulation of activity, apart from heterocyst differentiation, that is necessary to support symbiotic growth in any association is unknown.

4.2
Substage 5. Heterocyst Differentiation and Behavior

Physiologically, heterocysts are microoxic, heterotrophic, N_2-fixing cells whose differentiation from oxygenic photoautotrophic vegetative cells involves a hypothetically large, but unknown, extent of differential gene expression; estimates range from 140 (Wolk 2000) to 600 (Ehira et al. 2003), to more than 1,000 (Lynn et al. 1986) genes. A number of regulatory and structural genes for heterocyst differentiation and function have been identified in *Anabaena* sp. strain PCC 7120 (see Wolk 1996, 2000; Herrero et al. 2004) and *N. punctiforme* (Summers et al. 1995; Campbell et al. 1996, 1997; Wong and Meeks 2001) some of which are listed in Table 4 and will be elaborated below. Heterocysts are terminally differentiated, but their physiological life span is not precisely known. A small cluster of heterocyst-forming cyanobacteria, including *A. variabilis* strain ATCC 29413, express a unique vegetative cell nitrogenase under anoxic growth conditions (Thiel et al. 1995). However, in all other heterocyst-forming cyanobacteria (including *N. punctiforme* and *Anabaena* sp. strain PCC 7120, whose sequenced genomes lack homologous sequences), nitrogenase expression is confined to heterocysts and appears to be regulated by developmental, rather than environmental signals (Elhai and Wolk 1990). Consequently, nitrogen fixation is assumed to depend on heterocyst differentiation in symbiotic associations.

The working model of symbiotic interactions holds that the plant partners have evolved a means to co-opt the regulatory circuits that function to modulate heterocyst differentiation in the free-living growth state. Therefore, defining the mechanisms of enhanced symbiotic

Table 4. Heterocyst frequency and pattern as a consequence of gene inactivation (mutation) in positive-acting and negative-acting genes and when complemented with the respective gene in trans

Positive-acting gene	Mutation[a]	Gene in trans	Negative-acting gene	Mutation	Gene in trans
ntcA[b]	0	Wild-type	*patS*	MCH	0
hanA	0	Unknown	*hetN*	MCH	0
hetR	0	MCH	*patB*	MCH	Wild-type
hetF	0	MCH	*patU*	MCH	Unknown
patA	ca.1%	Wild-type	*patN*	MSH	Wild-type

[a]Mutation of positive genes results in no (or low *patA*) heterocyst differentiation. Mutation of negative genes results in differentiation of more (multiple) heterocysts than the parental culture. MCH is multiple contiguous heterocysts, MSH is multiple singular heterocysts

[b]Citations: *ntcA* is Herrero et al. (2001); *hanA* is Khudyakov and Wolk (1996); *hetR* is Buikema and Haselkorn (2001); *hetF* is Wong and Meeks (2001); *patA* is Liang et al. (1992); *patS* is Yoon and Golden (2001); *hetN* is Callahan and Buikema (2001); *patB* is Jones et al. (2003); *patU* and *patN* are Meeks et al. (2002)

heterocyst differentiation is both dependent on, and complementary to, understanding the free-living regulatory systems.

4.2.1
Differentiation in the Free-Living State

In the free-living state, heterocyst differentiation is repressed by the presence of a source of combined nitrogen. When the combined nitrogen is exhausted, heterocysts appear in the filaments in a nonrandom spacing pattern, typically as a single heterocyst flanked by 10 to 15 vegetative cells (Fig. 1B). There are two interrelated phenomena with respect to the presence and position of heterocysts in a filament. The first is the sequence of heterocyst appearance following exhaustion of combined nitrogen. Based on morphological observations and mutant analyses, the developmental sequence can be generally described in three stages: (i) initiation, (ii) transition and (iii) commitment.

Initiation. In the initiation stage no distinct morphological events are manifest. The vegetative cells of mutants arrested at this stage show no signs of differentiation and the cultures eventually die in the absence of NH_4^+. Genes encoding proteins we have classified as positive regulatory elements (Table 4), because mutants do not initiate heterocyst differentiation, are involved at some point in this stage (Meeks and Elhai 2002). These currently include *ntcA*, *hanA*, *hetR*, *hetF* and *patA*, but there are likely to be additional elements. NtcA senses the nitrogen limitation signal and activates genes involved in acquisition of nitrogen sources alternative to NH_4^+ (Herrero et al. 2001, 2004), including *hetR* transcriptional enhancement. HanA is a DNA binding protein with multiple cellular roles, not specifically heterocyst differentiation (Khudyakov and Wolk 1996). HetR is considered the primary positive element in the induction of heterocyst differentiation (Buikema and Haselkorn 2001; Huang et al. 2004). HetR enhances its own transcription (Black et al. 1993) and that of *ntcA* (Herrero et al. 2004), and is localized in the differentiating cells (Black et al. 1993; Wong and Meeks 2001). HetF is required for HetR autoinduced transcriptional enhancement and localization of HetR to developing heterocysts (Wong and Meeks 2001). PatA is required for intercalary vegetative cell differentiation into heterocysts; the mutant forms heterocysts only at the ends of the filaments (Liang et al. 1992). When overexpressed *in trans*, *patS*, and an exogenously supplied C-terminal pentapeptide of PatS, represses the initiation of heterocyst differentiation (Yoon and Golden 1998, 2001). These are the characteristics of a negative regulatory element.

Transition. Transition to the proheterocyst stage is accompanied by the degradation of photosynthetic pigments in specific cells and an enlargement of the same cell. Mutants cells arrested in the proheterocyst stage do not synthesize nitrogenase, but a pattern of nonfluorescent cells that are spaced in the filaments similar to the final pattern of heterocyst spacing can be detected. Proheterocysts will regress to a vegetative cell state if the culture is supplied with a source of combined nitrogen (Meeks and Elhai 2002). Candidate genes encoding proteins necessary for continued development of proheterocysts are *hetC* (Khudyakov and Wolk 1997) and *hetP* (Fernández-Piñas et al. 1994).

Commitment. Commitment of proheterocysts to terminal differentiation of mature heterocysts is accompanied by deposition of the unique polysaccharide and glycolipid envelope, decreased cellular O_2 tension, and induction of genes for synthesis, assembly and function of nitrogenase. There is growing list of genes encoding proteins essential for heterocyst function in an oxic environment (Herrero et al. 2004; Meeks 2005).

The second phenomenon relates to the spacing pattern. Three separate processes may be considered with regard to the spaced pattern of heterocysts in a filament: establishment, maintenance, and disruption of pattern. These processes and the initiation of differentiation are depicted in the schematic in Fig. 6.

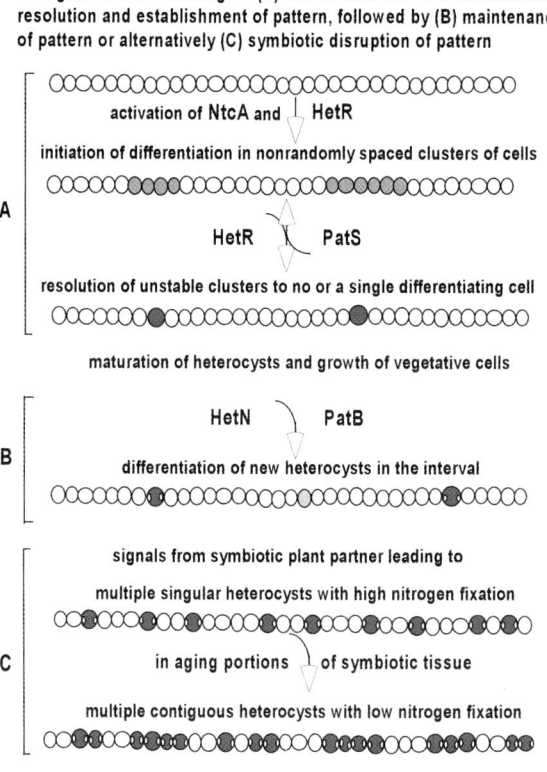

Fig. 6. Schematic of the sequence of events and major regulatory elements participating in: **A** Initiation of heterocyst differentiation and establishment of the pattern of spacing; **B** maintenance of the pattern; and **C** symbiotic disruption of the pattern

Establishment. Establishment of pattern involves those gene products that are required for initiation of the differentiation cascade as noted above. The factors involved in determining specifically which cells will initiate differentiation are currently unknown; we have speculated that position in the cell cycle when the environmental signal to differentiate is received is crucial (Meeks and Elhai 2002). Overexpression of the positive acting factors HetR and HetF results in clusters of heterocysts at single sites in the filaments (multiple contiguous heterocysts; MCH). Inactivation of four of the known negative regulatory elements also results in a MCH pattern. These mutant traits support the idea that initiation of differentiation occurs in a cluster of susceptible cells (biased initiation). The model for establishment of pattern holds that the cluster of differentiating cells are in an unstable developmental state and the interaction between positive and negative regulatory elements determines whether the cells will regress or continue to differentiate (resolution) (Meeks and Elhai 2002). HetR and PatS are projected to be the dominant positive and negative regulatory elements, respectively, with other elements acting in an ancillary role.

Maintenance. The spacing pattern is maintained as the vegetative cells grow and divide, using fixed nitrogen provided by the heterocysts. New heterocysts emerge at points approximately midway in the expanding vegetative cell interval between adjacent existing heterocysts. A distinct maintenance mechanism is implied by the observation that the *patS* mutant MCH phenotypic approaches the wild-type spacing pattern of single heterocysts at a site as the culture grows and new heterocysts differentiate (Yoon and Golden 2001). The negative regulatory elements HetN (Callahan and Buikema 2001) and PatB (Jones et al. 2003) appear to be involved more in maintenance than in establishment of the pattern. Controlled expression of *hetN* showed that the MCH phenotype of the mutant appeared only after the initial round of heterocyst differentiation following nitrogen deprivation had occurred (i.e., second generation heterocyst differentiation), paralleling the culture time when PatS apparently no longer influences the pattern.

Disruption. The pattern can be disrupted by certain chemical treatments (Meeks and Elhai 2002), in specific mutants (Table 4), and naturally by symbiotic association with a photosynthetic eukaryotic partner (see below). All chemical treatments and all but one of the mutations result in the differentiation of multiple heterocysts at a single site. The *patN* mutant yields multiple heterocysts, but they are present at single sites in the filaments with a decreased number of vegetative cells in the interval between heterocysts (termed multiple singular heterocysts; MSH) (Meeks et al. 2002). Symbiotic heterocyst frequencies in plant tissues range from 15 to 60% of the total cells and both MCH and MSH patterns can be observed (Adams 2000; Meeks and Elhai 2002; Rai et al. 2002).

4.2.2
Differentiation in Symbiosis

Symbiotic heterocyst differentiation differs from that in the free-living state in at least two ways: the signal that activates the differentiation cascade; and an enhanced frequency and altered spacing pattern.

Signal. Symbiotic heterocyst differentiation is hypothetically independent of the presence and concentration of combined nitrogen in the immediate environment (Meeks 1998, 2003; Meeks and Elhai 2002). Three lines of evidence support this hypothesis. First, approximately 10 µM NH_4^+ represses heterocyst differentiation in free-living cultures (Meeks et al. 1983). However, in the *A. punctatus* association, the concentration of N_2- derived NH_4^+ in the symbiotic cavity is calculated to be near to 550 µM (Meeks 2003), yet heterocysts continue to differentiate. Second, while heterocyst differentiation is repressed by nitrate in free-living wild-type cultures, it is not in mutants unable to assimilate nitrate. However, symbiotic heterocyst differentiation is repressed by nitrate in both wild-type and mutant strains (Campbell and Meeks 1992). Control experiments established that NO_3- derived NH_4^+ had not accumulated in the tissue. Even if it had, there is no reason to assume that the symbiotic *Nostoc* would response negatively to exogenous combined nitrogen when it apparently does not response to its own N_2-derived NH_4^+. These results implied that exogenous combined nitrogen was metabolized by the plant partner leading to repression of the plant signal to induce heterocyst differentiation in the symbiont. Third, the symbiotic vegetative cells are replete with nitrogen storage compounds, such as cyanophycin, phycobiliproteins and carboxysomes (Meeks 1998; Adams 2000). These compounds are typically catabolized to varying degrees in nitrogen limited and starved vegetative cells of free-living cultures, supporting the conclusion that symbiotic vegetative cells are not obviously nitrogen limited. Moreover, phycobiliproteins are typically lost during the differentiation of a vegetative cell into a heterocyst, most prominently at the prohetcrocyst stage, although they may reappear in older heterocysts (Thomas 1972). Conflicting reports have been made on the presence of phycobiliproteins in the heterocysts of *Nostoc* in association with *Anthoceros* (Rai et al. 1989; Meeks 1990). Phycobiliproteins are clearly present and functional in heterocysts of the symbiont in association with *Azolla* (Kaplan et al. 1986). Mutants blocked in phycobiliprotein degradation differentiate fully functional heterocysts containing the pigments (Baier et al. 2004), although data on their persistence with time were not given. If vegetative cells initiate differentiation in response to a plant signal, there is no reason to assume they would degrade their nitrogen stores, including phycobiliproteins, during the process. This is not to imply that phycobiliproteins continue to be synthesized in mature free-living or symbiotic heterocysts, only that they could well be present in the precursor vegetative cell. The observation that the GS antigen may be similar in concentration in symbiotic heterocysts and vegetative cells, in contrast to its 2-fold enhancement in heterocysts of free-living cultures (Rai et al. 1989) is also relevant to the inducing signal. GS transcription is elevated in heterocysts in a nitrogen limitation-activated NtcA dependent manner (Herrero et al. 2004). A lack of enhanced GS accumulation in symbiotic heterocysts is consistent with an environmental signal that is independent of nitrogen limitation.

If the regulatory target could be identified, perhaps the nature of the symbiotic signal could subsequently be deduced. Since we have established that *hetR* and *hetF* are essential for symbiotic heterocyst differentiation (Table 1) (Wong and Meeks 2002), attention shifts to genetic elements that function upstream of HetR; specifically NtcA activation and its interaction with *hetR*. Because NtcA is essential for infection (Table 1), its role in the subsequent step of symbiotic heterocyst differentiation cannot be genetically analyzed. Recent studies have indicated that NtcA activity in free-living cultures is modulated by the concentration of 2-oxoglutarate, with a high concentration signaling nitrogen limitation (Muro-Pastor et al. 2001; Vazquez-Bermudez et al. 2002). It is possible that the heterotrophic growth mode, combined with a decrease in glutamine formation, results in an elevation of the 2-oxoglutarate pool in the symbiotically associated *Nostoc*, thereby signaling apparent nitrogen limitation. However, if that were the case, the cells should then display a nitrogen limited morphology and physiology, which they appear not to do. Thus, the symbiotic signal and its regulatory target remain elusive.

Pattern. Planning experimental searches for the symbiotic signal and its regulatory target would be aided by knowledge of whether the symbiotic heterocyst spacing pattern reflects alterations in the mechanisms involved in establishment or maintenance of the free-living pattern. The enhanced symbiotic heterocyst frequency appears to be collectively manifest as singlets, doublets and higher ordered MCH. Unfortunately, due to the fragile cell-cell connections, symbiotic colony preparations typically squashed in a microscope slide to obtain an adequate microscopic focal plane yields single cells, bicells and very short filaments in which a spacing pattern is difficult to discern. However, a few meticulous light and scanning electron microscopic studies have very significantly shown that a gradient occurs from a low heterocyst frequency in the meristematic tip to higher frequencies in the more mature regions of the symbiotic structures in the *Azolla* (Peters and Mayne 1974; Hill 1975), cycad (Lindblad et al. 1985) and *Gunnera* (Söderbäk et al. 1990) associations. The spatially (thereby temporally) increasing numbers of heterocysts are first present singly in the filaments with a decreased number of vegetative cells in the intervals between heterocysts. Nitrogenase activity parallels the increase in heterocyst frequency up to a distance when multiple heterocysts begin to appear at sites in the filaments, whereupon nitrogenase activity declines with an increase in MCH, implying a predictable physiological senescence of older heterocysts, coupled with continued differentiation. This pattern is more consistent with a time-dependent increase in the frequency of functional heterocysts at single sites in the filament (i.e., MSH) as a consequence of disruption of the mechanisms involved in maintenance, rather than establishment, of pattern (Meeks and Elhai 2002). The time it takes for the gradient to establish and reach a senescent stage most likely varies with the plant partner. The *patN* mutant of *N. punctiforme* is the only current mutant that displays a MSH pattern in the free-living state (Meeks et al. 2002). However, in the *patN* mutant, the MSH pattern is established immediately following nitrogen limitation rather than sequentially as the pattern apparently arises in symbiosis. More information about the regulatory networks of establishment, and especially maintenance, of the heterocyst spacing pattern in the free-living state is required before a testable hypothesis of its disruption in plant symbioses can be formulated.

5
Future Directions – Genome and Genetic Analyses

The sequenced genome of *N. punctiforme* now makes many experimental approaches possible. Those include identification of all of the genes transcribed under a variety of specific growth conditions or environmental insults (i.e., the transcriptome), as well as correlation of transcription with the proteins present and their modification state (i.e., the proteome). Experiments are in progress in these two high throughput, systems level, approaches, with an emphasis on comparing symbiotic and free-living growth states. However, a lack of information on the molecular characteristics of the plant partner limits progress on understanding the complete system. Exogenous DNA has now been transferred into *Gunnera* seedlings via *Agrobacterium* mediated transformation and reporter genes expressed, establishing the potential for genetic manipulation (Wan-Ling Chiu, pers. comm.). *Agrobacterium* mediated transformations are also in progress with gametophyte tissue of the bryophytes *Anthoceros* and *Blasia*. As the international DNA sequencing capacity increases and costs decrease, it may soon be possible to obtain the genome sequence of one or more of the plant partners. Knowledge of the genetic potential of both partners of a symbiotic system would narrow the quest for metabolic systems synthesizing signals that initiate and control the interactions, and their sensory and regulatory targets. Genetic manipulation of both partners would allow for testing of working models, leading to biochemical identification of mechanisms, and ultimately manipulation of the associations.

Acknowledgements: The author gratefully acknowledges the inspiration of Sir William D.P. Stewart, in whose laboratory he was initially exposed to cyanobacterial symbiotic associations. Work in the author's laboratory is supported by grant 0317104 from the U.S.A. National Science Foundation.

References

Adams DG (2000) Symbiotic interactions. In: Whitton BA, Potts M (eds) The ecology of cyanobacteria, their diversity in time and space. Kluwer Academic Publ, Dordrecht, pp 523–561
Adams DG (2001) How do cyanobacteria glide? Microbiol Today 28:131–133
Armitage JP (1999) Bacterial tactic responses. Adv Microbial Physiol 41:229–289
Baier K, Lehmann H, Stephan DP, Lockau W (2004) NblA is essential for phycobilisome degradation in *Anabaena* sp. strain PCC 7120 but not for development of functional heterocysts. Microbiology 150:2739–2749
Bergersen FJ, Kennedy GS, Wittmann W (1965) Nitrogen fixation in the coralloid root of *Macrozamia communis* L Johnson. Aust J Biol Sci 18:1135–1142
Bergman B, Matveyev A, Rasmussen U (1996) Chemical signaling in cyanobacterial-plant symbioses Trends Plant Sci 1:191–197
Bhaya D, Bianco NR, Bryant D, Grossman A (2000) Type IV pilus biogenesis and motility in the cyanobacterium *Synechocystis* sp. PCC6803. Mol Microbiol 37:941–951
Bhaya D, Takahashi A, Grossman AR (2001) Light regulation of type IV pilus-dependent motility by chemotaxis-like elements in *Synehcocystis* PCC6803. Proc Natl Acad Sci USA 98:7540–7545
Black TA, Cai Y, Wolk CP (1993) Spatial expression and autoregulation of *hetR*, a gene involved in the control of heterocyst development in *Anabaena*. Mol Microbiol 9:77–84
Brahamsha B (1996) An abundant cell-surface polypeptide is required for swimming by the nonflagellated marine cyanobacterium *Synechococcus*. Proc Natl Acad Sci USA 93:6504–6509
Buikema WJ, Haselkorn R (2001) Expression of the *Anabaena hetR* gene from a copper-regulated promoter leads to heterocyst differentiation under repressing conditions. Proc Natl Acad Sci USA 98:2729–2734
Callahan SM, Buikema WJ (2001) The role of HetN in maintenance of the heterocyst pattern in *Anabaena* sp. PCC 7120. Mol Microbiol 40:941–950
Campbell EL, Meeks JC (1989) Characteristics of hormogonia formation by symbiotic *Nostoc* spp. in response to the presence of *Anthoceros punctatus* or its extracellular products. Appl Environ Microbiol 55:125–131
Campbell EL, Meeks JC (1992) Evidence for plant-mediated regulation of nitrogenase expression in the *Anthoceros-Nostoc* symbiotic association. J Gen Microbiol 138:473–480
Campbell EL, Hagen,KD, Cohen MF, Summers ML, Meeks JC (1996) The *devR* gene product is characteristic of receivers of two-component regulatory systems and is essential of heterocyst development in the filamentous cyanobacterium *Nostoc* sp. strain ATCC 29133. J Bacteriol 178:2037–2043
Campbell EL, Cohen MF, Meeks JC (1997) A polyketide-synthase-like gene is involved in the synthesis of heterocyst glycolipids in *Nostoc punctiforme* strain ATCC 29133. Arch Microbiol 167:251–258
Campbell EL, Brahamsha B, Meeks JC (1998) Mutation of an alternative sigma factor in the cyanobacterium *Nostoc punctiforme* results in increased infection of its symbiotic plant partner, *Anthoceros punctatus*. J Bacteriol 180:4938–4941

Campbell EL, Wong FCY, Meeks JC (2003) DNA binding properties of the HrmR protein of *Nostoc punctiforme* responsible for transcriptional regulation of genes involved in hormogonium differentiation. Mol Microbiol 47:573–582

Carpenter EJ, Foster RA (2002) Marine cyanobacterial symbioses. In: Rai A, Bergman B. Rasmussen U (eds) Cyanobacteria in symbiosis. Kluwer Academic Publ, Dordrecht, pp 11–17

Castenholz RW (2001) Phylum BX. Cyanobacteria oxygenic photosynthetic bacteria. In: Boone DR, Castenholz RW (eds) Bergey's manual of systematic bacteriology, 2nd edn, vol 1. The Archae and the deeply branching and phototrophic bacteria. Springer, Berlin Heidelberg New York, pp 473–599

Cohen MF, Meeks JC (1996) A hormogonium regulating locus, *hrmUA*, of the cyanobacterium *Nostoc punctiforme* strain ATCC 29133 and its response to an extract of a symbiotic plant partner *Anthoceros punctatus*. Mol Plant Microbe Interact 10:280–289

Cohen MF, Yamasaki H (2000) Flavonoid-induced expression of a symbiosis-related gene in the cyanobacterium *Nostoc punctiforme*. J Bacteriol 182:4644–4646

Cohen MF, Wallis JG, Campbell EL, Meeks JC (1994) Transposon mutagenesis of *Nostoc* sp strain ATCC 29133, a filamentous cyanobacterium with multiple cellular differentiation alternatives. Microbiology 140:3233–3240

Costa J-L, Lindblad P (2002) Cyanobacteria in symbiosis with cycads. In: Rai A, Bergman B. Rasmussen U (eds) Cyanobacteria in symbiosis. Kluwer Academic Publ, Dordrecht, pp 195–205

Costa J-L, Paulsrud P, Rikkinen J, Lindblad P (2001) Genetic diversity of *Nostoc* symbionts endophytically associated with two bryophyte species. Appl Environ Microbiol 67:4393–4396

Damerval T, Guglielmi G, Houmard J, Tandeau de Marsac N (1991) Hormogonium differentiation in the cyanobacterium *Calothrix*: a photoregulated developmental process. Plant Cell 3:191–201

DeBary A (1879) Die Erscheinung der Symbiose. Trubner, Strassburg

Doherty HM, Adams DG (1999) The organization and control of cell division genes expressed during differentiation in cyanobacteria. In: Peschek GA, Loffelhardt W, Schmetterer G (eds) The phototrophic prokaryotes. Kluwer Academic Publ, Dordrecht, pp 453–461

Douglas SE (1998) Plastid evolution: origins, diversity, trends. Curr Top Genet Dev 8:655–661

Ehira S, Amori M, Sato N (2003) Genome wide expression analysis of the responses to nitrogen deprivation in the heterocyst-forming cyanobacterium *Anabaena* sp. strain PCC 7120. DNA Res 10:97–113

Elhai J, Wolk CP (1990) Developmental regulation and spatial pattern of expression of the structural genes for nitrogenase in the cyanobacterium *Anabaena*. EMBO J 9:3379–3388

Enderlin CS, Meeks JC (1983) Pure culture and reconstitution of the *Anthoceros-Nostoc* symbiotic association. Planta 158:157–165

Fay P (1992) Oxygen relations of nitrogen fixation in cyanobacteria. Microbiol Rev 56:340–373

Fernández-Piñas F, Leganés F, Wolk CP (1994) A third genetic locus required for the formation of heterocysts in *Anabaena* sp. strain PCC 7120. J Bacteriol 176:5277–5283

Gallon JR (1992) Reconciling the incompatible: N_2 fixation and O_2. New Phytol 122:571–609

Gibson CE, Smith RV (1982) Freshwater plankton. In: Carr NG, Whitton BA (eds) The biology of cyanobacteria. Blackwell Scientific Publ, Oxford, pp 463–489

Hagen KD, Meeks JC (1999) Biochemical and genetic evidence for the participation of DevR in a phosphorelay signal transduction pathway essential for heterocyst maturation in *Nostoc punctiforme* ATCC 29133. J Bacteriol 181:4430–4434

Herdman M, Rippka R (1988) Cellular differentiation: hormogonia and baeocytes. Methods Enzymol 167:232–242

Herrero A, Muro-Pastor AM, Flores E (2001) Nitrogen control in cyanobacteria. J Bacteriol 183:411–425

Herrero A, Muro-Pastor AM, Valladares A, Flores E (2004) Cellular differentiation and the NtcA transcription factor in filamentous cyanobacteria. FEMS Microbiol Rev 28:469–487

Hill DJ (1975) The pattern of development of *Anabaena* in the *Azolla-Anabaena* symbiosis. Planta 122:179–184

Hoiczyk E, Baumeister W (1998) The junctional pore complex, a prokaryotic secretion organelle, is the molecular motor underlying gliding motility in cyanobacteria. Curr Biol 8:1161–1168

Huang X, Dong Y, Zhao J (2004) HetR homodimer is a DNA-binding protein required for heterocyst differentiation, and the DNA-binding activity is inhibited by PatS. Proc Natl Acad Sci USA 101:4848–4853

Johansson C, Bergman B (1994) Reconstitution of the symbiosis of *Gunnera mannicata* Linden: cyanobacterial specificity. New Phytolol 126:643–652

Joseph CM, Meeks JC (1987) Regulation of expression of glutamine synthetase in a symbiotic *Nostoc* strain associated with *Anthoceros punctatus*. J Bacteriol 169:2471–2475

Joseph NA, Adams DG (2000) Use of transposon-lux mutagenesis to study the *Nostoc-Blasia* symbiosis. In: Guerrero R (ed) 10th international symposium on phototrophic prokaryotes, book of abstracts. Univ Barcelona, Barcelona, p 125

Jones KM, Buikema WJ, Haselkorn R (2003) Heterocyst-specific expression of *patB*, a gene required for nitrogen fixation in *Anabaena* sp. strain PCC 7120. J Bacteriol 185:2306–2314

Kaplan D, Calvert HE, Peters GA (1986) The *Azolla-Anabaena azollae* relationship. XII. Nitrogenase activity and phycobiliproteins of the endophyte as a function of leaf age and cell type. Plant Physiol 80:884–890

Khudyakov I, Wolk CP (1996) Evidence that the *hanA* gene coding for HU protein is essential for heterocyst differentiation in, and cyanophage A-4(L) sensitivity of, *Anabaena* sp. strain PCC 7120. J Bacteriol 178:3572–3577

Khudyakov I, Wolk CP (1997) *hetC*, a gene coding for a protein similar to bacterial ABC protein exporters, is involved in early regulation of heterocyst differentiation in *Anabaena* sp. strain PCC 7120. J Bacteriol 179:6971–6978

Kimura J, Nakano T (1990) Reconstitution of a *Blasia-Nostoc* symbiotic association under axenic conditions. Nova Hedwigia 50:191–200

Kirby JR, Zusman DR (2003) Chemosensory regulation of developmental gene expression in *Myxococcus xanthus*. Proc Natl Acad Sci USA 100:2008–2013

Kluge M, Mollenhauer D, Wolf E, Schüâler A (2002) The *Nostoc-Geosiphon* endocytobiosis. In: Rai A, Bergman B, Rasmussen U (eds) Cyanobacteria in symbiosis. Kluwer Academic Publ, Dordrecht, pp 19–30

Knight CD, Adams DG (1996) A method for studying chemotaxis in nitrogen fixing cyanobacterium-plant symbioses. Physiol Mol Plant Path 49:73–77

Lazaroff N (1973) Photomorphogenesis and Nostocacean development. In: Carr NG, Whitton BA (eds) The biology of blue-green algae. Blackwell Scientific Publ, Oxford, pp 279–329

Liang J, Scappino L, Haselkorn RH (1992) The *patA* gene product, which contains a region similar to CheY of *Escherichia coli*, controls heterocyst pattern formation in the cyanobacterium *Anabaena* 7120. Proc Nat Acad Sci USA 89:5655–5659

Lin CC (1996) Dissimilatory pathways for sugars, polyols and carbohydrates. In: Neidhardt FC (ed) *Escherichia coli* and *Salmonella*, cellular and molecular biology. ASM Press, Washington, DC, pp 307–342

Lindblad P, Hällbom L, Bergman B (1985) The cyanobacterium-*Zamia* symbiosis: C_2H_2-reduction and heterocyst frequency. Symbiosis 1:19–28

Lynn ME, Battle JA, Ownby JD (1986) Estimation of gene expression in heterocysts of *Anabaena variabilis* by using DNA-RNA hybridization. J Bacteriol 136:1695–1699

Mague TH, Weare NM, Holm-Hansen O (1974) Nitrogen fixation in the North Pacific ocean. Marine Biol 24:109–119

McBride MJ (2001) Bacterial gliding motility: multiple mechanisms for cell movement over surfaces. Annu Rev Microbiol 55:49–75

Meeks JC (1990) Cyanobacterial-bryophyte associations. In: Rai AN (ed) Handbook of symbiotic cyanobacteria. CRC Press, Boca Raton, pp 43–63

Meeks JC (1998) Symbiosis between nitrogen-fixing cyanobacteria and plants. BioScience 48:266–276

Meeks JC (2003) Symbiotic interactions between *Nostoc punctiforme*, a multicellular cyanobacterium, and the hornwort *Anthoceros punctatus*. Symbiosis 34:55–71

Meeks JC (2005) The genome of the filamentous cyanobacterium *Nostoc punctiforme*, what can we learn from it about free-living and symbiotic nitrogen fixation? In: Palacios R, Newton WE (eds) Nitrogen fixation: 1888–2001, vol VI, Genomes and genomics of nitrogen-fixing organisms. Springer, Berlin Heidelberg New York, pp 27 70

Meeks JC, Elhai J (2002) Regulation of cellular differentiation in filamentous cyanobacteria in free-living and plant associated symbiotic growth states. Microbiol Mol Biol Rev 66:94–121

Meeks JC, Wycoff KL, Chapman JS, Enderlin CS (1983) Regulation of expression of nitrate and dinitrogen assimilation by *Anabaena* species. Appl Environ Microbiol 45:1351–1359

Meeks JC, Enderlin CS, Joseph CM, Chapman JS, Lollar MWL (1985) Fixation of $[^{13}N]N_2$ and transfer of fixed nitrogen in the *Anthoceros-Nostoc* association. Planta 164:406–414

Meeks JC, Steinberg N, Enderlin CS, Joseph CN, Peters GA (1987) *Azolla-Anabaena* relationship XIII. Fixation of $[^{13}N]N_2$. Plant Physiol 84:883–886

Meeks JC, Elhai J, Thiel T, Potts M, Larimer F, Lamerdin J, Predki P, Atlas R (2001) An overview of the genome of *Nostoc punctiforme*, a multicellular, symbiotic cyanobacterium. Photosyn Res 70:85–106

Meeks JC, Campbell EL, Summers ML, Wong FC (2002) Cellular differentiation in the cyanobacterium *Nostoc punctiforme*. Arch Microbiol 178:395–403

Mollenhauer D, Mollenhauer R, Kluge M (1996) Studies on initiation and development of the partner association in *Geosiphon pyriforme* (Kutz.) v. Wettstein, a unique encocytobiotic system of a fungus (Glomales) and the cyanobacterium *Nostoc punctiforme* (Kutz.). Hariot Protoplasma 193:3–9

Muro-Pastor MI, Reyes HC, Florencio FJ (2001) Cyanobacteria perceive nitrogen status by sensing intracellular 2-oxoglutarate levels. J Biol Chem 276:38320–38328

Ow MC, Gantar M, Elhai J (1999) Reconstitution of a cycad-cyanobacterial association. Symbiosis 27:125–134

Paerl HW (1992) Epi- and endobiotic interactions of cyanobacteria. In: Reisser W (ed) Algae and symbioses: plants, animals, fungi, viruses, interactions explored. Biopress, Bristol, pp 537–565

Parales RE, Ferradez A, Harwood CS (2004) Chemotaxis in Pseudomonads. In: Ramos J-L (ed) *Pseudomonas*, vol 1. Kluwer Academic Publ, Dordrecht, pp 793–815

Parkinson JS, Kofoid E (1992) Communication modules in bacterial signaling proteins. Annu Rev Genetics 26:71–112

Pate JS, Lindblad P, Atkins CA (1988) Pathways of assimilation and transfer of fixed nitrogen in coralloid roots of cycad-*Nostoc* symbioses. Planta 176:461–471

Perret X, Staehelin C, Broughton WJ (2000) Molecular basis of symbiotic promiscuity. Microbiol Mol Biol Rev 64:180–201

Peters GA, Mayne BC (1974) The *Azolla-Anabaena azollae* relationship. I. Initial characterization of the association. Plant Physiol 53:813–819

Potts M (2000) Nostoc. In: Whitton BA, Potts M (eds) The ecology of cyanobacteria, their diversity in time and space. Kluwer Academic Publ, Dordrecht, pp 465–504

Rai A, Borthakur M, Singh S, Bergman B (1989) *Anthoceros-Nostoc* symbiosis: immunoelectronmicroscopic localization of nitrogenase, glutamine synthetase, phycoerythrin and ribulose-1,5-bisphosphate carboxylase/oxygenase in the cyanobiont and the cultured (free-living) isolate *Nostoc* 7801. J Gen Microbiol 135:385–395

Rai A, Söderbäck E, Bergman B (2000) Cyanobacterium–plant symbioses. New Phytol 147:449–481

Rai A, Bergman B, Rasmussen U (2002) Cyanobacteria in symbiosis. Kluwer Academic Publ, Dordrecht

Rasmussen U, Nilsson M (2002) Cyanobacterial diversity and specificity in plant symbioses. In: Rai A, Bergman B, Rasmussen U (eds) Cyanobacteria in symbiosis. Kluwer Academic Publ, Dordrecht, pp 313–328

Rasmussen U, Johansson C, Bergman B (1994) Early communication in the *Gunnera-Nostoc* symbiosis: plant induced cell differentiation and protein synthesis in the cyanobacterium. Mol Plant Microbe Interact 7:696–702

Raven JA, Allen JF (2003) Genomics and chloroplast evolution: what did cyanobacteria do for plants? Genome Biol 4:209–213

Ridgeway JE (1967) The biotic relationship of *Anthoceros* and *Phaeoceros* to certain cyanophyta. Ann Missouri Bot Gdn 54:95–102

Rikkinen J (2002) Cyanoliches: an evolutionary overview. In: Rai A, Bergman B, Rasmussen U (eds) Cyanobacteria in symbiosis. Kluwer Academic Publ, Dordrecht, pp 31–72

Rippka R, Deruelles J, Waterbury JB, Herdman M, Stanier RY (1979) Generic assignments, strain histories and properties of pure cultures of cyanobacteria. J Gen Microbiol 111:1–61

Rippka R, Herdman M (1992) Pasteur culture collection of cyanobacteria in axenic culture. Institute Pasteur, Paris

Roberts RC, Mohr CD, Shapiro L (1996) Developmental programs in bacteria. Curr Top Dev Biol 34:207–257

Skimkets LJ, Brun YV (2000) Prokaryotic development: strategies to enhance survival. In: Brun YV, Skimkets LJ (eds) Prokaryotic development. ASM Press, Washington, DC, pp 1–7

Shulami S, Gat O, Sonenshein AL, Shoham Y (1999) The glucuronic acid utilization gene cluster from *Bacillus stearothermophilus* T-6. J Bacteriol 181:3695–3704

Silvester WB, Parsons R, Watt PW (1996) Direct measurement of release and assimilation of ammonia in the *Gunnera-Nostoc* symbiosis. New Phytol 132:617–625

Silvester WB, McNamara PJ (1976) The infection process and ultrastructure of the *Gunnera-Nostoc* symbiosis. New Phytol 77:135–141

Söderbäck E, Lindblad P, Bergman B (1990) Developmental patterns related to nitrogen fixation in the *Nostoc-Gunnera magellanica* Lam symbiosis. Planta 182:355–362

Spormann A (1999) Gliding motility in bacteria: Insights from studies of *Myxococcus xanthus*. Microbiol. Mol Biol Rev 63:621–641

Steinberg NA, Meeks JC (1989) Photosynthetic CO_2 fixation and ribulose bisphosphate carboxylase/oxygenase activity of *Nostoc* sp. strain UCD 7801 in symbiotic association with *Anthoceros punctatus*. J Bacteriol 171:6227–6233

Steinberg NA, Meeks JC (1991) Physiological sources of reductant for nitrogen fixation activity in *Nostoc* sp. strain UCD 7801 in symbiotic association with *Anthoceros punctatus*. J Bacteriol 173:7324–7329

Summers ML, Wallis JG, Campbell EL, Meeks JC (1995) Genetic evidence of a major role for glucose-6-phosphate dehydrogenase in nitrogen fixation and dark growth of the cyanobacterium *Nostoc* sp. strain ATCC 29133. J Bacteriol 177:6184–6194

Tandeau de Marsac N (1994) Differentiation of hormogonia and relationships with other biological processes. In: Bryant DA (ed) The molecular biology of cyanobacteria. Kluwer Academic Publ, Boston, pp 825–842

Thiel T, Lyons EM, Erker JC, Ernst A (1995) A second nitrogenase in vegetative cells of a heterocyst-forming cyanobacterium. Proc Natl Acad Sci USA 92:9358–9362

Thomas J (1972) Relationship between age of culture and occurrence of the pigments of photosystem II of photosynthesis in heterocysts of a blue-green alga. J Bacteriol 110:92–95

Van Rhijn P, Vanderleyden J (1995) The *Rhizobium*-plant symbiosis. Microbiol Rev 59:124–142

Vazquez-Bermudez MF, Herrero A, Flores E (2002) 2-oxoglutarate increases the binding affinity of the NtcA (nitrogen control) transcription factor for the *Synechococcus glnA* promoter. FEBS Lett 512:71–74

Villareal TA (1991) Nitrogen fixation by the cyanobacterial symbiont of the diatom genus *Hemiaulus*. Marine Ecol Prog Ser 76:201–204

Watts SD, Knight CD, Adams DG (1999) Characterisation of plant exudates inducing chemotaxis in nitrogen-fixing cyanobacteria. In: Peschek GA, Löffelhardt W,

Schmetterer G (eds) The phototrophic prokaryotes. Kluwer Academic Publ, New York, pp 679–684

West N, Adams DG (1997) Phenotypic and genotypic comparisons of symbiotic and free-living cyanobacteria from a single field site. Appl Environ Microbiol 63: 4479–4484

Whitton BA, Potts M (2000) The ecology of bacteria: their diversity in time and space. Kluwer Academic Publ, Dordrecht

Wolk CP (1996) Heterocyst formation. Annu Rev Genet 30:59–78

Wolk CP (2000) Heterocyst formation in *Anabaena*. In: Brun YV, Shimkets LJ (eds) Prokaryotic development. ASM, Washington, DC, pp 83–104

Wolk CP, Thomas J, Shaffer PW, Austin SM, Galonsky (1976) Pathway of nitrogen metabolism after fixation of ^{13}N-labeled nitrogen gas by the cyanobacterium *Anabaena cylindrica*. J Biol Chem 251:5027–5034

Wolk CP, Ernst A, Elhai J (1994) Heterocyst metabolism and development. In: Bryant DA (ed) The molecular biology of cyanobacteria. Kluwer Academic Publ, Dordrecht, pp 769–823

Wong FC, Meeks JC (2001) The *hetF* gene product is essential to heterocyst differentiation and affects HetR function in the cyanobacterium *Nostoc punctiforme*. J Bacteriol 183:26545–2661

Wong FC, Meeks JC (2002) Establishment of a functional symbiosis between the cyanobacterium *Nostoc punctiforme* and the bryophyte hornwort *Anthoceros punctatus* requires genes involved in nitrogen control and initiation of heterocyst differentiation. Microbiology 148:315–323

Yoon H-S, Golden JW (1998) Heterocyst pattern formation controlled by a diffusible peptide. Science 282:935–938

Yoon H-S, Golden JW (2001) PatS and products of nitrogen fixation control heterocyst pattern. J Bacteriol 183:2605–2613

Symbiosis of Thioautotrophic Bacteria with *Riftia pachyptila*

Frank J. Stewart, Colleen M. Cavanaugh

1
Introduction

The symbiosis between the giant vestimentiferan tubeworm *Riftia pachyptila* and an intracellular sulfur-oxidizing bacterium still fascinates researchers over 20 years after its discovery. This association, the first between a marine invertebrate and a chemoautotroph to be described, remains the best studied of the symbioses found at sulfide-rich hydrothermal vents. In the decade following the initial description of this symbiosis in 1981 (Cavanaugh et al. 1981; Felbeck 1981), many important studies have helped to characterize the physiological, biochemical, and anatomical adaptations that sustain this association (for other reviews see Fisher 1995; Nelson and Fisher 1995; van Dover 2000; Minic and Herve 2004; van Dover and Lutz 2004; Cavanaugh et al. 2005). Stable carbon isotope data and the absence of a mouth and gut strongly suggest that the adult *R. pachyptila* relies entirely on its bacterial symbionts for nutrition (see Fisher 1995; Nelson and Fisher 1995). These bacteria, which belong to the gamma subdivision of the Proteobacteria (Distel et al. 1988), oxidize reduced inorganic sulfur compounds to obtain energy and reducing power for autotrophic carbon fixation. Given their ability to synthesize C_3 compounds from a C_1 compound using chemical energy, *Riftia* symbionts are referred to as "chemosynthetic" (Cavanaugh et al. 2005).

Over the past 15 years, increasingly sophisticated experimental techniques (e.g., pressure chambers, vascular catheters) and new molecular technologies have dramatically increased our understanding of chemosynthetic symbioses. Specifically, for the *R. pachyptila* symbiosis, researchers provided new insights into the processes by which metabolites (e.g., carbon, sulfide, nitrogen) and waste products (e.g., protons) cycle among host, symbiont, and environment and identified some of the genes and corresponding enzymes involved in both host and symbiont metabolism. In addition, questions of host-

---

F.J. Stewart, C.M. Cavanaugh (e-mail: cavanaug@fas.harvard.edu)
Department of Organismic and Evolutionary Biology, Harvard University,
The Biological Laboratories, 16 Divinity Avenue, Cambridge MA 02138, USA

symbiont co-evolution and symbiont transmission were addressed in a number of studies, including recent work that successfully detected the free-living tubeworm protosymbiont using 16S rRNA probes for in situ hybridization (Harmer et al. 2005).

But many questions remain unanswered, particularly regarding the mechanisms of symbiont acquisition by the tubeworm, the spatio-temporal dynamics and processes of symbiont growth and metabolism, and the genetic structure and dispersal of symbiont populations. Our pursuit of answers to these questions has been hindered by the inability to culture the *Riftia* symbiont apart from its host. Fortunately, given the recent advances in genetic techniques and the substantial progress towards sequencing the *Riftia* symbiont genome (R. Feldman and R. Felbeck, pers. comm.), researchers are poised to reveal many of the genes and genetic interactions that guide the physiological, ecological, and evolutionary processes involved in this important, ecosystem-structuring symbiosis. This chapter presents an overview of the physiological ecology and evolution of the *Riftia* symbiosis, with mention of the genes thus far described for this association and the underlying questions that may guide future research.

1.1
Discovery of the *Riftia pachyptila* Symbiosis

Scientific understanding of chemosynthetic symbioses derives in large part from studies of the unique fauna associated with deep-sea hydrothermal vents. Early explorations revealed that, in contrast to common perceptions, the deep benthos was not a cold, food-limited habitat but instead contained flourishing ecosystems localized at hot springs emanating from mid-ocean spreading centers. First characterized along the Galapagos Rift and the East Pacific Rise in the eastern Pacific Ocean, hydrothermal vents were shown to support high concentrations of free-living microorganisms and dense aggregations of invertebrates, including the vestimentiferan tubeworm *Riftia pachyptila* (Fig. 1; Lonsdale 1977; Grassle 1985; van Dover 2000). Scientists first argued that suspended particulate organic matter and free-living chemoautotrophic bacteria were being filtered from the water column to support the abundant invertebrate populations (Lonsdale 1977; Corliss et al. 1979). But studies soon revealed that the adult *R. pachyptila* lacked a mouth and gut (Jones 1981) and was therefore incapable of suspension feeding. It appeared that tubeworm nutrition, and therefore the flux of energy through the vent food web, instead depended substantially on endosymbiotic chemosynthetic bacteria.

Fig. 1. *Riftia pachyptila* tubeworms on the East Pacific Rise. The branchial plume, a gill-like organ used for gas and metabolite exchange, protrudes from the white chitinous tube that protects the body of each worm

Initial evidence for a chemoautotrophic symbiosis in *R. pachyptila* came from microscopic and biochemical analyses showing Gram negative bacteria packed within the trophosome, a highly vascularized organ in the tubeworm trunk (Cavanaugh et al. 1981). Additional analyses involving stable isotope (Rau 1981), enzymatic (Felbeck 1981; Renosto et al. 1991), and physiological (Fisher et al. 1988) characterizations strongly suggested that the endosymbionts of *R. pachyptila* oxidize reduced sulfur compounds (e.g., hydrogen sulfide) to synthesize ATP for use in autotrophic carbon fixation via the Calvin cycle (i.e., "thioautotrophy"). The host tubeworm enables the uptake and transport of the substrates required for thioautotrophy (HS^-, O_2, and CO_2) and, in return, receives a portion of the organic matter synthesized by the symbiont population (Fig. 2). The bacterial population is the primary means of carbon acquisition for the symbiosis, and the adult tubeworm, given its inability to feed on particulate matter, is entirely dependent on its symbionts for nutrition.

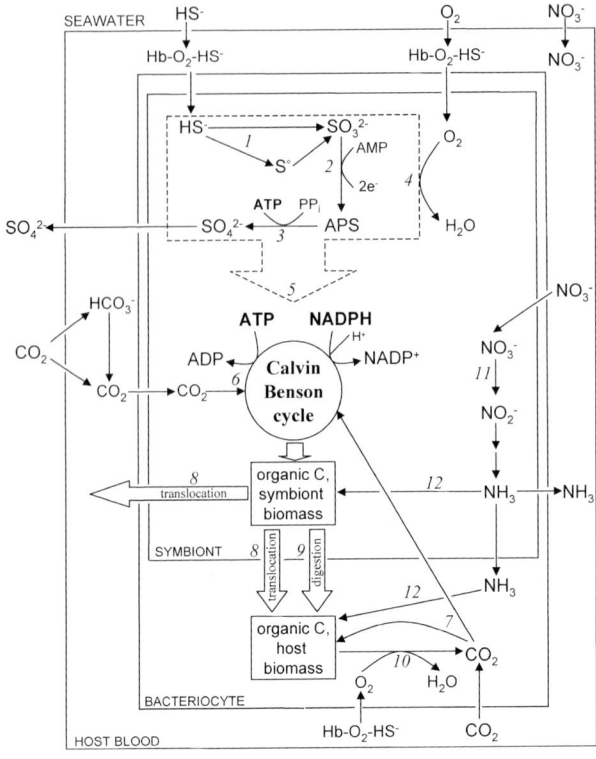

Fig. 2. Proposed model of metabolism in the symbiosis between *Riftia pachyptila* and a chemosynthetic sulfur-oxidizing bacterium. Reduced sulfur (primarily HS⁻) and NO_3^- enter the tubeworm blood from the environment through unidentified transport mechanisms. CO_2 and O_2 enter by diffusion. In the blood, HS⁻ and O_2 simultaneously and reversibly bind hemoglobin (Hb-O_2-HS⁻) for transport to the trophosome, where these substrates are used in symbiont sulfide oxidation (*dashed box*). HS⁻ is oxidized first to elemental sulfur (S^0) or directly to sulfite (SO_3^{2-}; *1*). SO_3^{2-} oxidation to sulfate (SO_4^{2-}) then proceeds through the APS pathway via the enzymes APS reductase (*2*) and ATP sulfurylase (*3*), yielding one ATP by substrate level phosphorylation. Electrons liberated during sulfur oxidation pass through an electron transport system, driving oxygen consumption (*4*) and the production of ATP and NADPH (*5*). Fixation of CO_2 occurs primarily via ribulose-1,5-bisphosphate oxygenase (RubisCO) in the Calvin Benson cycle (*6*), using ATP and NADPH generated from sulfur oxidation. Anaplerotic pathways in both host and symbiont (*7*) fix lesser amounts of CO_2. Transfer of organic matter from symbionts to host occurs via both translocation of simple nutritive compounds (e.g., amino acids) released by the bacteria (*8*) and direct digestion of symbiont cells (*9*). Host oxygen consumption (*10*) occurs in typical catabolic and anabolic pathways. Nitrate (NO_3^-), the dominant nitrogen source for the symbiosis, enters via an undescribed transport mechanism and is reduced to nitrite (NO_2^-) by the symbionts via an assimilatory nitrate reductase (*11*). NO_2^- is reduced via an uncharacterized pathway to yield ammonia (NH_3), which is used for biosynthesis by both symbiont and host (*12*). Abbreviations: APS, adenosine 5′-phosphosulfate. Reprinted with permission from TRENDS Microbiol (Stewart et al. 2005)

Discovery of this obligate mutualism prompted investigators to search for similar symbioses at vents and in other marine habitats (e.g., reducing sediments, hydrocarbon seeps). To date, chemosynthetic bacteria have been found in symbiosis with invertebrate hosts from six phyla as well as with ciliate protists (see review in Cavanaugh et al. 2005). Indeed, the presence of symbiotic bacteria is a defining characteristic of some taxa. For example, all members of the tubeworm family Siboglinidae examined to date, including the vestimentiferan (e.g., *R. pachyptila*) and the smaller pogonophoran tubeworms, contain intracellular symbionts. Members of this family occur not only at deep-sea hydrothermal vents (e.g., *Oasisia, Ridgeia, Riftia, Tevnia* sp.; McMullin et al. 2003) but also at hydrocarbon cold seeps (e.g., *Escarpia, Lamellibrachia* sp.; Sibuet and Olu 1998) and mud volcanoes (e.g., *Oligobrachia*; Pimenov et al. 2000; Gebruk et al. 2003). The symbionts of most of these worms are chemosynthetic sulfur oxidizers, but methane-oxidizing bacteria (methanotrophs) have been found in one host species, the pogonophoran tubeworm *Siboglinum poseidoni* (Schmaljohann and Flügel 1987; see following chapter).

Based on analyses comparing 16S rRNA gene sequences, symbionts of the vent tubeworms fall within the gamma Proteobacteria, a broad bacterial division that includes chemosynthetic symbionts from a wide diversity of host taxa (Fig. 3, adapted from McKiness 2004; Cavanaugh et al. 2005). When compared to other chemosynthetic symbionts of the gamma Proteobacteria and to epsilon Proteobacteria episymbionts of shrimp and alvinellid worms, the 16S rRNA gene sequence from the symbiont of *R. pachyptila* clusters with sequences from symbionts of other tubeworms, including another East Pacific Rise vent tubeworm, *Ridgeia piscesae*, and two species of seep tubeworms, *Escarpia spicata* and *Lamellibrachia columna* (Fig. 3). This result supports prior studies showing that vestimentiferan tubeworms from hydrothermal vents share a single, or very similar, symbiont phylotype (Feldman et al. 1997; Laue and Nelson 1997; di Meo et al. 2000; Nelson and Fisher 2000; McMullin et al. 2003). Outside of this "tubeworm group," the *Riftia* symbiont appears most closely related to chemosynthetic symbionts of tropical lucinid clams (Fig. 3). Interestingly, both lucinid clams and vestimentiferan vent tubeworms appear to acquire their symbionts from a pool of free-living bacteria (i.e., environmental transmission, see Gros et al. 1998, 2003 and Section 14.4 below). However, the extent to which a free-living symbiont stage facilitates the invasion of multiple hosts (e.g., tubeworms and clams) over evolutionary time remains equivocal.

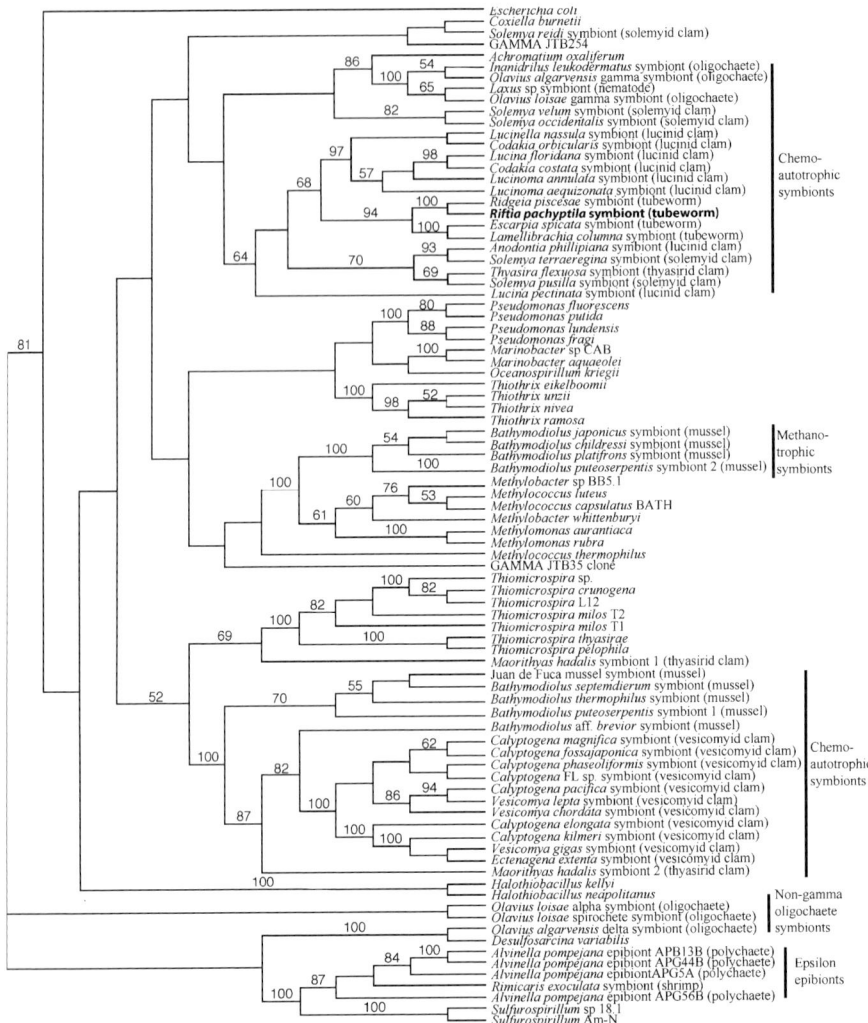

Fig. 3. Phylogeny showing the placement of the *Riftia* symbiont (*in bold*) relative to other chemosynthetic symbionts within the gamma Proteobacteria, episymbionts within the epsilon Proteobacteria, and free-living bacteria. The tree is a strict consensus of 46 trees obtained via parsimony analyses of 16S rRNA gene sequences (1,456 bp). Results greater than 50% from a 500 replicate bootstrap analysis are reported above respective branches. Symbionts are identified by host, with the common name of the host taxonomic group listed in parentheses. Adapted from McKiness (2004) and Cavanaugh et al. (2005)

For *R. pachyptila*, as for other marine chemosynthetic symbioses, much current research focuses on clarifying the mechanisms that mediate the environmental acquisition of symbionts as well as the transfer of carbon, nutrients, and sulfide within the symbiosis. These processes are vital to understanding the ecology and evolution of both the host and the symbiont. But elucidation of these mechanisms first requires knowledge of the physiochemical environment that determines the distribution and physiology of *R. pachyptila*.

1.2 Vent Habitat

Riftia pachyptila inhabits hydrothermal vent sites along the East Pacific Rise and the Galapagos Rift in the Eastern Pacific. The distribution of the tubeworm is intimately tied to the unique physiochemical characteristics of hydrothermal vents. Vent sites are typified by steep gradients between cold (~1.8°C), oxygen-rich (110 µM) bottom water and hot (up to 400°C), acidic (pH ~3 to 6) vent fluid. Hydrothermal fluids are usually laden with volcanic gases (e.g., methane and carbon dioxide) and reduced chemicals, including heavy metals and hydrogen sulfide (H_2S, HS^-, S^{2-}). Sulfide in vent fluids, produced by the geothermal reduction of seawater sulfate and the interaction of geothermally heated water with sulfur-containing rocks (e.g., basalt; Alt 1995; Elderfield and Schultz 1996; Rouxel et al. 2004), usually occurs at concentrations (3–12 mmol/kg) orders of magnitude higher than those in ambient seawater. It is at the interface (the chemocline) between the anoxic, reduced vent effluent and oxic bottom water that chemosynthetic vent symbioses thrive. Here, chemosynthetic symbionts access both the reduced compounds (e.g., sulfide) used as an energy source and the oxygen to which electrons are shuttled in aerobic energy metabolism.

While some vent symbioses, including those involving alvinellid polychaete worms and episymbiotic bacteria, congregate around sites where effluent (up to 400°C, pH ~3) directly exits the seafloor, *R. pachyptila* is typically clustered around diffuse or low flow vents. These vents, formed by ambient seawater mixing in the shallow subsurface with vent fluid, generally have a higher pH (~6), lower temperatures (1.8 to ~40°C), and, consequently, lower concentrations of reduced chemicals (e.g., sulfide up to 300 mM; Fisher 1995; van Dover 2000). But the physiochemical environment of diffuse vents is rarely stable; flow rate, temperature, and sulfide concentration may vary over timescales measured in seconds (Johnson et al. 1988a, 1994). To exploit this stochastic environment *R. pachyptila* relies on anatomical and physiological adaptations for sequestering sulfide, oxygen, inorganic carbon, and nitrogen from the chemocline and on the ability of its endosymbionts to use these substrates for energy metabolism and biomass synthesis.

2
Anatomy and Ultrastructure

Riftia pachyptila occurs in dense clumps attached to the seafloor substrate (e.g., basalt) at low flow vents. A narrow, elongate tube composed of chitin and scleroproteins and up to three meters in length protects the soft body of the worm, which is divided into four major regions. The branchial plume lies at the anterior of the worm in direct contact with the surrounding seawater. Infused with blood vessels, this gill-like organ allows an efficient exchange of metabolites (e.g., sulfide, oxygen, carbon dioxide, inorganic nitrogen) and waste products (e.g., ammonia, protons) between the worm and the surrounding seawater. Below the plume is the vestimentum, a circular muscle that houses the heart and brain of the worm as well as glands involved in tube secretion. The vestimentum also mediates the worm's position in its tube, enabling the animal to withdraw from predation or to extend its plume to access both sulfide-rich vent fluids and oxic bottom water.

The tubeworm trunk lies between the vestimentum and the segmented opisthosome that anchors the posterior of the worm to the tube. Encapsulated within the trunk wall in *R. pachyptila*, as in other vestimentiferan and pogonophoran tubeworm species, is a unique morphological adaptation designed specifically to house bacterial symbionts: the trophosome (Cavanaugh et al. 1981; Felbeck 1981; Jones 1981). The trophosome, which in *R. pachyptila* appears to develop from mesodermal tissue (Bright and Sorgo 2003) and replaces the transient gut present in larval and young juvenile tubeworms (Jones and Gardiner 1988), is a lobular organ consisting primarily of blood vessels, coelomic fluid, and specialized host cells called bacteriocytes. Bacteriocytes are packed with chemoautotrophic sulfide-oxidizing endosymbionts that are further encapsulated within a host-derived membrane bound vacuole (Fig. 4; Cavanaugh 1983, 1994; Fisher 1990). Bacterial abundance within this tissue is high, with cell density averaging 10^9 per gram of fresh trophosome (Hand 1987) and bacterial volume estimated to occupy between 15 and 35% of total trophosome volume (Powell and Somero 1986; Bright and Sorgo 2003).

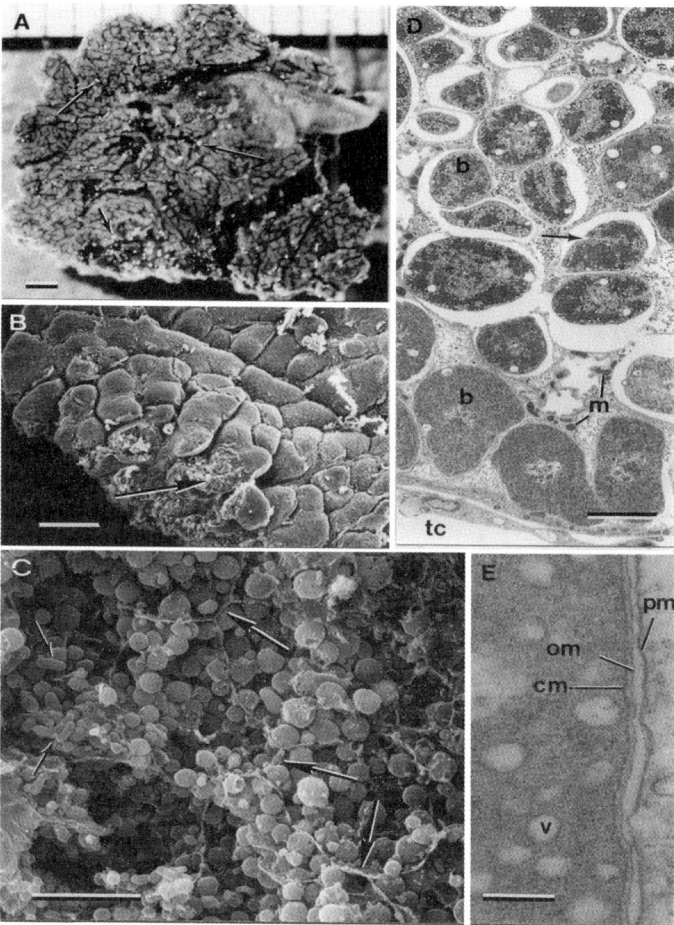

Fig. 4. *Riftia pachyptila* Jones (A-C: Galapagos Rift; D, E: 21°N, East Pacific Rise). A Photograph showing elemental sulfur crystals (*arrows*) scattered throughout trophosome; courtesy of M. L. Jones. B Scanning electron micrograph, showing lobules of trophosome; *arrow* indicates area of C (below) where surface epithelium was removed to reveal symbionts within trophosome. C Same, higher magnification, showing symbionts within trophosome; note spherical cells as well as rod-shaped cells (*small arrows*); *large arrows* indicate likely host cell membranes. D Cross section of portion of trophosome lobule, showing variable fine structure of symbionts, including membrane-bound vesicles in many cells; all symbionts contained within membrane-bound vacuoles, either singly or in groups or two or more; *arrow*, dividing bacterium; *b* bacteria; *m* mitochondria; *tc* trunk coelomic cavity. E Same, higher magnification, showing cell envelope of symbiont (resembling that of Gram-negative bacteria), intracytoplasmic vesicles, and peribacterial membrane; *v* vesicle; *cm* symbiont cytoplasmic membrane; *om* symbiont outer membrane; *pm* peribacterial membrane. *Scale bars*: A, 1 mm; B, 250 µm; C, 10 µm; D, 3 µm; E, 0.2 µm. Reprinted with permission from *Biol Soc Wash Bull* (Cavanaugh 1985)

In *R. pachyptila*, symbiont morphotype varies depending on location within the trophosome lobule (Fig. 5; Bosch and Grassé 1984a,b; Gardiner and Jones 1993; Bright et al. 2000). Bacteriocytes in the innermost (central) zone of the lobule primarily contain small, rod-shaped symbionts, while bacteriocytes nearer the periphery of the trophosome generally contain small and large cocci (1.6 to 10.7 µm diameter; Bright et al. 2000). Such pleomorphism may be caused either by intra-lobule biochemical gradients that impact symbiont metabolism, morphology, and growth or by differences in life cycle stage among symbiont cells (see below, Bosch and Grassé 1984a,b; Hand 1987; Bright et al. 2000).

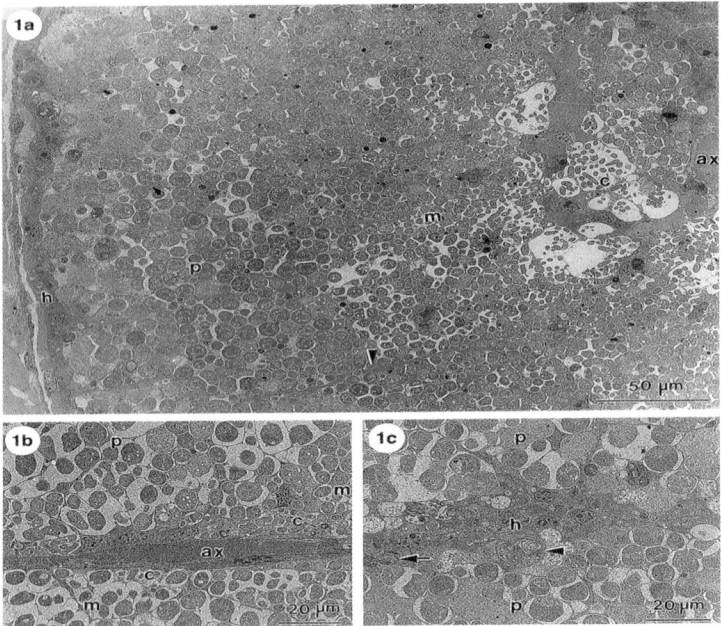

Fig. 5. Juvenile *Riftia pachyptila* transmission electron micrograph. Each trophosome lobule consists of non-symbiotic host tissue (*h*) surrounding the bacteriocytes of the peripheral zone (*p*) with large coccoid symbionts, the median zone (*m*) with small cocci, the central zone (*c*) with rods, and the non-symbiotic central axial blood vessel (*ax*). **a** Low magnification of lobule in cross section; note intracellular blood sinuses between bacteriocytes, one marked with an *arrowhead*. **b** Higher magnification of lobule center. **c** Higher magnification of periphery with degrading bacteria (*arrow*) and degrading bacteriocytes (*arrowhead*) in peripheral zone. Reprinted with permission from *Mar Biol* (Bright et al. 2000)

3
Nutritional Basis of the Symbiosis

3.1
Thioautotrophy

Nutrition in *Riftia pachyptila* depends on bacterial symbionts that fix carbon dioxide into organic matter using energy derived from the oxidation of reduced sulfur compounds. Figure 2 presents an overview of symbiont sulfur oxidation, as inferred from studies of free-living sulfur bacteria (see Kelly 1982; Nelson and Hagen 1995; Friedrich et al. 2001 for reviews of sulfur-based chemoautotrophy). The pathways by which the first half of this process, the oxidation of sulfide (H_2S, HS^-, S^{2-}) or thiosulfate ($S_2O_3^{2-}$) to sulfite (SO_3^{2-}), have not been clearly identified for the *Riftia* symbiosis and vary among different free-living thioautotrophs (Nelson and Hagen 1995). In contrast, considerable enzymatic and biochemical evidence (Felbeck 1981; Fisher et al. 1988; Renosto et al. 1991; Laue and Nelson 1994) suggests that tubeworm symbionts mediate sulfite oxidation via the energy-conserving adenosine 5'-phosphosulfate (APS) pathway (Fig. 2). In this pathway the enzyme APS reductase generates APS from adenosine monophosphate (AMP) and sulfite. In the presence of pyrophosphate, a second enzyme, ATP sulfurylase, then catalyzes the conversion of APS to ATP and sulfate (i.e., substrate level phosphorylation).

Activities of both APS reductase and ATP sulfurylase have been detected in the *R. pachyptila* trophosome (Felbeck 1981). Indeed, *Riftia* symbiont ATP sulfurylase was the first putative autotrophic ATP sulfurylase to be purified (Renosto et al. 1991). Unfortunately, neither enzyme is specifically diagnostic of thioautotrophy. ADP reductase and ATP sulfurylase operate in the reverse direction to catalyze the first steps of dissimilatory sulfate reduction by sulfate-reducing bacteria (see Peck and LeGall 1982). Also, ATP sulfurylase is ubiquitous in heterotrophic bacteria, fungi, and yeast, in which it functions in assimilatory biosynthetic pathways to incorporate sulfate into amino acids and other biomolecules (Segel et al. 1987). However, a probe specific to the gene encoding the *Riftia* symbiont ATP sulfurylase (*sopT*) was shown to specifically identify chemoautotrophic bacteria that use the APS pathway for sulfite oxidation (Laue and Nelson 1994). These results, along with the unique biochemical properties and notably high activity of *Riftia* symbiont ATP sulfurylase relative to ATP sulfurylase in photosynthetic and heterotrophic organisms (Renosto et al. 1991; Laue and Nelson 1994; Fisher 1995), strongly suggest that the *Riftia* symbiont uses this enzyme for sulfide-based chemoautotrophy. Further support for this hypothesis comes from recent crystallographic evidence showing that a key distinction between the *Riftia* symbiont ATP sulfurylase and the assimilatory enzymes from other organisms (e.g., fungi) lies in the orientation of the mobile loop that occupies the enzyme's active site (Beynon et al. 2001). The unique "open" loop position observed in the *Riftia* symbiont potentially lowers the affinity of the enzyme for sulfate and thereby drives the reaction in the direction of ATP synthesis (Beynon et al. 2001).

Ultimately, the electrons liberated during the entire oxidation of sulfide to sulfate are shuttled through an electron transport system, yielding a proton gradient that drives ATP production via oxidative phosphorylation. In most instances the terminal oxidant in electron transport is molecular oxygen. However, both nitrate and elemental sulfur have been shown to function in *R. pachyptila* as an electron acceptor during periods of anoxia (Hentschel and Felbeck 1993; Arndt et al. 2001). ATP from both oxidative phosphorylation and substrate level phosphorylation via the APS pathway is then available for CO_2 fixation in the Calvin cycle. Autotrophy in *R. pachyptila* is evidenced by the presence and activity of diagnostic Calvin cycle enzymes, namely, phosphoribulose kinase (PRK) and the CO_2-fixing enzyme ribulose 1,5-bisphosphate carboxylase/oxygenase (RubisCO; Felbeck 1981; Robinson et al. 1998), as well as by the uptake of radiolabeled CO_2 by live tubeworms (in pressure vessels), trophosome tissue homogenates, or isolated symbiont preparations (see Robinson and Cavanaugh 1995; Felbeck and Jarchow 1998; Bright et al. 2000). Indeed, the levels of carbon fixation (as RubisCO activity) in *R. pachyptila* are among the highest recorded for a chemosynthetic symbiosis (Felbeck 1981; Fisher 1995, and references therein). This is not surprising given the remarkably high growth rates observed for *R. pachyptila* – an increase of up to 1.4% of the tubeworm's total organic carbon per day (Childress et al. 1991; Lutz et al. 1994).

Thus, chemosynthesis by *Riftia* symbionts represents a significant input of fixed carbon into the vent ecosystem. To support this metabolism, the symbiosis must acquire all of the substrates necessary for both sulfide oxidation and carbon fixation: reduced sulfur, oxygen, dissolved inorganic carbon (DIC, as CO_2), and other nutrients (e.g., nitrogen and phosphorus) for use in biosynthesis. This requires that *R. pachyptila* access both oxic and anoxic environments, a task for which the tubeworm-symbiont association relies on specialized biochemistry and physiology.

3.2
Sulfide Acquisition

Thioautotrophy requires simultaneous access to both sulfide and oxygen. This dual requirement poses a unique challenge given that sulfide and oxygen occur in distinct habitat zones; in contrast to ambient seawater, the reduced vent fluid from which vent thioautotrophs obtain sulfide is typically anoxic or contains oxygen only at very low levels. In addition, sulfide spontaneously reacts with oxygen to form less-reduced sulfur compounds (S^0, $S_2O_3^{2-}$, or SO_4^{2-}; Zhang and Millero 1993). Such abiotic oxidation is typically several orders of magnitude slower than microbially-mediated sulfide oxidation (Millero et al. 1987; Johnson et al. 1988b) but nonetheless decreases the availability of these substrates, forcing thioautotrophs to compete with oxygen for free sulfide and reside at the interface between oxic-anoxic zones. Chemosynthetic symbionts have adapted to bridge this interface via association with a eukaryotic host (Cavanaugh 1985; Cavanaugh et al. 2005). Similar to free-living sulfur bacteria, hosts of thioautotrophic symbionts use specialized behavioral, anatomical, or physiological mechanisms to spatially or temporally span the oxic-anoxic interface and sequester both sulfide and oxygen (Cavanaugh 1994; Fisher 1996; Polz et al. 2000).

In *Riftia pachyptila* simultaneous acquisition of sulfide and oxygen occurs via a remarkable biochemical adaptation. In contrast to most invertebrate and vertebrate hemoglobins, with which sulfide interacts to inhibit oxygen-binding (Weber and Vinogradov 2001 and references therein), extracellular hemoglobins (two vascular and one coelomic) synthesized by *R. pachyptila* bind sulfide reversibly and independently of oxygen (Arp et al. 1985, 1987, Childress et al. 1991, Zal et al. 1996) and are then transported via the worm's circulatory system to the trophosome for use by the symbionts. Several studies have focused on the mechanism by which hydrogen sulfide (primarily as HS-; Goffredi et al. 1997a) binds to *Riftia* hemoglobin. Initially, two free cysteine residues, each located on a distinct globin subunit (Zal et al. 1997, Zal et al. 1998, Bailly et al. 2002), were proposed as sulfide binding sites. Indeed, Bailly et al. (2003) argue that sulfide-binding cysteine residues, which are well conserved in both symbiont-containing and symbiont-free annelids from sulfidic environments but are absent in annelids from sulfide-free habitats, are ancient features and that the annelid ancestor arose within a sulfur-rich environment. When annelid ancestors colonized sulfide-free environments, these cysteine residues, in the absence of their natural ligand (sulfide), may have reacted deleteriously with other blood components, leading eventually to the loss of the sulfide-binding function during annelid evolution (Bailly et al. 2003). But in *R. pachyptila*, which inhabits sulfide-rich vents, this specialized function of annelid hemoglobins presumably has been retained. Recently, however, Flores et al. (2005) have questioned the role of cysteine residues as the sole mechanism for sulfide binding. Using crystallographic data, these authors showed that in the coelomic hemoglobin free cysteine residues are located beneath the surface of the molecule and are therefore unlikely to serve as an efficient binding site. Rather, crystallographic data and ion chelation experiments suggest that >50% of bound sulfide may be explained by chelation to 12 Zn^+ ions near the poles of the coelomic hemoglobin molecule (Flores et al 2005). However, Zn^+ chelation accounted for considerably less (18%) of bound sulfide in experiments using the larger of the two vascular hemoglobins (Flores et al 2005). Clearly, sulfide binding may occur through multiple mechanisms in *Riftia* hemoglobins, and additional work is needed to fully understand the role of these molecules in providing symbionts access to energy substrates and circumventing the accumulation of sulfide in the blood.

3.3
Inorganic Carbon Acquisition

The *Riftia* symbiont uses the Calvin cycle for autotrophic carbon fixation and therefore requires access to dissolved organic carbon (DIC) in the form of CO_2, the DIC species that diffuses most readily through biological membranes (Fig. 2). While the majority of DIC in seawater (pH ~8.0;) is bicarbonate (HCO_3^-; pK_a of 6.1 at in situ temperature and pressure of ~10°C and 101.3 kPa; Dickson and Millero 1987), the lower pH associated with diffuse flow vents (pH ~6.0 around *R. pachyptila* plumes) generates higher concentrations of CO_2 and gives organisms that use the Calvin cycle a distinct advantage.

DIC uptake by *R. pachyptila* depends largely on gradients of pH and CO_2. To offset internal reactions that generate protons (e.g., symbiotic sulfide oxidation of HS^- to SO_4^{2-} and H^+), the tubeworm relies heavily on proton-equivalent ion export by energy-requiring H^+-ATPases (Goffredi et al. 1999; Goffredi and Childress 2001; Girguis et al. 2002). Indeed, proton elimination, which represents the worm's largest mass-specific metabolite flux, may be the single greatest energy cost incurred by *R. pachyptila* (Girguis et al. 2002). This important process helps maintain the extracellular vascular fluid at an alkaline pH of ~7.5 (Goffredi et al. 1997b). In contrast to the surrounding vent water in which a lower pH results in an elevated CO_2 concentration, the alkalinity of tubeworm blood favors the conversion of CO_2 to HCO_3^-, thereby establishing a CO_2 gradient across the tubeworm plume (Childress et al. 1993; Goffredi et al. 1997b; Scott 2003). This gradient, from higher external $[CO_2]$ to lower internal $[CO_2]$, drives the diffusion of CO_2 into the blood. DIC (as CO_2 and HCO_3^-) is then transported by the vascular system to the trophosome for use in symbiont CO_2 fixation via the Calvin cycle. The method by which HCO_3^- is converted back to CO_2 at the blood-bacteriocyte interface prior to incorporation by the symbionts is unclear; however, host-derived carbonic anhydrases, enzymes that catalyze the hydration of CO_2 into HCO_3^- in both prokaryotes and eukaryotes, may play a role in this process (Kochevar and Childress 1996; de Cian et al. 2003a,b).

The mechanism of DIC incorporation impacts the stable carbon isotope signature of the *Riftia*-chemoautotroph symbiosis. In particular, relative ^{13}C enrichment in *R. pachyptila* may be due in part to the form of symbiont RubisCO used for CO_2 fixation. The $\delta^{13}C$ values (expressed as ‰; $\delta^{13}C = [(R_{sample} - R_{std})/ R_{std})] * 10^3$, where $R = {}^{13}C / {}^{12}C$) reported for thioautotrophic vent symbioses typically cluster into two distinct groups: an isotopically light "–30‰ group" composed of vent bivalves with $\delta^{13}C$ values between –27 to –35‰ and a heavier "–11‰ group" including shrimp episymbionts, free-living vent bacterial mats, and vestimentiferan tubeworms with $\delta^{13}C$ values between –9 to –16‰ (Childress and Fisher 1992; van Dover and Fry 1994; Robinson and Cavanaugh 1995; Cavanaugh and Robinson 1996; Robinson et al. 2003). In general, these two groups correspond to the RubisCO form used by a given symbiont population, with form I RubisCO occurring in most members of the isotopically lighter group and form II in all members of the isotopically heavier group including the tubeworms (Robinson and Cavanaugh 1995; Cavanaugh and Robinson 1996). Robinson and Cavanaugh (1995) hypothesized that the difference in $\delta^{13}C$ values between these groups was due to variation in ^{13}C discrimination by the two structurally and kinetically distinct enzymes, as shown previously for form I from spinach (*Spinacia oleracea*) and form II from the proteobacterium *Rhodospirillum rubrum* (Roeske and O'Leary 1984). These authors therefore predicted that form II RubisCO of the *Riftia* symbiont discriminates less against ^{13}C than do typical form I enzymes. Indeed, Robinson et al. (2003) showed that the kinetic isotope effect (ϵ value), a measure of the extent to which $^{12}CO_2$ fixation is favored over $^{13}CO_2$ fixation (i.e., the degree of discrimination against ^{13}C), of purified form II RubisCO from *R. pachyptila* symbionts (ϵ = 19.5‰) is significantly lower than that of the form I enzyme (ϵ = 22 to 30‰; Guy et al. 1993). These data suggest that low ^{13}C discrimination by RubisCO may significantly impact the $\delta^{13}C$ values of tubeworm biomass.

Scott (2003) showed that rapid CO_2 fixation and steep gradients in CO_2 concentration may also contribute to ^{13}C enrichment in *R. pachyptila* biomass. RubisCO activity, by preferentially fixing $^{12}CO_2$ and leaving $^{13}CO_2$ behind, can temporarily enrich the cytoplasmic CO_2 pool. If carbon fixation is rapid, as would be expected given the high tubeworm growth rates and RubisCO activities recorded for this symbiosis, CO_2 equilibration between the isotopically heavier symbiont cytoplasm and the isotopically lighter bacteriocyte cytoplasm may not occur. The maintenance of this gradient forces RubisCO to draw from a more enriched $^{13}CO_2$ pool and therefore contributes to the relative abundance of ^{13}C in tubeworm biomass (Scott 2003).

3.4
Nitrogen Acquisition

Riftia pachyptila must also obtain all of the other macro- and micro-nutrients used in biosynthesis by both host and symbiont. While the pathways mediating assimilation and conversion of most metabolites are unknown for this symbiosis, researchers have identified some of the mechanisms of nitrogen metabolism in *R. pachyptila*. Laboratory incubations measuring the uptake of ^{15}N-nitrate ($^{15}NO_3^-$) showed that nitrate, which is abundant in situ (~40 µM; Johnson et al. 1988b), is the predominant inorganic nitrogen source for the symbiosis (Lee and Childress 1994). Indeed, activity of nitrate reductase, the bacterial enzyme mediating the reduction of nitrate to nitrite (NO_2^-), has been detected in *R. pachyptila* (Lee et al. 1999), indicating a role for nitrate reduction in either respiration or in the assimilatory pathway by which nitrate is ultimately converted to ammonia (NH_4^+) for use in biomass synthesis. Girguis et al. (2000), by measuring fluxes of inorganic nitrogen species into and out of *R. pachyptila* kept in pressurized chambers, ultimately showed that the symbiont population reduces nitrate to ammonia not for respiratory purposes but for incorporation into both symbiont and host biomass. In addition, glutamine synthetase (GS) and glutamate dehydrogenase (GDH), the primary enzymes mediating assimilation of ammonia into amino acids, have also been detected in *R. pachyptila* (Lee et al. 1999). Though both host and symbiont synthesize these enzymes, biochemical and molecular tests (e.g., protein characterization, Southern hybridization) showed that GS measured in the trophosome was of bacterial origin (Lee et al. 1999), implicating the symbiont population in inorganic nitrogen acquisition. However, subsequent detection of high GS activity in symbiont-free branchial plume tissue indicates that the tubeworm may also assimilate ammonia directly from the surrounding seawater (Minic et al. 2001).

But further enzymatic characterization of *R. pachyptila* tissues reinforces the hypothesis that the trophosome is a primary site for nitrogen assimilation and metabolism. Minic et al. (2001) demonstrated that only the bacterial symbiont has all of the enzymes required for the de novo synthesis of pyrimidines, implying that the tubeworm is entirely dependent on the bacterium for these nucleotides. Interestingly, while absent in the symbiont, the activities of at least three of the enzymes that mediate pyrimidine catabolism were detected in the tubeworm host, suggesting that pyrimidine degradation may represent an internal source of CO_2 and NH_3 for use in host biosynthesis (Minic et al. 2001). Similarly, the catabolic enzymes involved in the synthesis of polyamines from arginine appear to be present only in the symbiont (Minic and Herve 2003). Polyamines, which play important physiological roles in processes of growth, membrane structure, and nucleic acid synthesis and which represent a potential source of carbon and nitrogen upon degradation (Tabor and Tabor 1985), may therefore be available to *R. pachyptila* only via the metabolism of the bacterial symbiont (Minic and Herve 2003). These data further emphasize the extent to which the metabolism of the tubeworm host is intertwined with that of the symbiont. Undoubtedly, genomic characterization, in conjunction with enzymatic or gene expression analyses, will reveal additional interactions that sustain this symbiosis at the molecular level.

3.5
Organic Compound Transfer and Symbiont Growth

In chemosynthetic endosymbioses the transfer of fixed carbon from the symbiont to the host for use in biomass synthesis or energy metabolism can occur via two mechanisms: 1) the symbiont excretes autotrophically-fixed carbon in the form of soluble organic molecules that are then translocated to host cells, or 2) the host directly digests bacterial cells (Fig. 2).

Both translocation and digestion appear to contribute to host nutrition in *Riftia pachyptila*. Felbeck and Jarchow (1998), using radiotracer experiments with purified *Riftia* symbionts incubated in the presence ^{14}C-bicarbonate, showed that labeled sugars, organic acids, and amino acids (primarily succinate and glutamate) were excreted into the surrounding medium by the symbionts. These simple organic compounds, some of which had been previously identified in isolated trophosome during similar radiotracer experiments (Felbeck 1985), might be important intermediates in the transfer of fixed carbon from symbionts to host (Felbeck and Jarchow 1998). Recently, pulse-chase labeling analysis and autoradiography showed directly that a considerable fraction of the organic carbon fixed by the *Riftia* symbionts is released immediately (within 15 min) after fixation and assimilated into metabolically active tubeworm tissue (Bright et al. 2000). However, the appearance of radiolabel in host tissue and the concomitant loss of label from the trophosome during the chase period suggested that symbiont digestion also contributes to host nutrition (Bright et al. 2000). Indeed, these authors consistently observed symbionts being degraded near the periphery of the trophosome lobule (Bright et al. 2000); similar ultrastructural patterns had been reported previously (Bosch and Grassé 1984a; Hand 1987; Gardiner and Jones 1993). In addition, digestion of *Riftia* symbionts has been inferred from the relatively high lysozyme activity within the *R. pachyptila* trophosome (Boetius and Felbeck 1995).

The relative contribution of translocation and digestion to tubeworm nutrition is tied to the growth dynamics of the symbiont population. Several studies show that symbiont morphology varies predictably within a trophosome lobule, with the symbiont population dominated by small, actively dividing rods at the center of the lobule and large cocci in various stages of autolysis and digestion at the periphery (Fig. 5; Bosch and Grassé 1984a; Gardiner and Jones 1993; Bright and Sorgo 2003). Two hypotheses have been put forth to explain this variation in symbiont morphology and growth (reviewed in Bright and Sorgo 2003). First, blood flow from the periphery to the center of the trophosome forms biochemical gradients that influence the metabolism, and therefore the morphology and growth, of the symbionts (Hand 1987). Second, symbiont cells and host bacteriocytes both progress through a complex cell cycle involving cell proliferation at the lobule center, followed by migration toward the periphery, and subsequent lysis and degradation (Bosch and Grassé 1984a,b). Neither hypothesis is necessarily exclusive of the other. However, symbiont carbon fixation rates were shown to be roughly equivalent at the center and periphery of the lobule (Bright et al. 2000), implying that the substrates used in chemosynthesis may not occur over a gradient in the trophosome.

Support for a cell cycle hypothesis comes from quantitative ultrastructural analyses showing that the division by rods is in balance with the lysis of cocci and that intermediate stages between rods and cocci occur within the trophosome (Bright and Sorgo 2003). It therefore appears that in the central zone carbon fixation supports symbiont division whereas in the peripheral zone fixation supports an increase in per cell biomass, with excess carbon stored as glycogen (Sorgo et al. 2002). Recent evidence suggests that the symbiont-containing bacteriocytes undergo a similar cell cycle, with proliferation in the lobule center, terminal differentiation, and degradation in the periphery (Bright and Sorgo 2003). The life cycle hypothesis that appears to explain the gradation in symbiont size and shape within the trophosome implies a constraint on symbiont division outside the lobule center as well as a complex coordination of host and symbiont cell replication. The molecular mechanisms that regulate symbiont division, metabolism, and lysis likely involve complex symbiont-host signaling. Elucidation of these mechanisms, as well as the spatio-temporal dynamics of symbiont growth within the trophosome, remain as important goals in the study of the *Riftia* symbiosis.

4
Symbiont Transmission and Evolution

The evolutionary dynamics between symbiont and host depend largely on the symbiont transmission strategy. Symbiont transmission between successive host generations can occur environmentally (acquisition from a free-living population of symbiotic bacteria), horizontally (transfer between hosts sharing the same habitat), or vertically (transfer from parent to offspring via the egg). Vertically transmitted endosymbionts experience a unique selective regime within the host. These symbionts are effectively disconnected from their free-living counterparts and undergo a population bottleneck upon host colonization and another upon transmission (Mira and Moran 2002). These processes severely reduce the symbiont effective population size and thereby elevate the rate of fixation of slightly deleterious alleles via genetic drift (Ohta 1973; Wernegreen 2002). In addition, the asexuality and lack of recombination in endosymbionts exacerbates these genetic problems through what is known as Muller's ratchet (Muller 1964; Moran 1996). In Muller's ratchet wild type recombinants cannot be introduced into the endosymbiont population (Moran and Baumann 1994; Dale et al. 2003); genetic drift therefore occurs quickly, and the population cannot recover after fixation of deleterious alleles. Because the selective regime experienced by vertically transmitted endosymbionts is inextricably linked to the reproduction and dispersal of the host, co-speciation of symbiont and host is common and is used as a marker for vertical transmission (e.g., Chen et al. 1999; Thao et al. 2000; Degnan et al. 2004).

In contrast, environmentally transmitted symbiont populations have both a free-living and a symbiotic component. Exchange between these two pools presumably lessens the effects of Muller's ratchet by facilitating recombination between individual bacteria and leads to a larger effective population size, and therefore a potentially slower rate of nucleotide substitution, than for vertically transmitted symbionts (Moran 1996; Peek et al. 1998). Consequently, the link between symbiont and host evolution is weakened and co-speciation is not anticipated.

For the *Riftia* symbiosis, in which the adult host is completely dependent on its symbiont for sustenance, environmental transmission would seem an uncertain strategy, particularly given the erratic nature of the hydrothermal vent environment. Nonetheless, several lines of evidence strongly suggest that *R. pachyptila* acquires its symbiont de novo each generation from a pool of free-living bacteria. For example, tubeworm symbionts do not demonstrate co-speciation with their hosts, i.e., host-symbiont specificity and phylogenetic host-symbiont congruence are not evident in tubeworm symbioses. While repetitive-extragenic-palindrome PCR (REP-PCR) fingerprinting revealed that *Riftia* symbionts exhibit strain-level genetic variation that correlates with geographic location (di Meo et al. 2000), analysis of 16S rRNA gene sequences showed that vestimentiferan tubeworms belonging to the genera *Riftia, Tevnia,* and *Oasisia* appear to share a single, or very similar, symbiont phylotype (Feldman et al. 1997; Laue and Nelson 1997; di Meo et al. 2000; Nelson and Fisher 2000; McMullin et al. 2003). The distribution of a single phylotype among three vestimentiferan genera is strong evidence that vent tubeworms acquire their symbionts from a free-living pool of symbiont cells.

Other lines of evidence similarly suggest that *R. pachyptila* obtains its symbionts from the environment. PCR probing using universal eubacteria and *Riftia* symbiont-specific primers failed to detect bacterial 16 rRNA genes in DNA extracts from *R. pachyptila* eggs (Cary et al. 1993). These results suggest, but do not confirm, that the *Riftia* symbiont is not maternally transmitted via gonadal tissue. The apparent absence of symbiont cells in *R. pachyptila* eggs is consistent with the fact that tubeworm larvae and juveniles, while possessing a mouth and gut, lack symbiont-containing tissue. In contrast, adult tubeworms lack a mouth and gut and are therefore incapable of feeding autonomously. Rather, adults worms possess the bacteria-containing trophosome, the primary site for CO_2 fixation in this symbiosis.

The detection of specific functional genes in the *Riftia* symbiont also suggests environmental transmission. Millikan et al. (1999) used PCR to detect and amplify a *Riftia* symbiont gene with high sequence similarity to the flagellin gene, *fliC*, which encodes the primary subunits of the bacterial flagellum. This analysis, while showing that the *Riftia* symbiont has at least one of the genes required for flagellar synthesis, does not attest to the functional role of the flagellum in situ. Flagellar motility presumably would be unnecessary if the symbiont is always associated with the host (i.e., if the symbiont is vertically transmitted). Alternatively, if the symbiont is environmentally transmitted, a flagellum might mediate adhesion to and invasion of the tubeworm. Indeed, several other bacterial symbionts and pathogens have been shown to use flagellum-associated structures to colonize eukaryotic hosts (e.g., Chua et al. 2003; Gavin et al. 2003; Kirov 2003; Dons et al. 2004). But, while *Riftia* symbiont *fliC* induced the formation of flagella when the gene was cloned and expressed in *Escherichia coli* (Millikan et al. 1999), characterization of in situ flagellum synthesis is necessary to demonstrate that a flagellum mediates colonization of *R. pachyptila*. Similarly, symbiont genes whose products are homologous to known signal transduction proteins have been detected in the *Riftia* symbiont (Hughes et al. 1997). These genes, *rssS* and *rssB*, encode a histidine protein kinase (RssA) and a response regulator protein (RssB) and may facilitate communication of the *Riftia* symbiont with its external environment. Alternatively, *rssA* and *rssB* may regulate survival of the symbiont while within the tubeworm host and therefore be of no direct relevance to an environmental mode of transmission. As for *fliC*, in situ expression studies will be necessary to determine the functional significance of these signal transduction genes. Fortunately, expression analyses, such as those involving mRNA hybridization to genomic or proteomic arrays and in situ hybridization to mRNA (e.g., Pernthaler and Amann 2004), while previously daunting due to the inability to culture the symbiont apart from its host, are now possible given increasingly sensitive and specific molecular approaches and advances in maintaining vent organisms under in situ conditions (i.e., in pressure vessels; e.g., Girguis et al. 2000; Felbeck et al. 2004).

Finally, if the *Riftia* symbiont is obtained anew each generation, the symbiont must exist in a free-living form in the vent habitat. Indeed, the 16S rRNA phylotype of the *Riftia* symbiont has been detected in vent environments. Harmer et al. (2005), using both PCR and in situ hybridization with symbiont-specific probes, detected the free-living form of the *Riftia* symbiont on basalt blocks and settlement traps deployed in and around clumps of tubeworms at vent sites on the East Pacific Rise. These results provide the most encouraging evidence to date of environmental transmission in the *Riftia* symbiosis. This work may guide future ecological studies examining the impact of the tubeworm symbiosis on the free-living bacterial community at vents as well as complement ongoing efforts to determine the molecular mechanisms that mediate host-symbiont recognition and entry of the symbiont into the developing tubeworm.

5
Future Directions

In this review, we have attempted specifically to include those studies that use molecular or biochemical data to describe the physiology and evolution of the

Riftia pachyptila symbiosis. While this unique chemosynthetic association has been a primary focus of hydrothermal vent research (for additional reviews see Fisher 1995; Nelson and Fisher 1995; van Dover 2000; Cavanaugh et al. 2005; Minic and Herve 2004; van Dover and Lutz 2004), studies of the molecular biology of the *Riftia* endosymbiont have been hindered greatly by the inability to grow the bacterium in pure culture. However, with the *Riftia* symbiont genome sequence in process (R. Feldman and H. Felbeck, pers. comm.), we will soon have detailed knowledge of the genes that mediate symbiont growth, metabolism, and behavior both prior to and following invasion of the tubeworm host. Several important questions remain unanswered. For instance, through what mechanisms does symbiont sulfide oxidation proceed? How are biologically important elements (e.g., nitrogen, phosphorus, sulfur, iron) obtained by and cycled within the symbiont? Do these bacterial pathways provide important intermediates required for host metabolism? What transport processes and signal pathways are responsible for host-symbiont specificity and symbiont invasion of host cells? What factors (e.g., substrate gradients, cell cycle controls) regulate symbiont growth and division, and how are these factors coordinated with the regulation of the host cell cycle? How is the genetic structure of symbiont populations affected by environmental symbiont transmission, and how does population-level genetic diversity vary over spatial, habitat, and temporal gradients and between symbiotic and free-living populations? To answer these questions it will be particularly prudent to use genomic data not only to understand symbiont evolution but also to design in situ expression studies that can accurately assess the dynamics of both gene-gene and symbiont-host interactions. Doing so presents a challenge, but one for which researchers will be rewarded with even more fascinating details about this remarkable symbiosis.

Acknowledgements. We thank Monika Bright and Andrea Nussbaumer for stimulating discussions, the captains and crews of the R/V Atlantis and the R/V Atlantis II, the pilots and crews of the DSV Alvin, Chuck Fisher and all of the Chief Scientists and scientific teams of the deep-sea expeditions enabling research on the vent organisms. We gratefully acknowledge support from the Ocean Sciences Division of the National Science Foundation (NSF-OCE 9504257 and OCE 0002460), the NOAA Undersea Research Program through the West Coast and Polar Regions National Undersea Research Center at the University of Alaska at Fairbanks (UAF-WA # 98–06), the National Institutes of Health Genetics and Genomics Training Grant (to FJS) and a Hansewissenschaftkolleg Fellowship (to CMC).

References

Alt JC (1995) Subseafloor processes in mid-ocean ridge hydrothermal systems. In: Humphris SE, Zierenberg RA, Mullineaux LS, Thomson RE (eds) Seafloor hydrothermal systems: physical, chemical, biological, and geological Interactions. Geophysical monograph 91. Am Geophys Union, Washington, DC, pp 85–114

Arndt C, Gaill F, Felbeck H (2001) Anaerobic sulfur metabolism in thiotrophic symbioses. J Exp Biol 204:741–750

Arp AJ, Childress JJ, Fisher CR (1985) Blood gas transport in *Riftia pachyptila*. Bull Biol Soc Wash 6:289–300

Arp AJ, Childress JJ, Vetter RD (1987) The sulphide-binding protein in the blood of the vestimentiferan tube-worm, *Riftia Pachyptila*, is the extracellular haemoglobin. J Exp Biol 128:139–158

Bailly X, Jollivet D, Vanin S, Deutsch J, Zal F, Lallier F, Toulmond A (2002) Evolution of the sulfide-binding function within the globin multigenic family of the deep-sea hydrothermal vent tubeworm *Riftia pachyptila*. Mol Biol Evol 19:1421–1433

Bailly X, Leroy R, Carney S, Collin O, Zal F, Toulmond A, Jollivet D (2003) The loss of the hemoglobin H_2S-binding function in annelids from sulfide-free habitats reveals molecular adaptation driven by Darwinian positive selection. Proc Nat Acad Sci USA 100:5885–5890

Beynon JD, MacRae IJ, Huston SL, Nelson DC, Segel IH, Fisher AJ (2001) Crystal structure of ATP sulfurylase from the bacterial symbiont of the hydrothermal vent tubeworm *Riftia pachyptila*. Biochemistry 40:14509–14517

Boetius A, Felbeck H (1995) Digestive enzymes in marine-invertebrates from hydrothermal vents and other reducing environments. Mar Biol 122:105–113

Bosch C, Grassé PP (1984a) Cycle partiel des bactéries chimioautotrophes symbiotiques et eurs rapports avec les bactériocytes chez *Riftia pachyptila* Jones (Pogonophore Vestimentifère) I. Le trophosome et les bactériocytes. CR Acad Sci III Vie 299:371–376

Bosch C, Grassé PP (1984b) Cycle partiel des bactéries chimioautotrophes symbiotiques et eurs rapports avec les bactériocytes chez *Riftia pachyptila* Jones (Pogonophore Vestimentifère) II. L'évolution des bactéries symbotiques et des bactériocytes. CR Acad Sci III Vie 299:413–419

Bright M, Sorgo A (2003) Ultrastructural reinvestigation of the trophosome in adults of *Riftia pachyptila* (Annelida, Siboglinidae). Invert Biol 122:347–368

Bright M, Keckeis H, Fisher CR (2000) An autoradiographic examination of carbon fixation, transfer and utilization in the *Riftia pachyptila* symbiosis. Mar Biol 136:621–632

Cary SC, Warren W, Anderson E, Giovannoni SJ (1993) Identification and localization of bacterial endosymbionts in hydrothermal vent taxa with symbiont-specific polymerase chain reaction amplification and in situ hybridization techniques. Mol Mar Biol Biotech 2:51–62

Cavanaugh CM (1983) Symbiotic chemoautotrophic bacteria in marine invertebrates from sulfide-rich habitats. Nature 302:58–61

Cavanaugh CM (1985) Symbioses of chemoautotrophic bacteria and marine invertebrates from hydrothermal vents and reducing sediments. Bull Biol Soc Wash 6:373–388

Cavanaugh CM (1994) Microbial symbiosis: patterns of diversity in the marine environment. Am Zool 34:79–89
Cavanaugh CM, Robinson JJ (1996) CO_2 fixation in chemoautotroph-invertebrate symbioses: expression of Form I and Form II RubisCO. In: Lidstrom ME, Tabita FR (eds) Microbial growth on C_1 compounds. Kluwer Academic Publ Dordrecht, pp 285–292
Cavanaugh CM, Gardiner SL, Jones ML, Jannasch HW, Waterbury JB (1981) Prokaryotic cells in the hydrothermal vent tube worm *Riftia pachyptila* Jones: possible chemoautotrophic symbionts. Science 213:340–342
Cavanaugh CM, McKiness ZP, Newton ILG, Stewart FJ (2005) Marine chemosynthetic symbioses. In: Dworkin M, Falkow S, Rosenberg E, et al (eds) The Prokaryotes: a handbook on the biology of bacteria, 3rd edn. Springer, Berlin Heidelberg New York (in press)
Chen XA, Li S, Aksoy S (1999) Concordant evolution of a symbiont with its host insect species: molecular phylogeny of genus *Glossina* and its bacteriome-associated endosymbiont, *Wigglesworthia glossinidia*. J Mol Evol 48:49–58
Childress JJ, Fisher CR (1992) The biology of hydrothermal vent animals: physiology, biochemistry, and autotrophic processes. Oceanogr Mar Biol 30:337–441
Childress JJ, Fisher CR, Favuzzi JA, Kochevar RE, Sanders NK, Alayse AM (1991) Sulfide-driven autotrophic balance in the bacterial symbiont-containing hydrothermal vent tubeworm, *Riftia pachyptila* Jones. Biol Bull 180:135–153
Childress JJ, Lee RW, Sanders NK, Felbeck H, Oros DR, Toulmond A, Desbruyeres D, Kennicutt MC, Brooks J (1993) Inorganic carbon uptake in hydrothermal vent tubeworms facilitated by high environmental pCO_2. Nature 362:147–169
Chua KL, Chan YY, Gan YH (2003) Flagella are virulence determinants of *Burkholderia pseudomallei*. Infect Immun 71:1622–1629
Corliss JB, Dymond J, Gordon LI, Edmond JM, Herzen RPV, Ballard RD, Green K, Williams D, Bainbridge A, Crane K, van Andel TH (1979) Submarine thermal springs on the Galapagos Rift. Science 203:1073–1083
Dale C, Wang B, Moran N, Ochman H (2003) Loss of DNA recombinational repair enzymes in the initial stages of genome degeneration. Mol Biol Evol 20:1188–1194
De Cian MC, Andersen AC, Bailly X, Lallier FH (2003a) Expression and localization of carbonic anhydrase and ATPases in the symbiotic tubeworm *Riftia pachyptila*. J Exp Biol 206:399–409
De Cian MC, Bailly X, Morales J, Strub JM, Van Dorsselaer A, Lallier FH (2003b) Characterization of carbonic anhydrases from *Riftia pachyptila*, a symbiotic invertebrate from deep-sea hydrothermal vents. Proteins 51:327–339
Degnan PH, Lazarus AB, Brock CD, Wernegreen JJ (2004) Host-symbiont stability and fast evolutionary rates in an ant-bacterium association: cospeciation of *Camponotus* species and their endosymbionts, *Candidatus blochmannia*. Syst Biol 53: 95–110
Dickson AG, Millero FJ (1987) A comparison of the equilibrium constants for the dissociation of carbonic acid in seawater media. Deep Sea Res 34:1733–1743
Di Meo CA, Wilbur AE, Holben WE, Feldman RA, Vrijenhoek RC, Cary SC (2000) Genetic variation among endosymbionts of widely distributed vestimentiferan tubeworms. Appl Environ Microbiol 66:651–658

Distel DL, Lane DJ, Olsen GJ, Giovannoni SJ, Pace B, Pace NR, Stahl DA, Felbeck H (1988) Sulfur-oxidizing bacterial endosymbionts: analysis of phylogeny and specificity by 16S rRNA sequences. J Bacteriol 170:2506–2510

Dons L, Eriksson E, Jin YX, Rottenberg ME, Kristensson K, Larsen CN, Bresciani J, Olsen JE (2004) Role of flagellin and the two-component CheA/CheY system of *Listeria monocytogenes* in host cell invasion and virulence. Infect Immun 72: 3237–3244

Elderfield H, Schultz A (1996) Mid-ocean ridge hydrothermal fluxes and the chemical composition of the ocean. Annu Rev Earth Plant Sci 24:191–224

Felbeck H (1981) Chemoautotrophic potential of the hydrothermal vent tube worm, *Riftia pachyptila* Jones (Vestimentifera). Science 213:336–338

Felbeck H (1985) CO_2 fixation in the hydrothermal vent tube worm *Riftia pachyptila* (Jones). Physiol Zool 58:272–281

Felbeck H, Jarchow J (1998) Carbon release from purified chemoautotrophic bacterial symbionts of the hydrothermal vent tubeworm *Riftia pachyptila*. Physiol Zool 71:294–302

Felbeck H, Arndt C, Hentschel U, Childress JJ (2004) Experimental application of vascular and coelomic catheterization to identify vascular transport mechanisms for inorganic carbon in the vent tubeworm, *Riftia pachyptila*. Deep Sea Res 51: 401–411

Feldman R, Black M, Cary C, Lutz R, Vrijenhoek R (1997) Molecular phylogenetics of bacterial endosymbionts and their vestimentiferan hosts. Mol Mar Biol Biotech 6:268–277

Fisher CR (1990) Chemoautotrophic and methanotrophic symbioses in marine invertebrates. Rev Aquat Sci 2:399–436

Fisher CR (1995) Toward an appreciation of hydrothermal vent animals: their environment, physiological ecology, and tissue stable isotope values. In: Humphris SE, Zierenberg RA, Mullineaux LS, Thomson RE (eds) Seafloor hydrothermal systems: physical, chemical, biological, and geological Interactions. Geophysical monograph 91. Am Geophys Union, Washington, DC, pp 297–316

Fisher CR (1996) Ecophysiology of primary production at deep-sea vents and seeps. In: Uiblein R, Ott J, Stachowtish M (eds) Deep-sea and extreme shallow-water habitats: affinities and adaptations. Biosystematics and ecology series, vol 11. Austrian Academy of Sciences, Vienna, pp 311–334

Fisher, CR, Childress JJ, Arp AJ, Brooks JM, Distel D, Favuzzi JA, Macko SA, Newton A, Powell MA, Somero GN, Soto T (1988) Physiology, morphology, and biochemical composition of *Riftia pachyptila* at Rose Garden in 1985. Deep Sea Res 35:1745–1758

Flores JF, Fisher CR, Carney SL, Green BN, Freytag JK, Schaeffer SW, Royer JR. WE (2005) Sulfide binding is mediated by zinc ions discovered in the crystal structure of a hydrothermal vent tubeworm hemoglobin. Proc Nat Acad Sci USA 102:2713–2718

Friedrich CG, Rother D, Bardischewsky F, Quentmeier A, Fischer J (2001) Oxidation of reduced inorganic sulfur compounds by bacteria: emergence of a common mechanism? Appl Environ Microbiol 67:2873–2882

Gardiner SL, Jones ML (1993) Vestimentifera. In: Harrison FW, Gardiner SL (eds) Microscopic anatomy of invertebrates, vol 12. Onychophora, Chilopoda, and lesser Protostomata. Wiley-Liss, New York, pp 371–460

Gavin R, Merino S, Altarriba M, Canals R, Shaw JG, Tomas JM (2003) Lateral flagella are required for increased cell adherence, invasion and biofilm formation by *Aeromonas spp*. FEMS Microbiol Lett 224:77–83

Gebruk AV, Krylova EM, Lein AY, Vinogradov GM, Anderson E, Pimenov NV, Cherkashev GA, Crane K (2003) Methane seep community of the Hakon Mosby mud volcano (the Norwegian Sea): composition and trophic aspects. Sarsia 88:394–403

Girguis PR, Lee RW, Desaulniers N, Childress JJ, Pospesel M, Felbeck H, Zal F (2000) Fate of nitrate acquired by the tubeworm *Riftia pachyptila*. Appl Environ Microbiol 66:2783–2790

Girguis PR, Childress JJ, Freytag JK, Klose K, Stuber R (2002) Effects of metabolite uptake on proton-equivalent elimination by two species of deep-sea vestimentiferan tubeworm, *Riftia pachyptila* and *Lamellibrachia* cf *luymesi*: proton elimination is a necessary adaptation to sulfide-oxidizing chemoautotrophic symbionts. J Exp Biol 205:3055–3066

Goffredi SK, Childress JJ (2001) Activity and inhibitor sensitivity of ATPases in the hydrothermal vent tubeworm *Riftia pachyptila*: a comparative approach. Mar Biol 138:259–265

Goffredi SK, Childress JJ, Desaulniers NT, Lallier FH (1997a) Sulfide acquisition by the hydrothermal vent tubeworm *Riftia pachyptila* appears to be via uptake of HS^-, rather than H_2S. J Exp Biol 200:2069–2616

Goffredi SK, Childress JJ, Desaulniers NT, Lee RW, Lallier FH, Hammond D (1997b) Inorganic carbon acquisition by the hydrothermal vent tubeworm *Riftia pachyptila* depends upon high external pCO_2 and upon proton-equivalent ion transport by the worm. J Exp Biol 200:883–896

Goffredi SK, Childress JJ, Lallier FH, Desaulniers NT (1999) The ionic composition of the hydrothermal vent tube worm *Riftia pachyptila*: evidence for the elimination of SO_4^{2-} and H^+ and for a Cl^-/HCO_3^- shift. Physiol Biochem Zool 72:296–306

Grassle JF (1985) Hydrothermal vent animals – distribution and biology. Science 229:713–717

Gros O, De Wulf-Durand P, Frenkiel L, Moueza M (1998) Putative environmental transmission of sulfur-oxidizing bacterial symbionts in tropical lucinid bivalves inhabiting various environments. FEMS Microbiol Lett 160:257–262

Gros O, Liberge M, Heddi A, Khatchadourian C, Felbeck H (2003) Detection of the free-living forms of sulfide-oxidizing gill endosymbionts in the lucinid habitat (*Thalassia testudinum* environment). Appl Environ Microbiol 69:6264–6267

Guy RD, Fogel ML, Berry JA (1993) Photosynthetic fractionation of the stable isotopes of oxygen and carbon. Plant Physiol 101:37–47

Hand SC (1987) Trophosome ultrastructure and the characterization of isolated bacteriocytes from invertebrate-sulfur bacteria symbioses. Biol Bull 173:260–276

Harmer T, Nussbaumer A, Bright M, Cavanaugh CM (2005) Stalking the wild symbiont: free-living counterparts to tubeworm symbionts at deep-sea hydrothermal vents (in preparation)

Hentschel U, Felbeck H (1993) Nitrate respiration in the hydrothermal vent tubeworm *Riftia pachyptila*. Nature 366:338–340

Hughes DS, Felbeck H, Stein JL (1997) A histidine protein kinase homolog from the endosymbiont of the hydrothermal vent tubeworm *Riftia pachyptila*. Appl Environ Microbiol 63:3494–3498

Johnson KS, Childress JJ, Beehler CL (1988a) Short-term temperature variability in the Rose Garden hydrothermal vent field – an unstable deep-sea environment. Deep Sea Res 35:1711–1721

Johnson KS, Childress JJ, Hessler RR, Sakamoto-Arnold CM, Beehler CL (1988b) Chemical and biological interactions in the Rose Garden hydrothermal vent field, Galapagos spreading center. Deep Sea Res 35:1723–1744

Johnson KS, Childress JJ, Beehler CL, Sakamoto CM (1994) Biogeochemistry of hydrothermal vent mussel communities – the deep sea analog to the intertidal zone. Deep Sea Res 41:993–1011

Jones ML (1981) *Riftia pachyptila* Jones: observations on the vestimentiferan worm from the Galapagos Rift. Science 213:333–336

Jones ML, Gardiner SL (1988) Evidence for a transient digestive tract in Vestimentifera. Proc Biol Soc Wash 101:423–433

Kelly DP (1982) Biochemistry of the chemolithotrophic oxidation of inorganic sulphur. In: Postgate JR, Kelly DP (eds) Sulphur bacteria. R Soc Lond, pp 69–98

Kirov SM (2003) Bacteria that express lateral flagella enable dissection of the multifunctional roles of flagella in pathogenesis. FEMS Microbiol Lett 224:151–159

Kochevar RE, Childress JJ (1996) Carbonic anhydrase in deep-sea chemoautotrophic symbioses. Mar Biol 125:375–383

Laue BE, Nelson DC (1994) Characterization of the gene encoding the autotrophic ATP sulfurylase from the bacterial endosymbiont of the hydrothermal vent tubeworm *Riftia pachyptila*. J Bacteriol 176:3723–3729

Laue BE, Nelson DC (1997) Sulfur-oxidizing symbionts have not co-evolved with their hydrothermal vent tubeworm hosts: an RFLP analysis. Mol Mar Biol Biotech 6:180–188

Lee RW, Childress JJ (1994) Assimilation of inorganic nitrogen by marine invertebrates and their chemoautotrophic and methanotrophic symbionts. Appl Environ Microbiol 60:1852–1858

Lee RW, Robinson JJ, Cavanaugh CM (1999) Pathways of inorganic nitrogen assimilation in chemoautotrophic bacteria-marine invertebrate symbioses: expression of host and symbiont glutamine synthetase. J Exp Biol 202:289–300

Lonsdale P (1977) Clustering of suspension-feeding macrobenthos near abyssal hydrothermal vents at oceanic spreading centers. Deep Sea Res 24:857–863

Lutz RA, Shank TM, Fornari DJ, Haymon RM, Lilley MD, von Damm KL, Desbruyeres D (1994) Rapid growth at deep-sea vents. Nature 371:663–664

McKiness ZP (2004) Evolution of endosymbioses in deep-sea bathymodioline mussels (Mollusca:Bivalvia). PhD Thesis, Harvard University

McMullin ER, Hourdez S, Schaeffer SW, Fisher CR (2003) Phylogeny and biogeography of deep-sea vestimentiferan tubeworms and their bacterial symbionts. Symbiosis 34:1- 41

Millero FJ, Plese T, Fernandez M (1987) The dissociation of hydrogen-sulfide in seawater. Limnol Oceanogr 33:269–274

Millikan DS, Felbeck H, Stein JL (1999) Identification and characterization of a flagellin gene from the endosymbiont of the hydrothermal vent tubeworm *Riftia pachyptila*. Appl Environ Microbiol 65:3129–3133

Minic Z, Herve G (2003) Arginine metabolism in the deep sea tube worm *Riftia pachyptila* and its bacterial endosymbiont. J Biol Chem 278(42):40527–40533

Minic Z, Herve G (2004) Biochemical and enzymological aspects of the symbiosis between the deep-sea tubeworm *Riftia pachyptila* and its bacterial endosymbiont. Eur J Biochem 271:3093–3102

Minic Z, Simon V, Penverne B, Gaill F, Herve G (2001) Contribution of the bacterial endosymbiont to the biosynthesis of pyrimidine nucleotides in the deep-sea tubeworm *Riftia pachyptila*. J Biol Chem 276:23777–23784

Mira A, Moran NA (2002) Estimating population size and transmission bottlenecks in maternally transmitted endosymbiotic bacteria. Microbial Ecol 44:137–143

Moran NA (1996) Accelerated evolution and Muller's rachet in endosymbiotic bacteria. Proc Natl Acad Sci USA 93:2873–2878

Moran N, Baumann P (1994) Phylogenetics of cytoplasmically inherited microorganisms of arthropods. Trends Ecol Evol 9:15–20

Muller HJ (1964) The relation of recombination to mutational advance. Mutat Res 1:2–9

Nelson DC, Fisher CR (1995) Chemoautotrophic and methanotrophic endosymbiotic bacteria at deep-sea vents and seeps. In: Karl DM (ed) Microbiology of deep-sea hydrothermal vents. CRC Press, Boca Raton, pp 125–167

Nelson K, Fisher CR (2000) Absence of cospeciation in deep-sea vestimentiferan tubeworms and their bacterial endosymbionts. Symbiosis 28:1–15

Nelson DC, Hagen KD (1995) Physiology and biochemistry of symbiotic and free-living chemoautotrophic bacteria. Am Zool 35:91–101

Ohta T (1973) Slightly deleterious mutant substitutions in evolution. Nature 246:96–98

Peck HD Jr, LeGall J (1982) Biochemistry of dissimilatory sulphate reduction. In: Postgate JR, Kelly DP (eds) Sulphur bacteria. R Soc Lond, pp 13–36

Peek AS, Vrijenhoek RC, Gaut BS (1998) Accelerated evolutionary rate in sulfur-oxidizing endosymbiotic bacteria associated with the mode of symbiont transmission. Mol Biol Evol 15:1514–1523

Pernthaler A, Amann R (2004) Simultaneous fluorescence in situ hybridization of mRNA and rRNA in environmental bacteria. Appl Environ Microbiol 70:5426–5433

Pimenov NV, Savvichev AS, Rusanov II, Lein AY, Ivanov MV (2000) Microbiological processes of the carbon and sulfur cycles at cold methane seeps of the North Atlantic. Microbiology 69:709–720

Polz MF, Ott JA, Bright M, Cavanaugh CM (2000) When bacteria hitch a ride. ASM News 66:531–539

Powell MA, Somero GN (1986) Adaptations to sulfide by hydrothermal vent animals: sites and mechanisms of detoxification and metabolism. Biol Bull 171:274–290

Rau GH (1981) Hydrothermal vent clam and tube worm $^{13}C/^{12}C$: further evidence of non-photosynthetic food sources. Science 213:338–340

Renosto F, Martin RL, Borrell JL, Nelson DC, Segel IH (1991) ATP sulfurylase from trophosome tissue of *Riftia pachyptila* (hydrothermal vent tube worm). Arch Biochem Biophys 290:66–78

Robinson JJ, Cavanaugh CM (1995) expression of from I and form II Rubisco in chemoautotrophic symbioses: implications for the interpretation of stable isotope values. Limnol Oceanogr 40:1496–1502

Robinson JJ, Stein JL, Cavanaugh CM (1998) Cloning and sequencing of a form II ribulose-1,5-bisphosphate carboxylase/oxygenase from the bacterial symbiont of the hydrothermal vent tubeworm *Riftia pachyptila*. J Bacteriol 180:1596–1599

Robinson JJ, Scott KM, Swanson ST, O'Leary MH, Horken K, Tabita FR, Cavanaugh CM (2003) Kinetic isotope effect and characterization of form II RubisCO from the chemoautotrophic endosymbionts of the hydrothermal vent tubeworm *Riftia pachyptila*. Limnol Oceanogr 48:48–54

Roeske CA, O'Leary MH (1984) Carbon isotope effects on the enzyme-catalyzed carboxylation of ribulose bisphosphate. Biochemistry 23:6275–6284

Roeske CA, O'Leary MH (1985) Carbon isotope effect on carboxylation of ribulose bisphosphate catalyzed by ribulosebisphosphate carboxylase from *Rhodospirillum rubrum*. Biochemistry 24:1603–1607

Rouxel O, Fouquet Y, Ludden JN (2004) Subsurface processes at the Lucky Strike hydrothermal field, Mid-Atlantic Ridge: evidence from sulfur, selenium, and iron isotopes. Geochim Cosmochim Ac 68:2295–2311

Schmaljohann R, Flügel HJ (1987) Methane-oxidizing bacteria in pogonophora. Sarsia 72:91–98

Schulze A, Halanych KM (2003) Siboglinid evolution shaped by habitat preference and sulfide tolerance. Hydrobiologia 496(1–3):199–205

Scott KM (2003) A $d^{13}C$-based carbon flux model for the hydrothermal vent chemoautotrophic symbiosis *Riftia pachyptila* predicts sizeable CO_2 gradients at the host-symbiont interface. Environ Microbiol 5:424–432

Segel IH, Renosto F, PA Seubert (1987) Sulfate-activating enzymes. In: Jakoby WB, Griffith O (eds) Methods in enzymology, vol 143. Sulfur and sulfur amino acids. Academic Press, New York, pp 334–349

Sibuet M, Olu K (1998) Biogeography, biodiversity and fluid dependence of deep-sea cold-seep communities at active and passive margins. Deep Sea Res II 45:517–567

Sorgo A, Gaill F, Lechaire JP, Arndt C, Bright M (2002) Glycogen storage in the *Riftia pachyptila* trophosome: contribution of host and symbionts. Mar Ecol Prog Ser 231:115–120

Stewart FJ, Newton ILG, Cavanaugh CM (2005) Chemosynthetic endosymbioses: adaptations to oxic-anoxic interfaces. TRENDS Microbiol 13:439–448

Tabor CW, Tabor H (1985) Polyamines in microorganisms. Microbiol Rev 49:81–99

Thao ML, Moran NA, Abbot P, Brennan EB, Burckhardt DH, Baumann P (2000) Co-speciation of psyllids and their primary prokaryotic endosymbionts. Appl Environ Microbiol 66:2898–2905

Van Dover CL (2000) The ecology of deep-sea hydrothermal vents. Princeton Univ Press, Princeton, NJ

Van Dover CL, Fry B(1994) Microorganisms as food resources at deep-sea hydrothermal vents. Limnol Oceanog 39:51–57

Van Dover CL, Lutz RA (2004) Experimental ecology at deep-sea hydrothermal vents: a perspective. J Exp Mar Biol Ecol 300: 273–307

Weber RE, Vinogradov SN (2001) Nonvertebrate hemoglobins: functions and molecular adaptations. Physiol Rev 81:569–628

Wernegreen JJ (2002) Genome evolution in bacterial endosymbionts of insects. Nat Rev Genet 3:850–861

Zal F, Lallier FH, Wall JS, Vinogradov SN, Toulmond A (1996) The multi-hemoglobin system of the hydrothermal vent tube worm *Riftia pachyptila*.1. Reexamination of the number and masses of its constituents. J Biol Chem 271: 8869–8874

Zal F, Suzuki T, Kawasaki Y, Childress JJ, Lallier FH, Toulmond A (1997) Primary structure of the common polypeptide chain b from the multi-hemoglobin system of the hydrothermal vent tube worm *Riftia pachyptila*: an insight on the sulfide binding-site. Proteins 29:562–574

Zal F, Leize E, Lallier FH, Toulmond A, Van Dorsselaer A, Childress JJ (1998) S-sulfohemoglobin and disulfide exchange: the mechanisms of sulfide binding by *Riftia pachyptila* hemoglobins. Proc Nat Acad Sci USA 95:8997–9002

Zhang JZ, Millero FJ (1993) The products from the oxidation of H_2S in seawater. Geochim Cosmochim Ac 57:1705–1718

Symbioses of Methanotrophs and Deep-Sea Mussels (Mytilidae: Bathymodiolinae)

Eric G. DeChaine, Colleen M. Cavanaugh

1
Introduction

Symbioses between marine invertebrates and methanotrophs provide the bacteria with access to methane and oxygen and other substrates necessary for metabolism and the invertebrate host with a source of organic carbon. Methanotrophic bacteria utilize methane for generating ATP through oxidative phosphorylation and for the net synthesis of organic compounds used in cellular metabolism. Because methane is typically produced in anoxic environments, through either biological (the action of methanogenic Archaea) or inorganic processes, free-living aerobic methanotrophs are limited to the microaerophilic interface between oxic and anoxic zones (Anthony 1982). But in methanotrophic symbioses the invertebrate host acts as a "bridge" across the oxic-anoxic interface (as in chemoautotroph symbioses; see previous chapter), facilitating access to both oxygen and methane for the endosymbionts (Cavanaugh 1985; Cavanaugh et al. 2005). The methanotrophs in turn consume methane and provide the host with a sustainable carbon source not directly available to metazoans. The host derives other essential elements (e.g., N, P, S) from the symbiont and/or environmental sources.

The symbioses between mytilid mussels in the genus *Bathymodiolus* (family Mytilidae; subfamily Bathymodiolinae) and type I methanotrophic bacteria are the most prevalent, widespread, and best understood of the aerobic methane-based associations. Bathymodioline symbioses are globally distributed at deep-sea hydrothermal vents and cold seeps (Fig. 1, Table 1; von Cosel et al. 1994; O'Mullan et al. 2001; van Dover et al. 2001; Fiala-Médioni et al. 2002) and depending on the host mussel species, harbor either methanotrophs, chemoautotrophs (that oxidize reduced inorganic compounds accompanied by CO_2 fixation; see previous chapter), or both of these two metabolically and phylogenetically distinct gamma Proteobacteria

---

E.G. DeChaine, C.M. Cavanaugh (e-mail: cavanaug@fas.harvard.edu)
Department of Organismic and Evolutionary Biology, Havard University,
The Biological Laboratories, 16 Divinity Avenue, Cambridge MA 02138, USA

endosymbionts simultaneously (e.g., Distel et al. 1995; Dubilier et al. 1999). Methanotrophs and chemoautotrophs share the ability to synthesize organic compounds from C_1 compounds, utilizing both energy and carbon sources otherwise unavailable to their animal hosts. As these symbiotic bacteria have not yet been isolated in pure culture, much of the work has focused on characterizing these symbioses from physiological and phylogenetic perspectives. Recent advances in molecular techniques are permitting higher resolution examinations into symbiont distributions, evolution, and diversity.

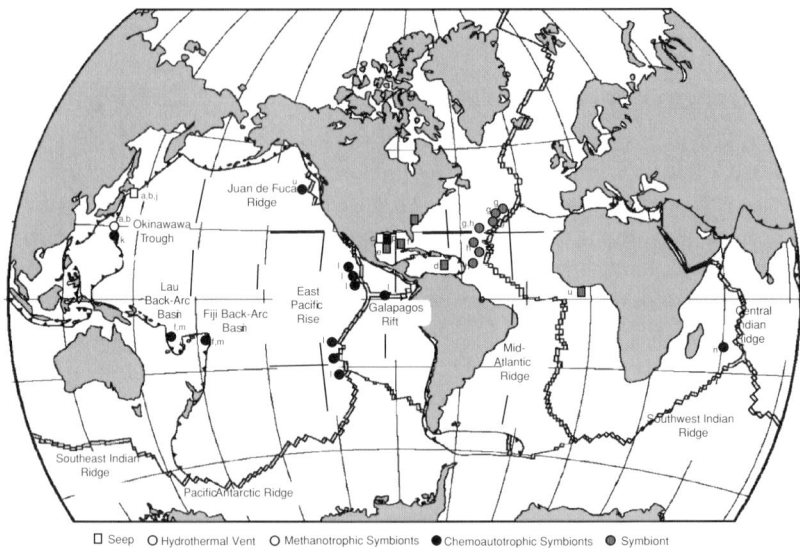

Fig. 1. Distribution of *Bathymodiolus* mussels and occurrence of methanotrophic and/or chemoautotrophic endosymbionts (modified from van Dover et al. 2002). Sites are shaded depending on the type of symbiont(s) hosted by *Bathymodiolus*. The biogeographic provinces are labeled for reference.

2
Methanotrophic Symbioses

Methanotrophic endosymbioses have only been characterized in a few marine invertebrate taxa and deep-sea habitats. Symbioses involving methanotrophs were first described in bathymodioline mussels inhabiting deep-sea cold seeps (Childress et al. 1986; Cavanaugh et al. 1987) and subsequently found in other bathymodioline mussels, a pogonophoran

Table 1. The taxonomic and biogeographic distribution of methanotroph-hosting invertebrates[a]

Host phylum / Family / Subfamily / Species	Type of symbiont[b]	Habitat	Collection sites[c]		References[d]
Mollusc					
Mytilidae					
Bathymodiolinae					
Bathymodiolus japonicus	M	Seep/vent	WP	Okinawa Trough, Sagami Bay	1, 2
B. platifrons	M	Seep/vent	WP	Okinawa Trough, Sagami Bay	2, 3
B. childressi	M	Seep	AG	Alaminos Canyon, Florida Escarpment	4, 5, 6
B. boomerang	M, C	Seep	A	Barbados Trench	7
B. brooksi	M, C	Seep	A	Alaminos Canyon	8, 9
B. heckerae	M, C	Seep	A, AG	Blake Ridge, Florida Escarpment	10–12
B. azoricus	M, C	Vent	A	Mid-Atlantic Ridge	13–15
B. puteoserpentis	M, C	Vent	A	Mid-Atlantic Ridge	15–18
B. sp. Gabon margin	M, C	Seep	A	Gabon Margin	19
B. thermophilus	C	Vent	EP	Eastern Pacific Rise	20, 21
B. sp. Juan de Fuca	C	Vent	EP	Juan de Fuca	22
B. aduloides	C	Seep	WP	Okinawa Trough	23
B. septemdierum	C	Seep	WP	Okinawa Trough	2, 23
B. elongates	C	Vent	WP	Lau and Fiji Basins	24
B. brevior	C	Vent	WP	Lau and Fiji Basins	25
B. aff. *brevior* (=*B. mariscindicus*)	C	Vent	I	Central Indian Ridge	26–28

Host phylum Family Subfamily Species	Type of symbiont[b]	Habitat	Collection sites[c]	References[d]
Tamu fisheri	C	Seep	AG	29
Annelida				
Siboglinidae				
Siboglinum poseidoni	M	Seep	A	30
Porifera				
Cladorhizidae				
Cladorhiza methanophila	M	Seep	A	31

[a] Known species of deep-sea vent and seep bathymodioline mussels are listed to highlight that these mussels can host methanotrophs, chemoautotrophs, or both types of bacteria: symbionts. The presence and characterization of symbionts is based on several techniques, including transmission electron microscopy, enzyme assays, radiolabeled uptake experiments, 16S rRNA sequence data, and/or in situ hybridization of symbiont-specific gene probes

[b] Type of symbiont: C chemoautotroph; M methanotroph

[c] Collection sites abbreviated as follows: A Atlantic; I Indian; AG Atlantic, Gulf of Mexico, EP Eastern Pacific, WP Western Pacific

[d] References – Hashimoto and Okutani (1994); [2]Fujiwara et al. (2000); [3]Barry et al. (2002); [4]Childress et al. (1986); [5]Fisher et al. (1987); [6]Kochevar et al. (1992); [7]von Cosel and Olu (1998); [8]Fisher et al. (1993); [9]Gustafson et al. (1998); [10]Cavanaugh et al. (1987); [11]Cary et al. (1988); [12]Cavanaugh (1992); [13]Trask and van Dover (1999); [14]Fiala-Médioni et al. (2002); [15]Pimenov et al. (2002); [16]Cavanaugh et al. (1992); [17]Distel et al. (1995); [18]Robinson et al. (1998); [19]Duperron et al. (2005a); [20]Belkin et al. (1986); [21]Nelson et al. (1995); [22]McKiness et al. (2005); [23]Yamanaka et al. (2000); [24]Pranal et al. (1997); [25]Dubilier et al. (1998); [26]van Dover et al. (2001); [27]Yamanaka et al. (2003); [28]McKiness and Cavanaugh (2005); [29]MacAvoy et al. (2005); [30]Schmaljohann (1991); [31]Vacelet et al. (1996)

tubeworm, and carnivorous sponges at seeps and hydrothermal vents (Table 1). The symbioses were identified based on the co-occurrence of bacteria containing intracytoplasmic membranes typical of type I methanotrophs and assays for enzymes diagnostic of methylotrophy (e.g., methanol dehydrogenase). *Bathymodiolus childressi* (Fig. 2a), from hydrocarbon seeps on the Louisiana Slope, Gulf of Mexico, was the first species found to house bacteria that exhibit intracytoplasmic membranes typical of type I methanotrophs, while the dual symbiosis was first described in *B. heckerae*, as observed through transmission electron microscopy (Childress et al. 1986; Cavanaugh et al. 1987). The methane oxidizing nature of the bacteria was confirmed through radiolabeled ^{14}C-methane uptake experiments that measure the amount of ^{14}CH$_4$ incorporated into acid-stable compounds and CO$_2$ and through methanol dehydrogenase (MeDH) assays that detect the activity of a key enzyme in the methane oxidation pathway (Cavanaugh et al. 1987; Fisher et al. 1987; Cavanaugh 1992). Subsequently, invertebrates from other methane-rich environments have been examined with a suite of techniques aimed at detecting symbiotic methanotrophs, such as additional enzyme assays, stable carbon isotope analysis (comparing $^{13}\delta$C values), phylogenetic analysis of 16S rRNA sequence data, and in situ hybridization of symbiont-specific genetic probes.

2.1
Other Invertebrate Hosts

While this chapter focuses on the association between bathymodioline mussels and their endosymbionts, because those are the best understood of methane-based symbioses, the two other known examples, a pogonophoran tubeworm and a carnivorous sponge, illustrate the diversity of invertebrate taxa that can harbor methanotrophs.

The pogonophoran *Siboglinum poseidoni* from methane-rich reducing sediments is the only tubeworm known to host methanotrophic endosymbionts (Schmaljohann 1991). *Siboglinum poseidoni* is abundant at methane- and sulfide-rich sites of the Skagerrak basin (~200–400 m depth) off the coast of Denmark (Dando et al. 1994). Endosymbiosis of methanotrophs in the pogonophoran tissue was detected through TEM showing symbionts containing stacked intracellular membranes typical of type I methanotrophs (Schmaljohann and Flügel 1987), enzyme assays indicating methanol dehydrogenase and hexulosephosphate synthetase activity, ^{14}CH$_4$ uptake experiments and stable carbon isotope analysis showing that pogonophoran cell carbon was derived from biogenic methane (Schmaljohann et al. 1990). No genetic analysis was performed, precluding any phylogenetic comparison with other methanotrophs.

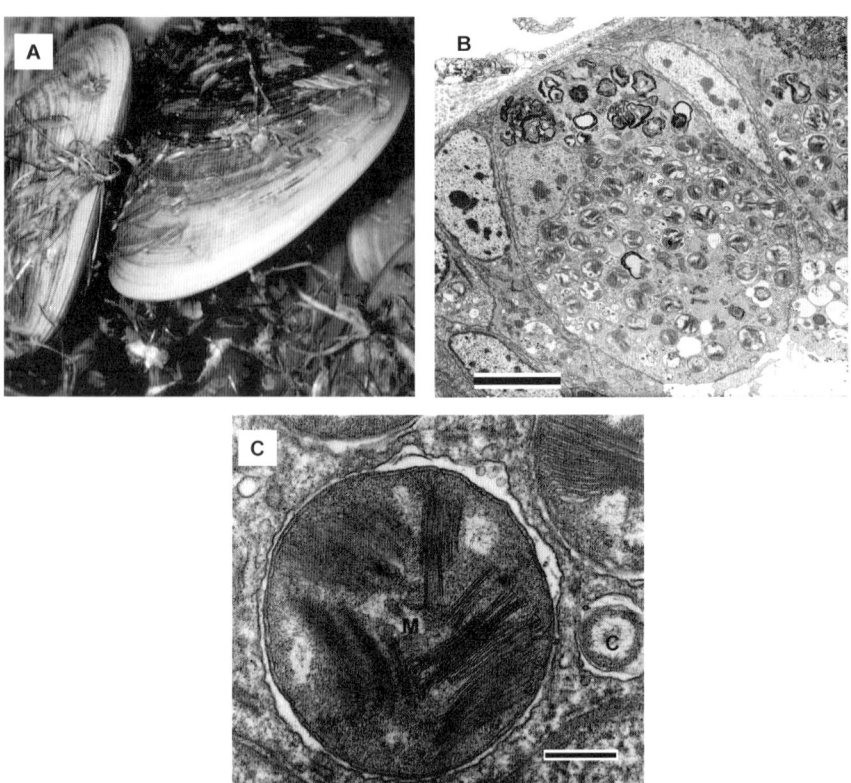

Fig. 2. Representative taxa involved in the mussel-chemosynthetic bacteria symbioses. **A** *Bathymodiolus childressi* mussels. B and C Transmission electron micrographs of *B. heckerae* gill epithelial tissue (from Cavanaugh et al. 1987). **B** Transverse section of host cells containing symbionts. *Scale bar* 5 µm. **C** Higher magnification of methanotropic (*M*) and chemoautotrophic (*C*) symbionts. *Scale bar* 0.3 µm

Methanotrophic bacteria are also found as endosymbionts associated with carnivorous sponges in the genus *Cladorhiza* (family Cladorhizidae), which inhabit deep-sea mud volcanoes near the Barbados accretionary prism (Vacelet et al. 1996). The presence of extracellular methanotrophs within the tissue of the sponge was revealed by TEM, with methanol dehydrogenase activity and $\delta^{13}C$ values consistent with the incorporation of carbon from biogenic methane into the sponge. As in the case of the pogonophoran, no genetic analysis was performed to determine the relationship among cladorhizid symbionts and other methanotrophs.

2.2
Methane-Utilizing Bacteria

The identification of symbionts, since none have been cultured, has depended on comparison with free-living aerobic methanotrophs. Methanotrophic bacteria, unique in their ability to use methane as a substrate, form a subset of methylotrophs, which utilize C_1 compounds for energy and carbon acquisition (Anthony 1982). Symbiotic methanotrophs are most closely related to free-living type I aerobic methanotrophs in the gamma Proteobacteria, based on their membrane organization, C_1 assimilation pathways, and phylogenetic relations (Hanson et al. 1991; Bratina et al. 1992; Bowman et al. 1993). While both type I and type II methanotrophs exhibit extensive internal membrane systems, in type I, the internal membranes are arranged in bundles of disc-shaped vesicles distributed throughout the entire cell (Fig. 2b,c), but paired membranes are restricted to the periphery of the cell in type II methanotrophs. Because anaerobic methanotrophs, which are responsible for much of global methane oxidation (Reeburgh 1980; Hinrichs et al. 1999; Boetius et al. 2000), and peat bog-inhabiting acidophilic methanotrophs in the alpha Proteobacteria (Dedysh 2002) are only distantly related to methanotrophic symbionts, our discussion concentrates on type I methanotrophs.

Details of methanotrophic symbiont metabolism are largely inferred from our knowledge of free-living type I and II aerobic methanotrophs. Through the oxidization of methane, electron transport and oxidative phosphorylation generate ATP and organic C_3 compounds are synthesized from formaldehyde (Fig. 3; Hanson and Hanson 1996). Methane monooxygenases (MMOs), either particulate (pMMO) bound to intracytoplasmic membranes or soluble (sMMO) depending on the species, initiate the oxidation of methane by introducing the oxygen to CH_4, thus forming H_2O and CH_3OH. Methanol is oxidized to formaldehyde (HCHO) by a periplasmic methanol dehydrogenase (MeDH). Formaldehyde is oxidized to formate through multiple enzyme systems, depending on the type of methanotroph. Finally, formate is oxidized to CO_2 by an NAD-dependent formate dehydrogenase. Though the overall methane oxidation pathway is the same for both types of methanotrophs, in type I and type II methanotrophs, C_1-utilization occurs through the ribulose monophosphate (RuMP) pathway and the serine pathway, respectively (Fig. 3).

2.3
Known Environments Inhabited by Methanotrophic Symbioses

Fluids rich in methane are released from hydrothermal vents, cold seeps, and mud volcanoes in the deep-sea. Hydrothermal vents, discovered in 1977,

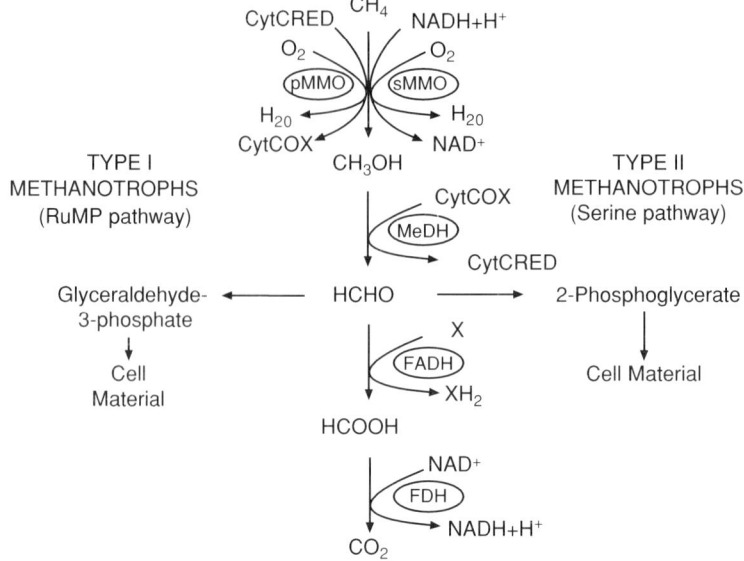

Fig. 3. Methane oxidation and carbon assimilation pathways for type I and type II methanotrophs via the RuMP (ribulose monophosphate) and serine pathways (compiled from Hanson and Hanson 1996). Enzymes involved in the oxidation of methane are abbreviated as follows: pMMO (particulate methane monooxygenase), sMMO (soluble methane monooxygenase), MeDH (methanol dehydrogenase), FADH (formaldehyde dehydrogenase), and FDH (formate dehydrogenase)

are distributed along deep-sea spreading ridges and back-arc basins throughout the world where volcanic activity associated with seafloor spreading allows seawater to circulate through the upper crust, becoming heated and enriched with reduced compounds (e.g., H_2S, Mn^{2+}, H_2, CO, and CH_4; Tunnicliffe et al. 1998; van Dover 2000). Hydrologic activity at cold seeps along continental margins and plate boundaries also releases sulfide-, methane-, and ammonia-rich fluids from the sediment (reviewed in van Dover 2000; Judd 2003). Mud volcanoes are formed when water, mud, and gas (usually dominated by CH_4, but may include CO_2 or nitrogen) are expelled from sedimentary sequences at zones of tectonic compression (Hedberg 1980; Brown 1990; Milkov 2000; Judd et al. 2002). The methane derived from these environmental sources forms a significant proportion of the global carbon budget (Hornafius et al. 1999; Judd et al. 2002).

Methane is generated via both biogenic and inorganic processes in seabed fluids (reviewed in Judd et al. 2002; Judd 2003). Anaerobic archaeal methanogens generate methane biologically (Huber et al. 1989; Shima et al. 2002). Thermocatalytic processes deep within sediments can also degrade organic matter to yield methane (Judd et al. 2002). Finally, the majority of methane at hydrothermal vents is believed to be of abiogenic origin, the result of degassing and cooling of mafic magmas and the serpentization of ultramafic rocks (Apps and van de Kamp 1993).

Methane concentrations, as well as overall fluid chemistry, vary greatly among sites inhabited by methanotrophic symbioses (Lupton et al. 1991; van Dover 2000; Kelley et al. 2001; Lilley et al. 2003). At each vent or seep, concentrations of methane (0.06 to 0.7 mmol kg^{-1}) are orders of magnitude higher than that of the ambient seawater (4×10^{-7} mmol kg^{-1}), but concentrations can also differ, among sites and over time (Kelley et al. 2001). Other environmental factors vary widely as well, including ion concentrations (e.g., Mg^{2+}, Ca^+, Na^+, Cl^-, SO_4^{2-}, H_2S, and H_2), pH (from ~2 to 10), and water temperature (from ~2 to 400°C). Overall, this translates into regional and fine-scale differences in chemistry that presumably influence the ability of methanotroph-hosting invertebrates to colonize, persist, and reproduce at a site.

3
Bathymodioline Symbioses

Thus far, symbiont-hosting mytilid mussels are restricted to deep-sea endemics that colonize whale and wood falls (e.g., mussels of the genus *Idas*) and species in the subfamily Bathymodiolinae (two genera: *Bathymodiolus* and *Tamu*) that inhabit vents and seeps (Kenk and Wilson 1985; Distel et al. 2000). Anatomical features of mussels in this subfamily, such as mantle fusion, a simple gut, and pronounced gills, distinguish them from other deep-sea mussels (Kenk and Wilson 1985; Gustafson et al. 1998). The first vent mytilid species to be described, *Bathymodiolus thermophilus*, was collected at the hydrothermal vent field of the Galapagos Rift Zone in 1977 (Lonsdale 1977; Grassle 1985). Since then, fifteen additional species, all of which harbor intracellular symbionts, have been described within this genus from deep-sea hydrothermal vents and cold seeps (see Table 1 for species and references), with new discoveries made with each exploratory dive to a novel vent or seep site (Miyazaki et al. 2004; Duperron et al. 2005a; McKiness and Cavanaugh 2005; McKiness et al. 2005). Indeed, mussel beds of *Bathymodiolus* species are designated as a habitat type for deep-sea chemosynthetic systems because they are the only known marine invertebrate symbiosis to be found in every explored biogeographic province of the deep-sea that contains hydrothermal vents and hydrocarbon seeps (Fig. 1; von Cosel et al. 1994; Gustafson et al. 1998; Sibuet and Olu 1998; van Dover 2000; Yamanaka et al. 2000; Hashimoto 2001; O'Mullan et al. 2001; van Dover et al. 2001, 2002; Fiala-Médioni et al. 2002; McKiness et al. 2005).

Bathymodioline mussels harbor bacterial endosymbionts in specialized epithelial cells, referred to as bacteriocytes, within the subfilamentar tissue of their gills (Fig. 2b, c; Cavanaugh et al. 1987, 1992; Fisher et al. 1987, 1993; Robinson et al. 1998; Fiala-Médioni et al. 2002). Development of this symbiont-containing gill tissue gives the mussels characteristically thick and opaque gills. Chemoautotrophs and methanotrophs have also been detected in the mantle and foot epithelia of *B. childressi*, but their role in the physiology, development, and evolution of the symbiosis remains unclear (Streams et al. 1997).

Mussels of the genus *Bathymodiolus* are unique among animals in that individuals can host both phylogenetically and physiologically distinct bacteria simultaneously, i.e., chemoautotrophic and methanotrophic endosymbionts, even within the same bacteriocyte (Fig. 2c; Cavanaugh et al. 1987, 1992; Kochevar et al. 1992; Distel et al. 1995; Nelson et al. 1995; Trask and van Dover 1999; Fiala-Médioni et al. 2002). Thus far, dual symbioses have only been described for mussels from the Atlantic and Gulf of Mexico (Table 1), but isotopic analyses suggest that an undescribed *Bathymodiolus* species from the Mariana Fore-Arc in the western Pacific might host dual symbionts as well (Yamanaka et al. 2003). The dual-symbiont condition allows the mussel environmental flexibility by expanding the resources, and thus chemical habitats, available to it.

Symbionts are housed within membrane-bound vacuoles, in bacteriocytes of the gill tissue of bathymodioline mussels. Methanotrophic symbionts (~1.5–2.0 μm in diameter), exhibiting the intracytoplasmic membranes typical of type I methanotrophs, are usually found individually within a vacuole, while the chemoautotrophs, coccoid cells ~0.3 μm in diameter, often occur multiply within a single vacuole (Fig. 2b, c). Though typically only one type of endosymbiont, either methanotroph or chemoautotroph, is found in a given vacuole, the symbionts can co-occur in the same vacuole.

3.1
Bacterial Symbionts

Phylogenetic analyses based on 16S rRNA have revealed that two types of gamma Proteobacteria are housed within the bacteriocytes of bathymodioline mussels. One type of symbiont includes lineages that cluster with free-living, type I methanotrophs, while the others form a monophyletic group of chemoautotrophic endosymbionts associated with mytilid mussels and vesicomyid clams (Fig. 4). Overall, six of the sixteen species of *Bathymodiolus* host dual symbionts, while an additional three harbor only methanotrophs and another seven, along with *Tamu fisheri*, are associated strictly with chemoautotrophs (Table 1). The association between bathymodioline mussels and these two distinct bacterial clades presents a unique situation to address questions about the origin and evolution of symbioses.

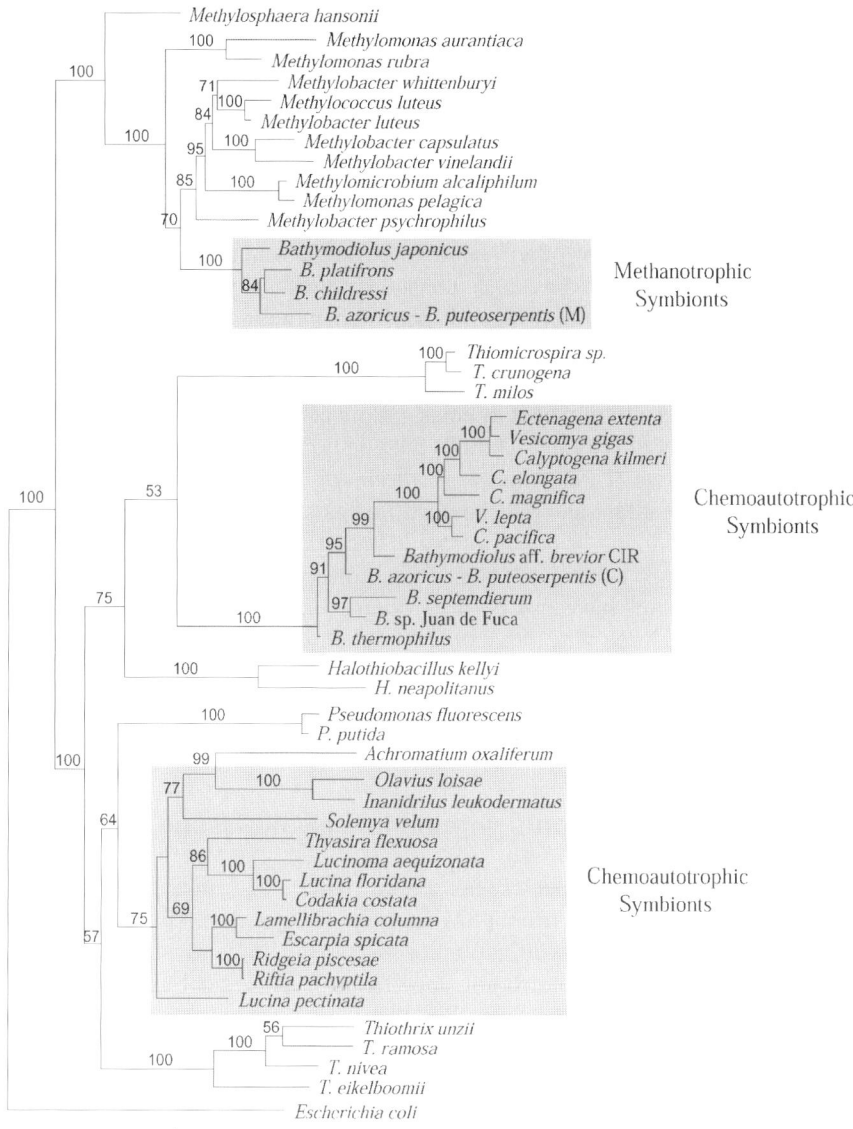

Fig. 4. Bayesian phylogram of chemoautotrophic and methanotrophic endosymbionts hosted by bathymodioline mussels, inferred from 16S rRNA gene sequences (modified from DeChaine et al. 2005). Posterior probabilities (>90) are shown above branches. Taxa include symbiotic and free-living gamma proteobacteria. Clades of symbiotic chemoautotrophs and methanotrophs are boxed in *gray*. *Bathymodiolus azoricus* and *B. puteoserpentis* host the same dual symbionts, which are labeled M and C for methanotroph and chemoautotroph, respectively.

To date, all methanotrophic endosymbionts of mussels form a monophyletic group nested within the type I free-living methanotroph clade (Fig. 4; DeChaine et al. 2005; Duperron et al. 2005a,b). Though the methanotrophs are generally host-specific, multiple host species can harbor the same methanotrophic symbiont phylotype and single mussel individuals can host multiple symbiont genotypes (DeChaine et al. 2005).

The bathymodioline chemoautotrophic symbionts cluster with symbionts of clams, to the exclusion of other chemoautotrophic vent symbionts (e.g., those of hydrothermal tubeworms and coastal mollusks; Fig. 4). As with the methanotrophs, diverse but related phylotypes have been uncovered for chemoautotrophic endosymbionts of mussels in the genus *Bathymodiolus*, including multiple phylotypes within an individual host as well as symbiont phylotypes shared among host species (Won et al. 2003a; DeChaine et al. 2005; Duperron et al. 2005b). Though the chemoautotrophic symbionts of the Eastern Pacific Rise (EPR) mussel, *B. thermophilus*, were shown to fix CO_2 and to utilize thiosulfate and hydrogen sulfide as energy sources and thus were labeled sulfur-oxidizing or thioautotrophic bacteria (Belkin et al. 1986; Nelson et al. 1995), energy substrates for chemoautotrophic endosymbionts of other mussels in the genus have not been determined. For a detailed discussion of chemoautotrophy, see the chapter on chemoautotrophic symbioses of vestimentiferan tubeworms (Stewart and Cavanaugh, this vol.) and the recent review by Cavanaugh et al. (2005).

3.2
Distribution of Symbionts within Mussel Gill Tissue

In most bathymodioline species, no spatial pattern of symbiont types in the bacteriocytes was detected from TEM analyses (Fisher et al. 1993; Distel et al. 1995; Fiala-Médioni et al. 2002; Won et al. 2003a). In contrast, hybridization of methano- and chemoautotrophic-specific probes to symbionts in the gill bacteriocytes of an undescribed *Bathymodiolus* species from a recently discovered hydrocarbon seep on the Gabon continental margin revealed a distinct distribution pattern for the two bacterial phylotypes (Duperron et al. 2005a). In situ hybridization (FISH) techniques revealed that methanotrophic symbionts occupied the basal region of the bacteriocyte while the chemoautotrophic symbionts occupied the apical end, in close proximity to the external, seawater environment. Duperron et al. (2005a) postulated that the high concentration of methane in the seep fluids could compensate for the diffusive loss of methane through the bacteriocyte, while the low concentration of sulfides might limit the distribution of the chemoautotrophs. Similar studies employing FISH probes on other species of *Bathymodiolus* from additional vent and seep sites are required to determine whether there are any general symbiont patterns or whether the distribution is host species- or habitat-specific.

4
Nutrient Assimilation

The metabolic needs of the host are met by the transfer of nutrients from endosymbionts. Our discussion focuses on assimilation of nutrients from the methanotrophic symbionts, since the previous chapter discusses nutrient acquisition by invertebrates hosting chemoautotrophs. Enzymatic and physiological assays have indicated that, in methanotrophic symbioses, the majority of carbon attained by the host is derived from utilization of methane by the symbiont. Evidence suggests that nitrogen, in the form of ammonium and nitrate, is assimilated by the mussel from symbiont and environmental sources. Future analyses aimed at elucidating the acquisition of other essential nutrients are needed to more fully understand the contribution of endosymbionts to the overall metabolism of the host.

4.1
Carbon Assimilation

A diverse array of enzymatic and physiological assays has shown that methane is the major source of carbon for methanotroph-hosting mussels. Because methane monooxygenases (MMOs) degrade rapidly (Cavanaugh et al. 1987) and methanol dehydrogenase (MeDH) is unique to methylotrophs, MeDH is commonly employed as a diagnostic enzyme for the detection of methanotroph activity (Cavanaugh et al. 1992; Fisher et al. 1993; Barry et al. 2002; Fiala-Médioni et al. 2002). Stable carbon isotope ratios, based on $^{13}\delta C$, provide support as to whether carbon is derived from methane of biogenic (depleted; –80‰) or thermogenic (–45 to –50‰) origin (Cavanaugh et al. 1987; Fisher 1990; Fisher et al. 1993; Barry et al. 2002). In addition to natural variation in $^{13}\delta C$ values due to the source methane, mussels that host dual symbionts exhibit $^{13}\delta C$ signatures reflecting a mix of methanotrophic and chemoautotrophic metabolisms, and thus values should be interpreted with caution (Fisher et al. 1994). The most definitive marker for the assimilation of carbon fixed by methanotrophs is $^{14}CH_4$ incubations followed by analyses of how much ^{14}C is incorporated into acid stable compounds and CO_2 (Cavanaugh 1992; Fisher and Childress 1992; Robinson et al. 1998). In growth experiments, shell growth rate in *B. childressi* was positively correlated with methane concentration, and the absence of methane inhibited growth altogether (Cary et al. 1988). Finally, the presence of methanotrophs in host tissue can be verified by using *pmoA* probes because the *pmoA* gene, which codes for a subunit of the particulate monooxygenase unique to methanotrophs, serves as an excellent marker for that group (Holmes et al. 1995; Pernthaler and Amann 2004).

The host can acquire biogenic carbon from the symbiont either through translocation of nutrients or direct digestion of the bacteria. Slow rates of carbon transfer from symbiont-bearing to symbiont-free tissue (Fisher and Childress 1992) and the degradation of methanotrophs in the basal region of bacteriocytes (Cavanaugh et al. 1992; Barry et al. 2002) suggest that hosts acquire much of their carbon by digesting their symbionts. But, by employing lysosomal enzyme cytochemistry and ^{14}C tissue autoradiography, Streams et al. (1997) demonstrated that hosts acquire nutrients through translocation of organic matter released by the symbionts as well as through direct digestion of symbionts.

4.2
Nitrogen and other Essential Nutrients

Both partners in the symbioses are dependent on environmental sources of nitrogen, typically available at deep-sea vents and seeps in the form of ammonium, nitrate, and/or organic nitrogen (Lee et al. 1992, 1999; Lee and Childress 1994). N_2 fixation, originally postulated for both chemoautotrophic and methanotrophic symbionts (based on depleted $\delta^{15}N$ values for mussel tissues), has not been detected. In retrospect, this is to be expected given the availability of other fixed nitrogen sources. Because endosymbionts are not in direct contact with the environment, where sediment and seep/vent effluent can be rich in ammonium and nitrate, the uptake of nitrogen is necessarily mediated by the host (Lee et al. 1999). The large variation in stable nitrogen isotope ratios, $\delta^{15}N$, among mussel populations suggests that some mussels acquire nitrogen containing compounds from their symbionts, while others must obtain it from the environment (Brooks et al. 1987; Page et al. 1990; Kennicutt et al. 1992; Fisher et al. 1994). Lee and Childress (1996) were unable to determine the relative contributions of the symbiont and host to overall nitrogen assimilation, but they did show that gill tissue from symbiotic mussels exhibited nitrate reductase activity (indicative of bacteria), and that the glutamine synthetase/glutamate dehydrogenase pathway (utilized by both host and bacteria) was probably responsible for ammonium assimilation.

Though bathymodioline mussels rely on their bacterial symbionts for the majority of their nutrition, their attenuated gut permits them to filter feed to a limited degree (Page et al. 1990). Indeed, suspension-feeding on ultraplankton provides *B. childressi* with supplemental nitrogen, essential for growth (Pile and Young 1999). The proportion of other nutrients (e.g., phosphorous, minerals) obtained through the symbionts and filter-feeding and the identity and concentration of these resources remain to be determined.

5
Evolution and Biogeography of Bathymodioline Symbioses

The historical biogeography of bathymodioline symbioses, inferred by placing portraits of relationships among individuals in a geographic context, provides a basis for understanding the current distribution and magnitude of diversity in the invertebrate host and symbionts. The disjunct distribution of hydrothermal vent and hydrocarbon seep communities, due to factors such as topography, physical oceanography, and tectonic activity, could promote genetic differentiation, local adaptation, and speciation by inhibiting gene flow among populations (Tunnicliffe 1991; Tunnicliffe et al. 1998; Kim and Mullineaux 1998; van Dover 2002; DeChaine et al. 2005). The global distribution of bathymodioline symbioses at deep-sea vents and seeps makes these mussels excellent model systems for addressing questions of deep-sea historical biogeography.

Vent endemic mussels in the genus *Bathymodiolus* have the best-understood biogeographic history of all hydrothermal vent fauna (Tunnicliffe and Fowler 1996; Vrijenhoek 1997; Vrijenhoek et al. 1998; O'Mullan et al. 2001; Won et al. 2003a,b; McKiness et al. 2005; DeChaine et al. 2005). Bathymodioline mussels are globally distributed at deep-sea hydrothermal vents and cold seeps (Fig. 1) and fossil evidence suggests that they existed with chemosynthetic symbionts as early as the Jurassic (~150 Mya; Campbell and Bottjer 1993; Taviani 1994). Phylogenetic inferences suggest that the common ancestor of the vent and seep endemic Bathymodiolinae was derived from mytilids that first inhabited whale and wood falls (Distel et al. 2000). At the species level, mussel populations are genetically differentiated along the northern Mid-Atlantic Ridge (MAR; Won et al. 2003a), while those on the fast-spreading and relatively continuous Eastern Pacific Rise (EPR) are more homogeneous (Won et al. 2003b). Differences in the degree of isolation among species and populations are presumably due to inter-ridge variations in spreading rate, discontinuity of the ridge segments, and deep-ocean currents (van Dover et al. 2002).

Populations of bathymodioline symbionts are geographically structured as well (Won et al. 2003a; DeChaine et al. 2005). Genetic analyses of the ITS region from the chemoautotrophic symbiont populations showed that populations hosted by mussels at the Lost City and MAR were genetically isolated from one another and have experienced independent demographic histories (DeChaine et al. 2005). The degree to which symbiont population structure is affected by host distribution and/or environmental factors remains elusive due to uncertainties in how symbionts are transmitted between hosts from one generation to the next and the limited number of populations studied (Le Pennec and Beninger 1997; Eckelbarger and Young 1999; Won et al. 2003a; DeChaine et al. 2005).

Continued sampling of *Bathymodiolus* species from novel vent and seep sites, characterization of symbioses, and additional population-level studies will provide the basis for understanding how dispersal barriers have influenced gene flow and genetic differentiation in hosts and symbionts, and thus the process of speciation and evolution of the symbiosis.

6
Summary and Conclusions

The symbioses between invertebrates and chemosynthetic bacteria allow both host and symbiont to colonize and thrive in otherwise inhospitable deep-sea habitats. Given the global distribution of the bathymodioline symbioses, this association is an excellent model for evaluating co-speciation and evolution of symbioses. Thus far, the methanotroph and chemoautotroph endosymbionts of mussels are tightly clustered within two independent clades of gamma Proteobacteria, respectively. Further physiological and genomic studies will elucidate the ecological and evolutionary roles that these bacterial clades play in the symbiosis and chemosynthetic community. Due to the overall abundance of the methanotrophic symbioses at hydrothermal vents and hydrocarbon seeps, they likely play a significant, but as of yet unquantified, role in the biogeochemical cycling of methane. With this in mind, the search for methanotrophic symbioses should not be restricted to these known deep-sea habitats, but rather should be expanded to include methane-rich coastal marine and freshwater environments inhabited by methanotrophs and bivalves. Our current understanding of the bathymodioline symbioses provides a strong foundation for future explorations into the origin, ecology, and evolution of methanotroph symbioses, which are now becoming possible through a combination of classical and advanced molecular techniques.

Acknowledgements. We thank Nicole Dubilier and Rudi Amann for hosting us at the MPI for Marine Microbiology (Bremen) and, along with Annelie Pernthaler and Sebastien Duperron, for stimulating discussions on methanotrophic symbioses. Without the Chief Scientists, Captains and crews of the research vessels (including R/V Atlantis II, R/V Atlantis, and R/V Knorr), and the Expedition Leaders and crews of DSV Alvin and ROV Jason, we could not explore the vast unknown deep sea – to them we are grateful.

Research in my laboratory (CMC) on methanotrophic symbioses has been supported by grants from NSF (Biological Oceanography, RIDGE, Ecosystems), NOAA National Undersea Research Center for the West Coast and Polar Regions, and the Office of Naval Research, and by an NSF Postdoctoral Fellowship in Microbial Biology (DBI-0400591 to EGD) and a Hansewissenschaftkolleg Fellowship (to CMC), which we gratefully acknowledge.

References

Anthony C (1982) The biochemistry of methylotrophs. Academic Press, New York
Apps JA, van de Kamp PC (1993) Energy gases of abiogenic origin in the Earth's crust. In: Howell DG, Wiese K, Fanelli M, Zink L, Cole F (eds) The future of energy gases. US Geol Surv Prof Pap 1570:81–132
Barry JP, Buck KR, Kochevar RK, Nelson DC, Fujiwara Y, Goffredi SK, Hashimoto J (2002) Methane-based symbiosis in a mussel, *Bathymodiolus platifrons*, from cold seeps in Sagami Bay, Japan. Invert Biol 121:47–54
Belkin S, Nelson DC, Jannasch HW (1986) Symbiotic assimilation of CO_2 in two hydrothermal vent animals, the mussel *Bathymodiolus thermophilus* and the tubeworm *Riftia pachyptila*. Biol Bull 170:110–121
Boetius A, Ravenschlag K, Schubert CJ, Rickert D, Widdel F, Gieseke A, Amann R, Jorgensen BB, Witte U, Pfannkuche O (2000) A marine microbial consortium apparently mediating anaerobic oxidation of methane. Nature 407:623–626
Bowman JP, Jimenez L, Rosario I, Hazen TC, Sayler GS (1993) Characterization of the methanotrophic bacterial community present in a trichloroethylene-contaminated groundwater site. Appl Environ Microbiol 59:2380–2387
Bratina BJ, Brusseau GA, Hanson RS (1992) Use of 16S rRNA analysis to investigate phylogeny of methylotrophic bacteria. Int J Syst Bacteriol 42:645–648
Brooks JM, Kennicutt MC II, Fisher CR, Macko SA, Cole K, Childress JJ, Bidigare RR, Vetter RD (1987) Deep-sea hydrocarbon seep communities: evidence for energy and nutritional carbon sources. Science 20:1138–1142
Brown KM (1990) The nature and hydrogeologic significance of mud diapirism and diatremes from accretionary systems. J Geophys Res 95:8969–8982
Campbell KA, Bottjer DJ (1993) Fossil cold seeps. Res Exp 9:326–343
Cary SC, Fisher CR, Felbeck H (1988) Mussel growth supported by methane as sole carbon and energy source. Science 240:78–80
Cavanaugh CM (1985) Symbioses of chemoautotrophic bacteria and marine invertebrates from hydrothermal vents and reducing sediments. Biol Soc Wash Bull 6:373–388
Cavanaugh CM (1992) Methanotroph-invertebrate symbioses in the marine environment: ultrastructural, biochemical, and molecular studies. In: Murrell JC, Kelly DP (eds) Microbial growth on C1 compounds. Intercept, Andover, UK, pp 315–328
Cavanaugh CM, Levering PR, Maki JS, Mitchell R, Lidstrom ME (1987) Symbiosis of methylotrophic bacteria and deep-sea mussels. Nature 325:346–347

Cavanaugh CM, Wirsen C, Jannasch HJ (1992) Evidence for methylotrophic symbionts in a hydrothermal vent mussel (Bivalvia: Mytilidae) from the Mid-Atlantic Ridge. Appl Environ Microbiol 58:3799–3803

Cavanaugh CM, McKiness ZP, Newton ILG, Stewart FJ (2005) Marine chemosynthetic symbioses. In: Dworkin M, Falkow S, Rosenberg E, et al (eds) The Prokaryotes: a handbook on the biology of bacteria, 3rd edn. Springer, Berlin Heidelberg New York (in press)

Childress JJ, Fisher CR, Brooks JM, Kennicutt MC II, Bidigare R, Anderson AE (1986) A methanotrophic marine molluscan (Bivalvia, Mytilidae) symbiosis: mussels fueled by gas. Science 233:1306–1308

Dando PR, Bussman I, Niven SJ, O'Hara SCM, Schmaljohann R, Taylor LJ (1994) A methane seep area in the Skagerrak, the habitat of the pogonophore *Siboglinum poseidoni* and the bivalve mollusc *Thyasira sarsi*. Mar Ecol Prog Ser 107:157–167

DeChaine EG, Bates A, Shank TM, Cavanaugh CM (2005) Off-axis symbiosis found: characterization and biogeography of bacterial symbionts of *Bathymodolus* mussels from Lost City hydrothermal vents, In review

Dedysh SN (2002) Methanotrophic bacteria of acidic sphagnum peat bogs. Microbiology 71:638–650

Distel D, Lee HK, Cavanaugh CM (1995) Intracellular coexistence of methano- and thioautotrophic bacteria in a hydrothermal vent mussel. Proc Natl Acad Sci USA 92:9598–9602

Distel D, Baco AR, Chuang E, Morrill W, Cavanaugh C, Smith CR (2000) Do mussels take wooden steps to deep-sea vents? Science 403:725–726

Dubilier N, Windoffer R, Giere O (1998) Ultrastructure and stable carbon isotope composition of the hydrothermal vent mussels *Bathymodiolus brevior* and *B.* sp. affinis *brevior* from the North Fiji Basin, western Pacific. Mar Ecol Prog Ser 165:187–193

Dubilier N, Amann R, Erseus C, Muyzer G, Park SY, Giere O, Cavanaugh CM (1999) Phylogenetic diversity of bacterial endosymbionts in the gutless marine oligochete *Olavius loisae* (Annelida). Mar Ecol Prog Ser 178:271–280

Duperron S, Nadalig T, Caprais J-C, Sibuet M, Fiala-Medioni A, Amann R, Dubilier N (2005a) Dual symbiosis in a *Bathymodiolus* mussel from a methane seep on the Gabon continental margin (South East Atlantic): 16S rRNA phylogeny and distribution of the symbionts in the gills (in press)

Duperron ST, Bergen C, Zielinski F, McKiness ZP, DeChaine EG, Sibuet M, Cavanaugh CM, Dubilier N (2005b) A dual symbiosis shared by two Mid-Atlantic Ridge bathymodioline mussels (Bivalvia: Mytilidae) Env. Microbiol. In press.

Eckelbarger KJ, Young CM (1999) Ultrastructure of gametogenesis in a chemosynthetic mytilid bivalve (*Bathymodiolus childressi*) from a bathyal, methane seep environment (northern Gulf of Mexico). Mar Biol 135:635–646

Fiala-Médioni A, McKiness Z, Dando P, Boulegue J, Mariotti A, Alayse-Danet A, Robinson J, Cavanaugh C (2002) Ultrastructural, biogeochemical, and immunological characterization of two populations of a new species of Mytilid mussel, *Bathymodiolus azoricus*, from the Mid-Atlantic Ridge: evidence for a dual symbiosis. Mar Biol 141:1035–1043

Fisher CR (1990) Chemoautotrophic and methanotrophic symbioses in marine invertebrates. Rev Aquat Sci 2:399–436

Fisher CR, Childress JJ (1992) Organic carbon transfer from methanotrophic symbionts to the host hydrocarbon-seep mussel. Symbiosis 12:221–235

Fisher CR, Childress JJ, Oremland RS, Bidigare RR (1987) The importance of methane and thiosulfate in the metabolism of the bacterial symbionts of two deep-sea mussels. Mar Biol 96:59–71

Fisher CR, Brooks JM, Vodenichar JS, Zande JM, Childress JJ, Burke RA Jr (1993) The co-occurrence of methanotrophic and chemoautotrophic sulfur-oxidizing bacterial symbionts in a deep-sea mussel. Mar Ecol 14:277–289

Fisher CR, Childress JJ, Macko SA, Brooks JM (1994) Nutritional interaction in Galapagos Rift hydrothermal vent communities: inferences from stable carbon and nitrogen isotope analyses. Mar Ecol Prog Ser 103:45–55

Fujiwara Y, Takai K, Uematsu K, Tsuchida S, Hunt JC, Hashimoto J (2000) Phylogenetic characterization of endosymbionts in three hydrothermal vent mussels: influence on host distributions. Mar Ecol Prog Ser 208:147–155

Grassle JF (1985) Hydrothermal vent animals – distribution and biology. Science 229:713–717

Gustafson R, Turner R, Lutz R, Vrijenhoek R (1998) A new genus and five new species of mussels (Bivalvia, Mytilidae) from deep-sea sulfide/hydrocarbon seeps in the Gulf of Mexico. Malacologia 40:63–112

Hanson RS, Hanson TE (1996) Methanotrophic bacteria. Microbiol Rev 60:439–471

Hanson RS, Netrusov AI, Tsuji K (1991) The obligate methanotrophic bacteria Methylococcus, Methylomonas, Methylosinus and related bacteria. In: Balows A, Truper HG, Dworkin M, Harder W, Schleifer KH (eds) The prokaryotes. Springer, Berlin Heidelberg New York, pp 2350–2365

Hashimoto J (2001) A new species of *Bathymodiolus* (Bivalvia: Mytilidae) from hydrothermal vent communities in the Indian Ocean. Venus Jpn J Malacol 60: 141–149

Hashimoto J, Okutani T (1994) Four new mytilid mussels associated with deep sea chemosynthetic communities around Japan. Venus Jpn J Malacol 53:61–83

Hedberg HD (1980) Methane generation and petroleum migration. In: Roberts WH III, Cordell RJ (eds) Problems of petroleum migration. American Association of Petroleum Geologists. Studies in geology no 10. Am Assoc Petrol Geol, Tulsa, pp 179–206

Hinrichs K-U, Hayes JM, Sylva SP, Brewer PG, DeLong EF (1999) Methane-consuming archaebacteria in marine sediments. Nature 398:802–805

Holmes A, Owens N, Murrell J (1995) Detection of novel marine methanotrophs using phylogenetic and functional gene probes after methane enrichment. Microbiology 141:1947–1955

Hornafius JS, Quigley D, Luyendyk BP (1999) The world's most spectacular marine hydrocarbon seeps (Coal Oil Point, Santa Barbara Channel, California): quantification of emissions. J Geophys Res 104:20703–20711

Huber R, Kurr M, Jannasch HW, Stetter KO (1989) A novel group of abyssal methnogenic archaebacteria (*Methanopyrus*) growing at 110-degrees-C. Nature 342:833–834

Judd AG (2003) The global importance and context of methane escape from the seabed. Geo Mar Lett 23:147–154

Judd AG, Hovland M, Dimitrov LI. Garcia Gil S, Jukes V (2002) The geological methane budget at continental margins and its influence on climate change. Geofluids 2:109–126

Kelley DS, Karson JA, Blackman DK, Früh-Green GL, Butterfield DA, Lilley MD, Olson EJ, Schrenk MO, Roe KK, Lebon GT, Rivizzigno P, and the AT3–60 Shipboard Party (2001) An off-axis hydrothermal vent field near the Mid-Atlantic Ridge at 30°N. Nature 412:145–149

Kenk VC, Wilson BR (1985) A new mussel (Bivalvia, Mytilidae) from hydrothermal vents in the Galapagos Rift zone. Malacologia 26:253–271

Kennicutt MC, Burke RA, MacDonald IR, Brooks JM, Denoux GJ, Macko SA (1992) Stable isotope partitioning in seep and vent organisms: chemical and ecological significance. Chem Geol 101:293–310

Kim S, Mullineaux LS (1998) Distribution and near-bottom transport of larvae and other plankton at hydrothermal vents. Deep Sea Res II 45:423–440

Kochevar RE, Childress JJ, Fisher CR, Minnich E (1992) The methane mussel: roles of symbiont and host in the metabolic utilization of methane. Mar Biol 112:389–401

Lee RW, Childress JJ (1994) Assimilation of inorganic nitrogen by chemoautotrophic and methanotrophic symbioses. Appl Environ Microbiol 60:1852–1858

Lee RW, Childress JJ (1996) Inorganic N assimilation and ammonium pools in a deep-sea mussel containing methanotrophic endosymbionts. Biol Bull 190:373–384

Lee RW, Thuesen EV, Childress JJ, Fisher CR (1992) Ammonium and free amino acid uptake by a deep-sea mussel containing methanotrophic bacterial symbionts. Mar Biol 113:99–106

Lee R, Robinson JJ, Cavanaugh CM (1999) Pathways of inorganic nitrogen assimilation in chemoautotrophic bacteria-marine invertebrate symbioses: expression of host and symbiont glutamine synthetase. J Exp Biol 202:289–300

Le Pennec M, Beninger PG (1997) Ultrastructural characteristics of spermatogenesis in three species of deep-sea hydrothermal vent mytilids. Can J Zool 75:308–316

Lilley MD, Butterfield DA, Lupton JE, Olson EJ (2003) Magmatic events can produce rapid changes in hydrothermal vent chemistry. Nature 472:878–881

Lonsdale P (1977) Clustering of suspension-feeding macrobenthos near abyssal hydrothermal vents at oceanic spreading centers. Deep Sea Res 24:857

Lupton JE, Lilley MD, Olson EJ, von Damm KL (1991) Gas chemistry of vent fluids from 9°–10°N on the East Pacific Rise. EOS 72:F481

MacAvoy SE, Fisher CR, Carney RS, Macko SA (2005) Nutritional associations among fauna at hydrocarbon seep communities in the Gulf of Mexico. Mar Ecol Prog Ser (in press)

McKiness ZP, Cavanaugh CM (2005) The ubiquitous mussel: *Bathymodiolus* aff. *brevior* symbiosis discovered at the Central Indian Ridge hydrothermal vents. Mar Ecol Prog Ser (in press)

McKiness ZP, McMullin ER, Fisher CR, Cavanaugh CM (2005) A new bathymodioline mussel symbiosis at the Juan de Fuca hydrothermal vents (submitted)

Milkov AV (2000) Worldwide distribution of submarine mud volcanoes and associated gas hydrates. Mar Geol 167:29–42

Miyazaki J-I, Shintaku M, Kyuno A, Fujiwara Y, Hashimoto J, Iwasaki H (2004) Phylogenetic relationships of deep-sea mussels of the genus *Bathymodiolus* (Bivalvia: Mytilidae). Mar Biol 144:527–535

Nelson DC, Hagan KD, Edwards DB (1995) The gill symbiont of the hydrothermal vent mussel *Bathymodiolus thermophilus* is a psychrophilic, chemoautotrophic, sulfur bacterium. Mar Biol 121:487–495

O'Mullan GD, Maas PAY, Lutz RA, Vrijenhoek RC (2001) A hybrid zone between hydrothermal vent mussels (Bivalvia: Mytilidae) from the Mid-Atlantic Ridge. Mol Ecol 10:2819–2831

Page HM, Fisher CR, Childress JJ (1990) Role of filter-feeding in the nutritional biology of a deep-sea mussel with methanotrophic symbionts. Mar Biol 104: 251–257

Peek AS, Vrijenhoek RC, Gaut BS (1998) Accelerated evolutionary rate in sulfur-oxidizing endosymbiotic bacteria associated with the mode of symbiont transmission. Mol Biol Evol 15:1514–1523

Pernthaler A, Amann R (2004) Simultaneous fluorescence in situ hybridization of mRNA and rRNA in environmental bacteria. Appl Environ Microbiol 70: 5426–5433

Pile AJ, Young CM (1999) Plankton availability and retention efficiencies of cold-seep symbiotic mussels. Limnol Oceanogr 44:1833–1839

Pimenov NV, Kalyuzhnaya MG, Khmelenina VN, Mityushina LL, Trotsenko YA (2002) Utilization of methane and carbon dioxide by symbiotrophic bacteria in gills of Mytilidae (*Bathymodiolus*) from the Rainbow and Logatchev hydrothermal fields on the Mid-Atlantic Ridge. Microbiology 71:587–594

Pranal V, Fiala-Médioni A, Guezennec J (1997) Fatty acid characteristics in two symbiont-bearing mussels from deep-sea hydrothermal vents of the south-western Pacific. J Mar Biol Assoc UK 77:473–492

Reeburgh WS (1980) Anaerobic methane oxidation: rate distributions in Skan Bay sediments. Earth Planet Sci Lett 47:345–352

Robinson JJ, Polz MF, Fiala-Médioni A, Cavanaugh CM (1998) Physiological and immunological evidence for two distinct C_1-utilizing pathways in *Bathymodiolus puteoserpentis* (Bivalvia: Mytilidae), a dual endosymbiotic mussel from the Mid-Atlantic Ridge. Mar Biol 132:625–633

Schmaljohan R (1991) Oxidation of various potential energy sources by the methanotrophic endosymbionts of *Siboglinum poseidoni* (Pogonophora). Mar Ecol Prog Ser 76:143–148

Schmaljohan R, Flügel HJ (1987) Methane-oxidizing bacteria in Pogonophora. Sarsia 72:91–98

Schmaljohan R, Faber E, Whiticar MJ, Dando PR (1990) Co-existence of methane-and sulphur-based endosymbioses between bacteria and invertebrates at a site in the Skagerrak. Mar Ecol Prog Ser 61:119–124

Shima S, Warkentin E, Thauer RK, Ermler U (2002) Structure and function of enzymes involved in the methanogenic pathway utilizing carbon dioxide and molecular hydrogen. J Biosci Bioeng 93:519–530

Sibuet M, Olu K (1998) Biogeography, biodiversity and fluid dependence of deep-sea cold-seep communities at active and passive margins. Deep Sea Res II 45:517–567

Streams ME, Fisher CR, Fiala-Médioni A (1997) Methanotrophic symbiont location and fate of carbon incorporated from methane in a hydrocarbon seep mussel. Mar Biol 129:465–476

Taviani M (1994) The Calcari-A-Lucina macrofauna reconsidered – deep-sea faunal eases from Miocene age cold vents in the Romagna Appenine, Italy. Geo Mar Lett 14:185–191

Trask JL, van Dover CL (1999) Site-specific and ontogenetic variations in nutrition of mussels (*Bathymodiolus* sp.) from the Lucky Strike hydrothermal vent field, Mid-Atlantic Ridge. Limnol Oceanogr 44:334–343

Tunnicliffe V (1991) The biology of hydrothermal vents – ecology and evolution. Oceanogr Mar Biol 29:319–407

Tunnicliffe V, Fowler C (1996) Influence of sea-floor spreading on the global hydrothermal vent fauna. Nature 379:531–533

Tunnicliffe V, McArthur AG, Mchugh D (1998) A biogeographical perspective of the deep-sea hydrothermal vent fauna. Adv Mar Biol 34:353–442

Vacelet J, Fiala-Médioni A, Fisher CR, Boury-Esnault N (1996) Symbiosis between methane-oxidizing bacteria and a deep-sea carnivorous cladorhizid sponge. Mar Ecol Prog Ser 145:77–85

Van Dover CL (2000) The ecology of deep-sea hydrothermal vents. Princeton Univ Press, Princeton, NJ, 424 pp

Van Dover CL (2002) Trophic relationships among invertebrates at the Kairei hydrothermal vent field (Central Indian Ridge). Mar Biol 141:761–772

Van Dover CL, Humphris SE, Fornari D, Cavanaugh CM, Collier R, Goffredi SK, Hashimoto J, Lilley M, Reysenbach AL, Shank TM, von Damm KL, Banta A, Gallant RM, Götz D, Green D, Hall J, Harmer TL, Hurtado LA, Johnson P, McKiness ZP, Meredith C, Olson E, Pan IL, Turnipseed M, Won Y, Young CR III, Vrijenhoek RC (2001) Biogeography and ecological setting of Indian Ocean hydrothermal vents. Science 294:818–823

Van Dover CL, German C. R, Speer K. G, Parson L. M, and Vrijenhoek R. C (2002. Evolution and biogeography of deep-sea vent and seep invertebrates. Science 295: 1253–1257

Von Cosel R, Olu K (1998) Gigantism in Mytilidae. A new Bathymodiolus from cold seep areas on the Barbados accretionary Prism. CR Acad Sci Paris Sci Vie 321:655–663

Von Cosel R, Métivier B, Hashimoto J (1994) Three new species of *Bathymodiolus* (Bivalvia: Mytilidae) from hydrothermal vents in the Lau Basin and the North Fiji Basin, Western Pacific, and the Snake Pit area, Mid-Atlantic Ridge. Veliger 37:374–392

Vrijenhoek RC (1997) Gene flow and genetic diversity in naturally fragmented metapopulations of deep-sea hydrothermal vent animals. J Heredity 88:285–293

Vrijenhoek RC, Shank T, Lutz R (1998) Gene flow and dispersal in deep-sea hydrothermal vent animals. Cah Biol Mar 39:363–366

Won Y, Hallam SJ, O'Mullan GD, Pan IL, Buck KR, Vrijenhoek RC (2003a) Environmental acquisition of thiotrophic endosymbionts by deep-sea mussels of the genus *Bathymodiolus*. Appl Environ Microbiol 69:6785–6792

Won Y, Young CR, Lutz RA, Vrijenhoek RC (2003b) Dispersal barriers and isolation among deep-sea mussel populations (Mytilidae: *Bathymodiolus*) from eastern Pacific hydrothermal vents. Mol Ecol 12:169–184

Yamanaka T, Mizota C, Maki Y, Fujikura K, Chiba H (2000) Sulfur isotope composition of soft tissues of deep-sea mussels, *Bathymodiolus* spp, in Japanese waters. Benthos Res 55:63–68

Yamanaka T, Mizota C, Fujiwara Y, Chiba H, Hashimoto J, Gamo T, Okudaira T (2003) Sulphur-isotopic composition of the deep-sea mussel *Bathymodiolus marisindicus* from currently active hydrothermal vents in the Indian Ocean. J Mar Biol Assoc UK 83:841–848

Symbioses between Bacteria and Gutless Marine Oligochaetes

Nicole Dubilier, Anna Blazejak, Caroline Rühland

1
Introduction

The first description of a gutless marine oligochaete was published in 1977 (Jamieson 1977), but it was not until 1979 that the reduction of a mouth and gut in this species and several other marine tubificid worms was recognized (Erséus 1979a,b; Giere 1979). At the time, the only other free-living worms known to lack a mouth or gut were pogonophores (now called Frenulata) that were most commonly found buried deep in the reducing sediments of continental slopes. Extensive studies on these very long and thin pogonophore worms indicated that their high surface areas enabled them to gain their nutrition from the uptake of dissolved organic compounds from the environment (Southward and Southward 1980). It was therefore assumed that the gutless oligochaetes that are also quite thin (0.1–0.2 mm) and relatively long (up to 2–3 cm), also gain their nutrition through the diffusive uptake of organic compounds from the sediment pore waters.

The discovery of the giant tube worm, *Riftia pachyptila*, at hydrothermal vents in the late 1970s revolutionized our understanding of the nutrition of gutless marine worms. These worms without a mouth or gut were clearly too thick and the concentration of organic compounds at the vents too low to explain the high biomass of these worms through diffusive uptake of nutrients alone. Indeed, very soon after their discovery Cavanaugh et al. (1981) and Felbeck (1981) showed that *R. pachyptila* lives in symbiosis with chemoautotrophic sulfur-oxidizing bacteria that use reduced sulfur compounds from vent fluids as electron donors to gain energy, and fix CO_2 into organic compounds that are passed on to the worms.

In the wake of the discovery of the *Riftia* symbiosis it became quickly clear that other marine invertebrates with reduced guts could harbor

N. Dubilier (e-mail: ndubilie@mpi-bremen.de), A. Blazejak, C. Rühland
Max-Planck-Institut für Marine Mikrobiologie, Celsiusstr. 1, 28359 Bremen, Germany

similar symbionts. The reexamination of the morphology of gutless oligochaetes revealed the presence of a thick layer of bacteria just below the cuticle (outer most layer of the body wall) of the worms (Giere 1981; Richards et al. 1982). Enzyme assays and uptake experiments with inorganic carbon indicated that the bacterial symbionts are thiotrophic (i.e., CO_2-fixing sulfur oxidizers) (Felbeck et al. 1983), although we now know that this is only part of the story, as only some of the symbionts in these hosts are sulfur-oxidizing bacteria (see Sect. 12.5 and 12.6).

Since the early 1980s a wealth of morphological and ecophysiological studies by Olav Giere (University of Hamburg, Germany) and taxonomical studies of the hosts by Christer Erséus (University of Göteborg, Sweden) have laid the basis for research on these symbioses with molecular analyses of the symbionts by the authors of this review adding a new dimension since the mid 1990s. In the following, we will describe these molecular investigations in detail but also review earlier studies, to provide a comprehensive understanding of what is known about the symbioses between bacteria and gutless oligochaetes (see also a recent review by Bright and Giere 2005).

2
Biogeography of the Hosts

Gutless oligochaetes are an ideal host group for studying the biogeography and evolution of marine symbioses as they occur throughout the world in a wide range of different habitats with some species widely distributed and others highly endemic (Erséus 1992). They are also one of the few marine groups in which such a large number of host species (>100) are so closely related to each other, forming a monophyletic group with all gutless oligochaetes descendents from a single common ancestor (Nylander et al. 1999).

2.1
Geographic Distribution

The first gutless oligochaetes were found in coral reef sediments in the Pacific (Jamieson 1977), and North Atlantic (Erséus 1979b; Giere 1979), and the highest diversity of oligochaete species is still regularly found in shallow water calcareous sediments (Erséus 1984, 1990). For example, as many as 18 species have been described from sediments around a small island in the Bahamas (Erséus 2003). While tropical and subtropical coral reef sediments clearly represent "hotspots" for gutless oligochaete diversity, these worms also occur at greater depths and more temperate regions such as the Northern West Atlantic (Davis 1985) and continental shelves off the coast of California (Erséus 1991) and Peru (Finegenova 1986). Given the small size of these worms and the limited ability of most taxonomists to recognize these species, they may often be overlooked in benthic surveys.

Of particular interest for studies of biogeography and cospeciation between host and symbiont is the disjunct distribution of gutless oligochaetes. For example, the two host species that co-occur in the Mediterranean, *Olavius algarvensis* and *O. ilvae* are more closely related to other non-Mediterranean species than to each other (Giere and Erséus 2002). This is particularly intriguing in view of the restricted possibilities for dispersal in marine oligochaetes. A planktonic egg or larval stage does not occur and the worms prefer deeper sediments, where they are not directly exposed to ocean currents.

2.2
Phylogeny of the Hosts

Over 80 gutless oligochaete species have been described, and there are numerous undescribed species, mainly from the Indo-Pacific region and the East Atlantic Ocean, still available for identification. Given that a collection trip to a novel region can yield as many as 9 new species (Erséus 2003), the diversity of gutless oligochaetes may be even higher. There is a broad range from primitive to highly derived species, indicating that the symbiotic condition has led to a radiative evolution in these worms. Despite this high diversity, all gutless oligochaetes belong to a single monophyletic group within the marine tubificids, based on both morphological (Erséus 1984, 1992) and molecular analyses (Nylander et al. 1999; Erséus et al. 2000). Within this group, two sister genera have been identified, *Inanidrilus* and *Olavius* (Erséus 1984), with the monophyly of *Inanidrilus* confirmed in all analyses, in contrast to the genus *Olavius* that may be paraphyletic (Erséus 2003). (Note that in earlier publications, the genus name *Phallodrilus* was used for some gutless oligochaetes but these were later renamed as either *Inanidrilus* or *Olavius* species by Erséus 1984).

3
Environment

Until recently, it was assumed that gutless oligochaetes only occur in large numbers in shallow water calcareous sediments, living interstitially in the pore waters. The discovery of high abundances of gutless oligochaetes in silicate sediments near sea-grass beds off the coast of Elba in the Mediterranean (Perner 2003) and in 100–400 m water depth in muds off the coast of Chile and Peru (Levin et al. 2002, 2003) indicate that gutless oligochaetes may be much more widespread and occur in many more environments than presently known.

3.1
Reduced Sulfur Compounds

The close timing between the discovery of sulfur-oxidizing bacteria in the gutless tube worms from hydrothermal vents and the symbiotic bacteria in gutless oligochaetes, led to the assumption that the oligochaete symbionts were also dependent on reduced sulfur compounds such as sulfide and thiosulfate from the environment. Most ecological studies have centered on a single species, *Inanidrilus leukodermatus*, from a single site in Bermuda (Giere et al. 1982). Sulfide concentrations at this site ranged between 2–32 µM in 5 and 10 cm sediment depth where the worms were most abundant (Giere et al. 1982). While these concentrations may seem low to microbiologists that often use much higher sulfide concentrations to isolate sulfur-oxidizing bacteria, they are comparable to those that hydrothermal vent fauna with sulfide-oxidizing bacteria are exposed to (Johnson et al. 1988; Le Bris et al. 2003). The concentration of other reduced sulfur compounds, such as thiosulfate, has not been determined in the habitat of gutless oligochaetes. Experiments on the uptake of CO_2 in *I. leukodermatus* indicate that thiosulfate is preferred over sulfide, but the use of environmentally unrealistic concentrations of thiosulfate in the mM range and unusually high CO_2 uptake rates in the absence of either sulfide or thiosulfate make the interpretation of these results difficult (Giere et al. 1988).

The presence of reduced sulfur compounds has been assumed to be one of the most important environmental factors for gutless oligochaetes, as the lack of these electron donors would lead to the starvation of the sulfur-oxidizing symbionts and in turn, their hosts. This perception has changed with the discovery of gutless oligochaetes from the island of Elba (Mediterranean) that live in sediments in which sulfide concentrations are usually in the nM range (Dubilier et al. 2001), and only occasionally in the low µM range (Perner 2003). However, low sulfide concentrations need not be limiting if the flux of sulfide from production by free-living sulfate-reducing bacteria is high enough. Sulfate reduction rates in these sediments range between 100–300 nmol cm^{-3} per day, showing that sulfide is produced at rates comparable to those of other sandy sediments (Perner 2003). However, this sulfide is apparently oxidized very quickly, as free sulfide concentrations are extremely low in the pore waters. It is not known if the flux of free sulfide in these sediments is sufficient to support the growth of these worms. Since these worms also harbor sulfate-reducing symbionts that produce sulfide internally (see Sect. 12.6), they may be independent of reduced sulfur compounds from the environment.

3.2
Oxygen and Other Electron Acceptors

Gutless oligochaetes require oxygen, not only for their own respiration, but also for their sulfur-oxidizing symbionts, that use oxygen or other oxidized compounds such as nitrate as an electron acceptor. While the worms can survive short periods in the absence of oxygen, presumably by switching to an anaerobic metabolism as observed in many other marine invertebrates (Grieshaber et al. 1992; Dubilier et al. 1995b), longer periods of anoxia for several days lead to massive mortalities (Dubilier, unpubl. obs.). As all tubificids, the worms contain well developed blood vessels with a red respiratory pigment (presumably hemoglobin), enabling them to store oxygen for limited time periods.

At almost all sites investigated, gutless oligochaetes are most dominant in the deeper sediment layers at 5–15 cm below the sediment surface. In most sediments, including those of coral reefs, oxygen penetrates at most only a few cm into the sediment (Falter and Sansone 2000). At the Elba site in the Mediterranean, oxygen can penetrate as deep as 3 cm below the sediment surface (D. de Beer, unpubl. data), but not to where the worms occur at 10–15 cm sediment depth. At the up-welling site off the coast of Peru, where *O. crassitunicatus* was the dominant member of the infauna, oxygen concentrations were extremely low just above the sediment surface (<1 µM) and it is very likely that such low concentrations persist for longer time periods (Levin et al. 2002, 2003). Interestingly, at this site the worms occur in the upper sediment layers between 1–5 cm sediment depth, indicating that the worms migrate upwards when oxygen becomes limiting.

In all known habitats of gutless oligochaetes there is no overlap between oxygen in the upper sediment layers and reduced sulfur compounds in the deeper sediment layers. Since the worms do not build tubes or burrows, it is assumed that they migrate between the lower sulfidic and the upper oxygenated sediments. This model implies that sulfide is taken up in the anoxic deeper sediment layers, oxidized to sulfur by either nitrate from the environment or oxygen from the worm's hemoglobin, and the sulfur stored in the bacteria until the worms migrate to upper sediment layers where the sulfur could be fully oxidized to sulfate. Experiments show that the worms can migrate actively through the sediment (Giere et al. 1991). However, nothing is known about the time periods spent by individual worms in oxygenated or sulfidic sediments, whether nitrate is used as an electron acceptor, and whether the symbiotic bacteria can oxidize sulfide using oxygen stored in the worm's hemoglobin.

3.3
Other Environmental Factors

As described above, when gutless oligochaetes were first discovered it was assumed that they gained their nutrition by uptake of dissolved organic compounds from the environment. At the Bermuda site where *I. leukodermatus* occurs in high abundances, concentrations of dissolved free amino acids and organic carbon, as well as total carbohydrates were higher than in other Bermudian carbonate sediments (Giere et al. 1982). However, in uptake experiments with radiolabeled substrates, organic carbon was taken up much more slowly than inorganic carbon by *I. leukodermatus*. In contrast, Liebezeit et al. (1983) argued that dissolved organic carbon may contribute significantly to the nutrition of *I. leukodermatus*. At the Elba site in the Mediterranean, the worms are most abundant in sediments around sea grass beds, indicating that organic substrates from the sea grasses may play a role in the distribution of these worms. Clearly, well-designed uptake experiments are needed in which net uptake (and not only total uptake as in previous studies) of dissolved organic compounds and CO_2 are examined in more detail including a differentiation between uptake by the symbiotic bacteria and the host itself.

4
Structural Aspects

4.1
Morphology of the Symbiosis

In the first two descriptions of gutless oligochaetes, namely *Inanidrilus albidus* (albus = Latin: white) and *I. leukodermatus* (leukos = Greek: white), the intensive white color of the worms, now known to be common to all gutless oligochaetes, was mentioned but not further investigated (Fig. 4a) (Jamieson 1977; Giere 1979). This white coloring comes from the refraction of light by the symbiotic sulfur-oxidizing bacteria just below the cuticle of the worm that are filled with sulfur and the storage compound polyhydroxybutyric acid (PHB). The white color of the worms distinguishes them from other marine oligochaetes, making their identification in the field as symbiont-bearing worms easy, although only in live specimens (when preserved in alcohol or formaldehyde the white color is lost).

In all gutless oligochaetes the morphology of the symbiosis is remarkably similar. The worms have no mouth, gut, or anus, and unique among all gutless worms, they are the only host group that completely lack nephridia, excretory organs used to remove nitrogenous waste compounds and for osmoregulation. The symbiotic bacteria occur in a multicellular layer just below the thin cuticle of the worms, called the symbiotic region (Fig. 1). Calculations based on visual examinations of transmission electron micrographs indicate that an average sized worm harbors at least 10^6 bacterial cells, corresponding to roughly 25% of the worm's volume (Giere et al. 1995).

The bacteria in the apical part of the symbiotic region are extracellular, and sit between extensions of the epidermal cells. Diffusion experiments with fluorescein-labeled dextrane showed that the cuticle is permeable for substrates as large as 70 kDa as well as small negatively charged molecules such as fluorescein (J. Krieger and N. Dubilier, unpubl. data). This indicates that the symbiotic bacteria have free access to most substrates in the pore waters that the worms live in. In the basal area of the symbiotic region, the bacteria are regularly enclosed in vacuoles of the epidermal cells and appear to be in various stages of lysis. It is not known how important lysis of the bacteria is for the nutrition of the hosts, or whether "milking" of the bacteria, in which organic compounds are transferred to the host, may be the main mode of nutrient transfer. Lysis may also play a role in the regulation of bacterial growth (Giere et al. 1995).

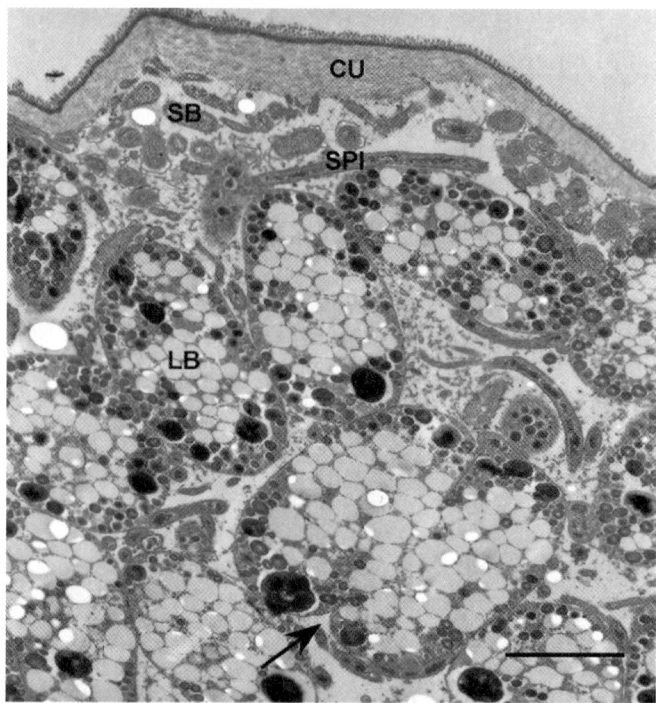

Fig. 1. Transmission electron micrograph of symbiotic region below the cuticle (*CU*) of the gutless oligochaete *O. crassitunicatus*. Three bacterial morphotypes are clearly visible: large oval-shaped bacteria (*LB*) with sulfur and polyhydroxybutyric acid vesicles in the cytoplasm, small cocci- to rod-shaped bacteria (*SB*), and long, thin bacteria (*SPI*). In this host, the large bacteria have been identified as Gamma 1 symbionts, the small bacteria as delta proteobacterial symbionts, and the long thin bacteria as spirochetes (Fig. 2). The *arrow* shows a Gamma 1 symbiont in the final stages of division

4.2
Multiple Bacterial Morphotypes

The first ultrastructural studies of oligochaete symbionts described two bacterial morphotypes (Giere 1981, 1985): large oval bacteria 2–7 µm in length filled with sulfur and PHB vesicles and smaller cocci- to rod-shaped bacteria 0.7–1.5 µm in length without any conspicuous inclusions or structures in their cytoplasm (Fig. 1) (also see Bright and Giere 2005 for a more detailed review of the ultrastructure of the symbiosis). The large oval bacteria are now known to be sulfur-oxidizing gamma *Proteobacteria* (Gamma 1 symbionts in Fig. 2). In contrast, the small bacteria can not be identified based on their ultrastructure alone, and depending on the host species they occur in, belong to the alpha, gamma, or delta *Proteobacteria* (see Sect. 5.2). A third very thin and long morphotype (0.3×10 µm), was first described in *O. crassitunicatus* from the Peru margin (Giere and Krieger 2001), and has now been identified as a spirochete (Fig. 1 and Sect. 5.2.4). The identity of a fourth bacterial morphotype of intermediate size between the large and small bacteria and described in *O. loisae* (Dubilier et al. 1999) and *O. algarvensis* (Giere and Erséus 2002) remains unclear. Overall, the ultrastructure of the bacterial symbionts can only be used for identification of the sulfur-oxidizing Gamma 1 symbionts with their obvious sulfur and PHB vesicles and in some cases for the long and thin spirochaete symbionts. The morphology of the other bacterial morphotypes, in particular the small rod- and cocci-shaped bacteria, does not yield enough characters to allow a clear taxonomic identification based on ultrastructure alone.

4.3
Transmission of the Symbionts

When gutless oligochaetes become fully mature they develop so-called genital pads, a pair of sack-like pockets on the ventral side of the worms close to the oviporus, the opening through which eggs are deposited. These genital pads are packed with symbiotic bacteria and are only separated from the environment by a very thin cuticle layer of the worm (Giere and Langheld 1987). While all genital organs and the maturing eggs inside of the worm appear to be free of bacteria, the freshly laid egg is surrounded by bacteria that are enclosed in an extracellular space between the inner egg membrane and the outer egg integument (Krieger 2000). It is assumed that these bacteria are transmitted from the parent to the egg via the genital pads, which are believed to rupture during egg deposition (Giere and Langheld 1987). Several hours after egg deposition, the bacteria accumulate at one pole of the egg and penetrate the egg membrane (Krieger 2000).

In contrast to all other oligochaetes that deposit their eggs in cocoons, the eggs of gutless oligochaetes are deposited singly and immediately adhere to the surrounding sediment because of a sticky mucous covering on the outside of the egg integument (Giere and Langheld 1987). In addition to the vertical transmission of symbionts via the genital pads, the

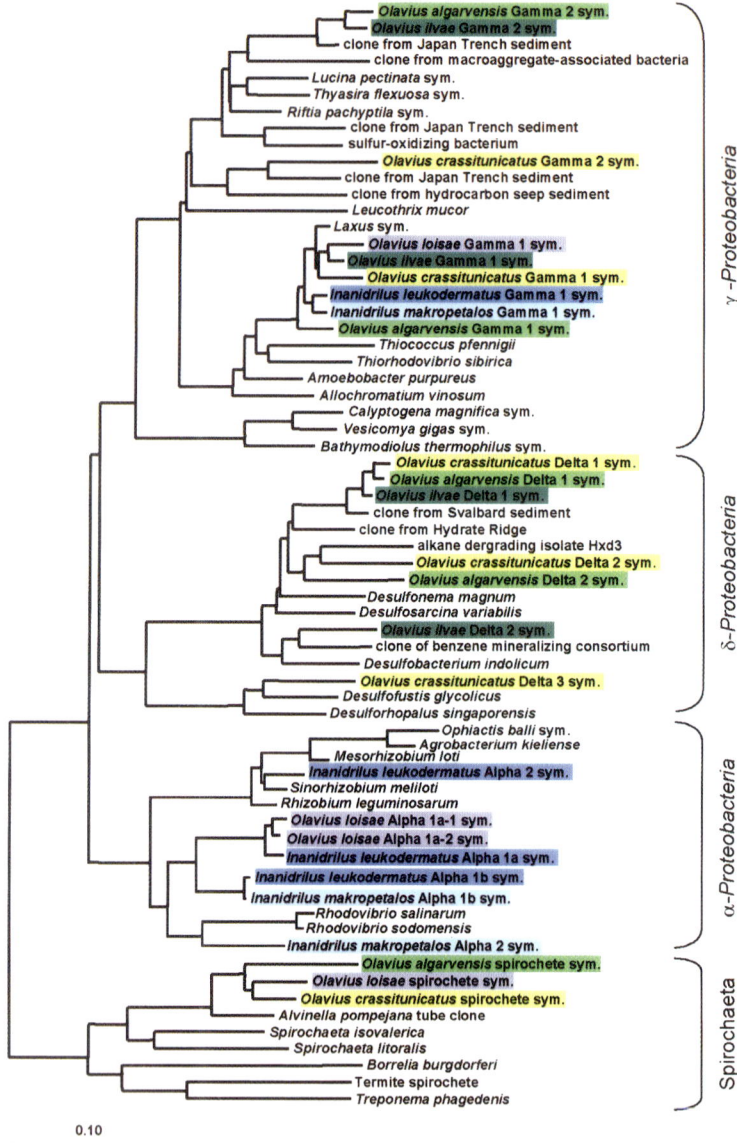

Fig. 2. Phylogenetic placement of the symbiotic bacteria in gutless oligochaetes, based on parsimony analyses of 16S rRNA sequences. The host species are from Bermuda (*I. leukodermatus* in *dark blue*), the Bahamas (*I. makropetalos* in *light blue*), the island of Elba in the Mediterranean (*O. algarvensis* in *light green* and *O. ilvae* in *dark green*), the Australian Great Barrier Reef (*O. loisae* in *purple*), and the continental margin of Peru (*O. crassitunicatus* in *yellow*). As many as six bacterial phylotypes can co-occur in a single host (e.g., *O. crassitunicatus*). *Bar* 10% estimated sequence divergence

deposition of the eggs into the surrounding sediment would offer free-living bacteria from the environment an opportunity to invade the egg. Thus, it is conceivable that some of the symbionts are inherited vertically from the parents, and some horizontally from the environment. Even if all symbiont phylotypes are inherited vertically in recent oligochaetes, the first steps in the development of these multiple symbioses may have originally evolved through horizontal infection of the egg, followed by a successful establishment as a symbiont and transfer to a vertical mode of infection.

5
Molecular Identification and Phylogeny of the Symbionts

5.1
Detection of Multiple Symbiont Phylotypes

The gene of choice for studying the phylogeny of the bacterial symbionts has been the 16S ribosomal RNA (rRNA) gene. Crucial to the study of these hosts has been the full cycle rRNA approach, in which comparative 16S rRNA sequence analysis is coupled with fluorescence in situ hybridization (FISH) (Amann et al. 1995). Oligonucleotide probes of 18–22 bp length are designed based on the 16S rRNA sequences isolated from the hosts, labeled with a fluorescent dye, and hybridized to sections of the worms to ensure that the 16S rRNA sequences originated from the symbionts and not from bacterial contaminants.

In the first phylogenetic study of oligochaete symbionts, on *I. leukodermatus* from Bermuda, DNA was isolated from a pooled sample of 100 worms (Dubilier et al. 1995a). Direct sequencing of PCR products without cloning yielded a single unambiguous 16S rRNA sequence (*I. leukodermatus* Gamma 1 symbiont in Fig. 2). FISH confirmed that this sequence originated from the large, oval-shaped morphotype with sulfur and PHB vesicles (Dubilier et al. 1995a). With the FISH methods used at that time (2 step detection with biotinylated probes and avidin-fluorescein) background fluorescence was too high to distinguish if the smaller symbiont morphotype was labeled by the probe for the Gamma 1 symbiont. We therefore were not able to resolve whether the different bacterial morphotypes represented structural dimorphism of a single bacterial 16S rRNA phylotype or 2 distinct phylotypes.

We now know that all gutless oligochaetes examined (at present nine host species) harbor at least three bacterial 16S rRNA phylotypes, and in some hosts, as many as six phylogenetically distinct symbiont phyloypes can co-occur in the symbiotic region (Blazejak et al. 2005). As our molecular techniques have improved, we have been able to detect a higher number of symbiont phylotypes in the worms, including species such as *I. leukodermatus*, in which we originally found only a single phylotype but have now identified at least four co-occurring bacterial symbionts (Fig. 2) (Blazejak et al., submitted). Important for the detection of multiple phylotypes in a single host is the amplification of the 16S rRNA gene with a low number of PCR cycles to decrease PCR bias (Polz and Cavanaugh 1998). By combining up to 10 parallel PCR reactions (amplified from the same DNA sample), the number of PCR cycles can be decreased to as few as 15 cycles, with a reconditioning step recommended, to decrease heteroduplex formation (Thompson et al. 2002). Furthermore, at least 100 clones per individual should be screened, and at least three individuals per host species examined, to ensure that rare phylotypes are identified. Finally, improved FISH techniques have been crucial for the discovery of multiple symbionts. These include probes directly labeled with fluorescent dyes to decrease background labeling (Dubilier et al. 1999) and enhanced detection techniques such as catalyzed reporter deposition (CARD)-FISH (Pernthaler et al. 2002) for symbionts with low rRNA concentrations (Blazejak et al. 2005). Furthermore, confocal studies using dual and triple hybridizations to compare the signals from specific symbiont probes with those from general bacterial probes have been useful in ensuring that all bacteria in the symbiotic region are identified (Blazejak et al. 2005).

5.2
Phylogeny

The bacterial symbionts of gutless oligochaetes are phylogenetically diverse, and belong to the gamma, delta, and alpha *Proteobacteria* as well as the *Spirochaeta* (Fig. 2). Despite this phylogenetic diversity, the associations are highly specific, with each host species harboring between 3–6 symbiont phylotypes that are specific to the species and are not found in other host species.

5.2.1
Gamma Proteobacterial Symbionts

In all gutless oligochaete species examined to date, the large oval-shaped symbiont morphotype with sulfur and PHB vesicles (see Sect. 4.2) has been identified as a Gamma 1 phylotype (Fig. 2 and 3) that falls within a closely related cluster of sequences in the gamma subdivision of the *Proteobacteria*. The oligochaete Gamma 1 symbionts are most closely related to the ectosymbionts of the marine nematode *Laxus* (Polz et al. 1994), and build a monophyletic group with these in all phylogenetic analyses. Such a close relationship between these bacteria is surprising given their different lifestyles as endo- and ectosymbionts and the large evolutionary distance between oligochaetes and nematodes. This distance rules out cospeciation as an explanation for the monophyly of oligochaete and nematode symbionts. As these two host groups co-occur in similar geographic locations and environments, it is likely that their shared biogeographical distribution influenced the establishment of these symbioses.

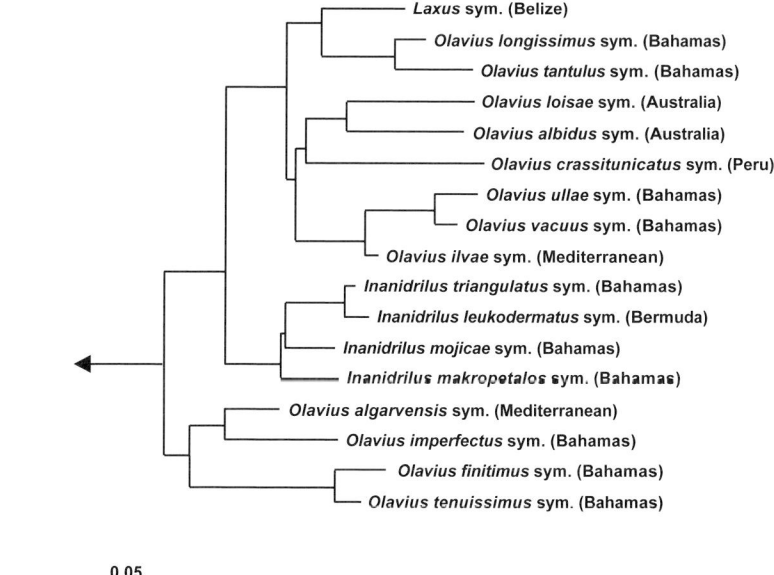

Fig. 3. Phylogenetic tree of the chemoautotrophic, sulfur-oxidizing Gamma 1 symbionts of gutless oligochaetes, based on maximum-likelihood analyses of the 16S rRNA gene. Symbionts from *Inanidrilus* species form a monophyletic group, while those from *Olavius* species fall into two clades that are phylogenetically separate (i.e., paraphyletic). *Bar* 5% estimated sequence divergence

The Gamma 1 oligochaete symbionts are closely related to each other (95.1–99.7% sequence similarity), but their 16S rRNA sequences differ clearly between host species (Fig. 3). The differences between host species are consistent, even between the two most closely related symbionts from *I. leukodermatus* and *I. triangulatus*, and are found in all individuals of the same species. In all phylogenetic analyses (parsimony, distance, and maximum likelihood) the symbionts from hosts belonging to the *Inanidrilus* genus fall in a monophyletic group, while those from *Olavius* hosts fall into two separate clades (Fig. 3). This corresponds well with the general phylogeny of the hosts, with *Inanidrilus* species belonging to a monophyletic and *Olavius* species to a paraphyletic group (see Sect. 2.2). A more detailed comparison of Gamma 1 symbionts and their hosts is currently in progress and will reveal if cospeciation played a role in the establishment of these symbioses. At this point it is clear that the phylogeny of these symbionts cannot be explained solely by their geographic location, as symbionts from co-occurring host species, such as *O. algarvensis* and *O. ilvae* from the Mediterranean island of Elba, are more closely related to symbionts from the Bahamas than to each other (Fig. 3).

In addition to the Gamma 1 bacteria that occur in all gutless oligochaetes, some host species have established associations with a second gamma proteobacterial symbiont, called Gamma 2 symbionts (Fig. 2). These fall into two distinct lineages, one that includes the symbionts from the Mediterranean host species *O. algarvensis* and *O. ilvae* (Rühland et al., in prep.) and one with the Gamma 2 symbiont of *O. crassitunicatus* from Peru that falls on a separate branch (Blazejak et al. 2005). FISH analyses show that in all three host species, these bacteria are much smaller than the Gamma 1 symbionts, and correspond to the small symbiont morphotype described in Sect. 4.2. The closest relatives of these symbionts are clone sequences from cold-seep communities in the Japan Trench (Li et al. 1999), as well as a chemoautotrophic, sulfur-oxidizing bacterium from a shallow hydrothermal vent in the Mediterranean (Sievert et al. 1999).

5.2.2
Delta Proteobacterial Symbionts

Symbionts belonging to the delta *Proteobacteria* were first found in *O. algarvensis* from the Mediterranean (Delta 1 symbiont in Fig. 2) (Dubilier et al. 2001). In addition to this Delta 1 symbiont, we have recently identified a second delta proteobacterial phylotype in *O. algarvensis* (Delta 2 symbiont in Fig. 2) (Rühland et al., in prep.). It is now clear that at least two further host species also harbor delta proteobacterial symbionts, *O. ilvae* from the Mediterranean (Rühland et al., in prep.), and *O. crassitunicatus* from Peru (Blazejak et al. 2005). The Delta 1 symbionts of these three hosts build a closely related monophyletic group, indicating that these symbionts have descended from a single common ancestor. In contrast, the associations with the Delta 2 symbionts, and in *O. crassitunicatus* the Delta 3 symbionts, were established multiple times in convergent evolution, as these symbionts are only distantly related to each other and do not fall in monophyletic groups.

The delta proteobacterial symbionts are small and cocci-shaped in all three host species, but their distribution in the symbiotic region is very different in the Mediterranean and Peruvian hosts (Fig. 4b,c). In *O. algarvensis* and *O. ilvae*, these symbionts are distributed throughout the entire symbiotic region and are in close contact to the Gamma 1 and 2 symbionts. In contrast, the delta proteobacterial symbionts in *O. crassitunicatus* from Peru are concentrated in a thin ring just below the cuticle of the worm and have little contact to the gamma proteobacterial symbionts. This suggests that the delta proteobacterial symbionts of *O. crassitunicatus* are more dependent on substrates from the environment, while those in the Mediterranean hosts may be more important for providing reduced sulfur compounds to the sulfur-oxidizing symbionts (see Sect. 12.6).

5.2.3
Alpha Proteobacterial Symbionts

To date, alpha proteobacterial symbionts have been found in three gutless oligochaete species that are all from similar environments, shallow water coral reef sediments, but from different geographic locations (*I. leukodermatus* from Bermuda, *I. makropetalos* from the Bahamas, and *O. loisae* from the Australian Great Barrier Reef). All three host species harbor symbionts belonging to a clade of relatively closely related (92.6% sequence similarity) bacteria called Alpha 1 symbionts (Fig. 2). Within this clade, different phylotypes can co-occur in the same host species (Fig. 2), e.g., the Alpha 1a and b symbionts of *I. leukodermatus* (Blazejak et al., submitted) and the Alpha 1a-1 and 1a-2 phylotypes in *O. loisae* (Dubilier et al. 1999). In addition to the Alpha 1 clade of symbionts, *I. makropetalos* and *I. leukodermatus* harbor additional lineages of alpha proteobacterial symbionts (Alpha 2 symbionts in Fig. 2) (Blazejak et al., submitted). All alpha proteobacterial symbionts have been identified as the small, rod- to cocci-shaped morphotype described in (12.) 4, and they appear to be evenly distributed throughout the symbiotic region (Blazejak et al., submitted).

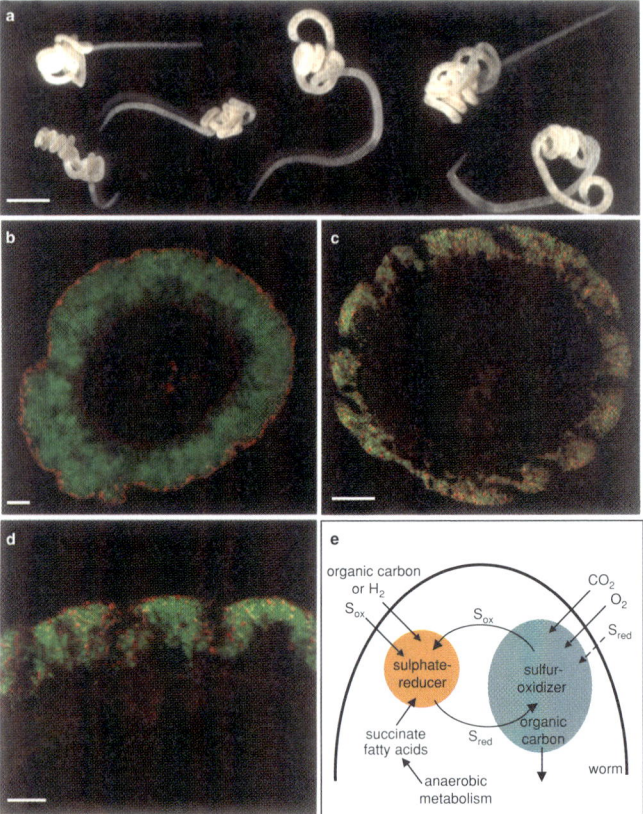

Fig. 4. a Live gutless oligochaetes. Note the characteristic white color of the worms, due to the bacterial symbionts just below the cuticle. The extended front ends of the worms have less symbionts, and are therefore more transparent. **b – d** Fluorescence in situ hybridization (FISH) identification of bacterial symbionts in *O. crassitunicatus* (**b**) and *O. algarvensis* (**c, d**). Dual hybridization with GAM42a and DSS658/DSR651 probes, showing gamma proteobacterial symbionts in *green*, and delta proteobacterial symbionts in *red*. Note the different distribution of the symbionts in the two host species, with the delta proteobacterial symbionts of *O. crassitunicatus* directly below the cuticle, and those of *O. algarvensis* more evenly distributed throughout the symbiotic region. *Bars* b and d 10 µm, c 20 µm. **e** Model of the syntrophic sulfur cycle between the sulfur-oxidizing and sulfate-reducing symbionts. The bacteria exchange oxidized (S_{ox}) and reduced (Sr_{ed}) sulfur compounds such as sulfate and sulfide. The sulfur oxidizer fixes CO_2 into organic compounds which are passed on to the host. In environments with low sulfide concentrations, the dominant flow of electrons is through the sulfate reducer in the form of organic carbon or H_2 if the sulfate reducer is autotrophic. Anaerobic metabolites such as succinate and the fatty acids proprionate and acetate that the worm produces under oxygen-limiting conditions and normally excretes, can be recycled by the sulfate reducer

5.2.4
Spirochete Symbionts

A 16S rRNA phylotype belonging to the spirochetes was first found in *O. loisae* from the Australian Great Barrier Reef (Dubilier et al. 1999). However, it was not possible to show that this phylotype originated from a symbiont until the enhanced detection method CARD-FISH was used on spirochetes in *O. crassitunicatus* from Peru (Blazejak et al. 2005). These spirochete symbionts correspond to the long and thin morphotype described in this host species (Giere and Krieger 2001). In addition to *O. loisae* and *O. crassitunicatus*, spirochete symbionts have also been found in *O. algarvensis* from the Mediterranean (Fig. 2). These symbionts are closely related to each other (95.4%) and form a monophyletic group in all phylogenetic analyses. The close relationship of these symbionts despite the large geographic distances between their hosts and the differences in their habitats (deep-water slope sediments (*O. crassitunicatus*), shallow coral reef sediments (*O. loisae*), and silicate sea-grass sediments (*O. algarvensis*) indicates that the spirochete symbiosis is integral to these oligochaete hosts and independent of geographic or environmental factors.

6
Functional Aspects

The phylogenetic diversity of the oligochaete symbionts is mirrored in their physiological diversity. Before the discovery that gutless oligochaetes harbor multiple symbionts, it was assumed that these associations are driven solely by chemoautotrophy. Evidence for a symbiosis fueled by sulfur-oxidizing, CO_2-fixing bacteria was based on rapid uptake and incorporation of radiolabeled bicarbonate as well as the presence of characteristic enzymes for the fixation of CO_2, such as ribulose-1,5-bisphosphate carboxylase/oxygenase (RubisCO) and the oxidation of sulfur (Felbeck et al. 1983). However, these studies were conducted on whole animal homogenates so that it remained unresolved whether all or only some of the bacteria that coexist in gutless oligochaetes are thiotrophic.

It is now clear that the Gamma 1 symbionts, which occur in all host species and are the dominant symbionts in the worms in terms of biomass, are responsible for most if not all of the observed thiotrophic activity. The thiotrophic nature of these symbionts is suggested by their close evolutionary relationship to a clade of purple sulfur bacteria that includes *Allochromatium vinosum* and *Thiococcus pfennigii* (Fig. 2). Immunocytochemistry analyses with an antiserum directed against Form I RubisCO showed consistent labeling of the Gamma 1 symbionts, and large deposits of sulfur in intracellular globules of these symbionts were shown using electron microscopic spectroscopy (Krieger 2000; Dubilier et al. 2001). It is assumed that the Gamma 1 symbionts provide the oligochaetes with a source of nutrition, either through regular transfer of carbon compounds, or through their lysis and digestion in host vacuoles (Giere and Langheld 1987). However, this has not been confirmed experimentally.

The physiological nature of the Gamma 2 symbionts is not currently known. They do not appear to be essential to the hosts, as they only occur in some, but not in all species (Fig. 2). The Gamma 2 symbionts are related to phylogenetic groups that include bacteria from cold seeps and a thiotrophic vent isolate (Sect. 5.2.1), suggesting that these symbionts might also participate in chemosynthethic pathways. It is not clear if the Gamma 2 bacteria express RubisCO, as the immunocytochemistry analyses described above were conducted prior to the discovery of these symbionts. We are currently using immunofluorescence studies with antisera against Form I and II RubisCO in combination with 16S rRNA FISH to understand more about the metabolism of these symbionts.

The metabolism of the alpha proteobacterial symbionts is another as yet unsolved puzzle. The Alpha 1 symbionts are most closely related to *Rhodovibrio salinarum* and *R. sodomensis*, halophilic bacteria that are photoheterotrophic under anoxic conditions but can also grow in the dark heterotrophically under aerobic conditions. It is intriguing that in all gutless oligochaetes, either Alpha 1 or Delta 1 symbionts coexist with the Gamma 1 symbionts, and it is tempting to speculate that the Alpha 1 symbionts might also play a role in recycling of anaerobic waste products of the worms as suggested for the delta proteobacterial symbionts (see below).

The Alpha 2 symbiont of *I. leukodermatus* is most closely related to nitrogen-fixing symbionts of leguminous plants (Fig. 2), indicating that these symbionts might also fix N_2. However, numerous attempts to amplify a gene characteristic of nitrogen-fixing bacteria, the *nif* H gene (Zehr and Ward 2002), were unsuccessful (N. Dubilier and J.P. Zehr, unpubl. data). Since this symbiont is only present in *I. leukodermatus*, it does not appear to be essential for nitrogen uptake, as other oligochaete hosts are clearly able to acquire nitrogen without Alpha 2 symbionts.

The spirochete symbionts fall on a neighboring branch with a sequence from the tubes of the hydrothermal vent polychaete, *Alvinella pompejana*, obtained from an enrichment culture grown on a very rich medium (M.A.Cambon-Bonavita, unpubl. data). The free-living, marine spirochetes *Spirochaeta isovalerica* and *S. litoralis* consistently form a neighboring clade of the oligochaete spirochetes. These bacteria were isolated from sulfidic muddy sediments and are obligate anaerobes that ferment carbohydrates mainly to acetate, ethanol, CO_2, and H_2 (Hespell and Canale-Parola 1973; Harwood and Canale-Parola 1983). While fermentation is one possible metabolic pathway of the oligochaete spirochetes, they could also have a completely different metabolism, just as the spirochete symbionts in termites do not possess properties common to their closest free-living relatives within the genera *Treponema*. Instead, termite spirochetes were recently discovered to be chemoautotrophic, using H_2 and CO_2 to produce acetate (Leadbetter et al. 1999), and were also shown to be able to fix nitrogen (Lilburn et al. 2001). These types of metabolism would clearly be beneficial to the oligochaete hosts, providing them with additional sources of reduced carbon and nitrogen.

In contrast to the uncertainties in the function of the alpha proteobacterial and spirochete symbionts, the metabolism of the delta proteobacterial symbionts is clear. These symbionts are sulfate-reducing bacteria based on their close phylogenetic relationship to free-living sulfate-reducers, the presence of a gene (*dsr*AB) encoding an enzyme used for dissimilatory sulfate reduction, and the detection of sulfate reduction in the worms at rates comparable to those of free-living sulfate reducers (Dubilier et al. 2001). The coexistence of sulfate-reducing and sulfide-oxidizing bacteria as endosymbionts in oligochaete hosts suggests that these are engaged in a syntrophic sulfur cycle in which oxidized and reduced sulfur compounds are cycled between the two symbionts (Fig. 4e). The sulfate reducer produces reduced sulfur compounds (sulfide or sulfur) using organic carbon (or H_2 if it is autotrophic) from the environment. The reduced sulfur compounds are used by the sulfide-oxidizer as an electron donor for the autotrophic fixation of CO_2.

This syntrophic association provides several benefits for the hosts. The cycling of reduced and oxidized sulfur compounds between the symbionts increases their energy yields, as shown for continuous co-cultures of sulfate-reducing and sulfide-oxidizing bacteria (van den Ende et al. 1997). Another advantage for the worm is that the sulfate-reducers could take up anaerobic metabolites from the worm such as succinate and fatty acids that are produced when oxygen becomes limiting and are normally excreted. This would allow the worms to recycle these energy rich fermentation products. Finally, the presence of an internal sulfide producer allows these hosts to colonize habitats in which sulfide is present at very low concentrations as in Elba, or only intermittently available, or even completely absent. However, the presence of sulfate-reducing symbionts in *O. crassitunicatus* that occurs in sediments that appear to be well supplied with sulfide (see Sect. 3.1), shows that the role of these bacteria is not restricted to supplying sulfide. For these worms, the benefits of internal sulfur cycling and the reuse of fermentation products appears to provide a sufficient selective advantage.

7
Outlook

The field of symbiosis has entered an incredibly exciting period because new techniques, in particular in molecular biology, are providing us with a wealth of novel tools to study symbiotic associations. These tools are particularly useful for studying associations in which the partners cannot be separated from each other. As in many other obligate symbioses, the bacterial symbionts of gutless oligochaetes remain as yet uncultivable despite numerous attempts by the authors and others. The lack of a gut or mouth in the oligochaete worms makes it highly unlikely that these hosts can be cultured aposymbiotically (free of symbionts).

Polymerase chain reaction (PCR) approaches have been extremely valuable in the past for assessing the diversity of uncultivable organisms such as bacterial symbionts. However, the study of functional genes has been slow using PCR based methods. There are only a limited number of primers available for genes of interest, and the development of new primers that can successfully amplify a functional gene of unknown sequence in the target organisms takes time. Finally, the inability to amplify a given gene does not necessarily mean that the gene is not present. It is therefore particularly exhilarating to experience the breakthrough of an era in which high throughput DNA sequencing and computational genomics can provide information on entire genomes and help discover novel proteins and metabolic strategies previously not even conceived to play a role in the organisms of interest (DeLong 2004).

We have recently begun analyzing the genomes of the *O. algarvensis* symbionts using bacterial artificial chromosome (BAC) libraries that we are end-sequencing and screening for functional genes. This approach has led to the identification of genes encoding enzymes involved in the oxidation of reduced inorganic sulfur compounds (*dsr*ABFKNO, *apr*AB, *sox*B, etc.), fixation of CO_2 (Form I RubisCO genes), and the dissimilatory reduction of nitrate (*nir*K, *nap*AGF), as well as the discovery of genes involved in antimicrobial processes (*nor*M, *acr*BDF) and virulence (*vap*BC, *pil*T).

A complete analysis of the genomes of the 5 coexisting symbionts in *O. algarvensis* is currently in progress in collaboration with the Joint Genome Institute in California, USA (www.jgi.doe.Gov/sequencing/cspseqplans.html). This project will not only offer the opportunity to identify the genomic differences and similarities of the multiple symbionts and how they interact with their host, but also to examine the metabolic and genetic interactions within the symbiotic bacterial consortium as well as the interactions of these symbionts with the environment.

Despite the invaluable contribution that molecular methods and genomics have made in advancing our knowledge of symbiotic associations, these methods can clearly not give us all the answers. It is essential that the information we gain through studying the genes involved in these symbioses is used to return in a full cycle to morphological, physiological, and ecological analyses to investigate under which conditions genes are expressed and how their expression affects the biology and evolution of the symbiosis. Only such an integrated and interdisciplinary approach can provide a full understanding of the interactions between these prokaryotic symbionts and their eukaryotic hosts.

Acknowledgements. We are grateful to Silke Wetzel for her excellent technical work, to Mirjam Perner and Christina Zaluski for contributions made to this research as part of their Diplom Thesis, and to Rudolf Amann for his continuous enthusiasm for and financial support of this research. We are indebted to Olav Giere and Christer Erséus for their collaboration in the "Gutless Oligochaete Project" and many fruitful discussions. This work was supported by the Max Planck Society and the German Research Foundation.

References

Amann RI, Ludwig W, Schleifer K-H (1995) Phylogenetic identification and in situ detection of individual microbial cells without cultivation. Microbiol Rev 59:143–169

Blazejak A, Erséus C, Amann R, Dubilier N (2005) Coexistence of bacterial sulfide-oxidizers, sulfate-reducers, and spirochetes in a gutless worm (Oligochaeta) from the Peru margin. Appl Environ Microbiol 71:1553–1561

Blazejak A, Kuever J, Amann R, Dubilier N Phylogeny of 16S rRNA, RubisCO, and APS reductase genes from gamma and alpha proteobacterial symbionts in gutless marine worms (Oligochaeta) from Bermuda and Bahamas (submitted)

Bright M, Giere O (2005) Microbial symbiosis in Annelida. Symbiosis 38:1–45
Cavanaugh CM, Gardiner SL, Jones ML, Jannasch HW, Waterbury JB (1981) Prokaryotic cells in the hydrothermal vent tube worm *Riftia pachyptila* Jones: possible chemoautotrophic symbionts. Science 213:340–342
Davis D (1985) The Oligochaeta of Georges Bank. Proc Biol Soc Wash 98:158–176
DeLong EF (2004) Microbial population genomics and ecology: the road ahead. Environ Microbiol 6:875–878
Dubilier N, Giere O, Distel DL, Cavanaugh CM (1995a) Characterization of chemoautotrophic bacterial symbionts in a gutless marine worm (Oligochaeta, Annelida) by phylogenetic 16S rRNA sequence analysis and in situ hybridization. Appl Environ Microbiol 61:2346–2350
Dubilier N, Giere O, Grieshaber MK (1995b) Morphological and ecophysiological adaptations of the marine oligochaete *Tubificoides benedii* to sulfidic sediments. Amer Zool 35:163–173
Dubilier N, Amann R, Erséus C, Muyzer G, Park S, Giere O, Cavanaugh CM (1999) Phylogenetic diversity of bacterial endosymbionts in the gutless marine oligochaete *Olavius loisae* (Annelida). Mar Ecol Prog Ser 178:271–280
Dubilier N, Mülders C, Ferdelman T, de Beer D, Pernthaler A, Klein M, Wagner M, Erséus C, Thiermann F, Krieger J, Giere O, Amann R (2001) Endosymbiotic sulphate-reducing and sulphide-oxidizing bacteria in an oligochaete worm. Nature 411:298–302
Erséus C (1979a) *Inanidrilus bulbosus* gen. et sp. n., a marine tubificid (Oligochaeta) from Florida, USA. Zool Scr 8:209–210
Erséus C (1979b) Taxonomic revision of the marine genus *Phallodrilus* Pierantoni (Oligochaeta, Tubificidae), with descriptions of thirteen new species. Zool Scr 8:187–208
Erséus C (1984) Taxonomy and phylogeny of the gutless Phallodrilinae (Oligochaeta, Tubificidae), with descriptions of one new genus and twenty-two new species. Zool Scr 13:239–272
Erséus C (1990) The marine Tubificidae (Oliogchaeta) of the barrier reef ecosystems at Carrie Bow Cay, Belize, and other parts of the Caribbean Sea, with descriptions of twenty-seven new species and revision of *Heterodrilus*, *Thalassodrilides* and *Smithsonidrilus*. Zool Scr 19:243–303
Erséus C (1991) Two new deep-water species of the gutless genus *Olavius* (Oligochaeta: Tubificidae) from both sides of North America. Proc Biol Soc Wash 104:627–630
Erséus C (1992) A generic revision of the Phallodrilinae (Oligochaeta, Tubificidae). Zool Scr 21:5–48
Erséus C (2003) The gutless Tubificidae (Annelida: Oligochaeta) of the Bahamas. Meiofauna Marina 12:59–84
Erséus C, Prestegaard T, Källersjö M (2000) Phylogenetic analysis of Tubificidae (Annelida, Clitellata) based on 18S rDNA sequences. Mol Phylogen Evol 15:381–389
Falter JL, Sansone FJ (2000) Hydraulic control of porewater geochemistry within the oxic-suboxic zone of a permeable sediment. Limnol Oceanogr 45:550–557
Felbeck H (1981) Chemoautotrophic potential of the hydrothermal vent tube worm *Riftia pachyptila* Jones (Vestimentifera). Science 213:336–338

Felbeck H, Liebezeit G, Dawson R, Giere O (1983) CO_2 fixation in tissues of marine oligochaetes (*Phallodrilus leukodermatus* and *P. planus*) containing symbiotic, chemoautotrophic bacteria. Mar Biol 75:187–191

Finegenova N (1986) Six new species of marine Tubificidae (Oligochaeta) from the continental shelf off Peru. Zool Scr 15:45–51

Giere O (1979) Studies on marine Oligochaeta from Bermuda, with emphasis on new *Phallodrilus* species (Tubificidae). Cah Biol Mar 20:301–314

Giere O (1981) The gutless marine oligochaete *Phallodrilus leukodermatus*. Structural studies on an aberrant tubificid associated with bacteria. Mar Ecol Prog Ser 5:353–357

Giere O (1985) The gutless marine tubificid *Phallodrilus planus*, a flattened oligochaete with symbiotic bacteria. Results from morphologcial and ecological studies. Zool Scr 14:279–286

Giere O, Erséus C (2002) Taxonomy and new bacterial symbioses of gutless marine Tubificidae (Annelida, Oligochaeta) from the Island of Elba (Italy). Org Divers Evol 2:289–297

Giere O, Krieger J (2001) A triple bacterial endsymbiosis in a gutless oligochaete (Annelida): ultrastructural and immunocytochemical evidence. Invert Biol 120:41–49

Giere O, Langheld C (1987) Structural organisation, transfer and biological fate of endosymbiotic bacteria in gutless oligochaetes. Mar Biol 93:641–650

Giere O, Liebezeit G, Dawson R (1982) Habitat conditions and distribution pattern of the gutless oligochaete *Phallodrilus leukodermatus*. Mar Ecol Prog Ser 8:291–299

Giere O, Wirsen CO, Schmidt C, Jannasch HW (1988) Contrasting effects of sulfide and thiosulfate on symbiotic CO_2-assimilation of *Phallodrilus leukodermatus* (Annelida). Mar Biol 97:413–419

Giere O, Conway NM, Gastrock G, Schmidt C (1991) 'Regulation' of gutless annelid ecology by endosymbiotic bacteria. Mar Ecol Prog Ser 68:287–299

Giere O, Nieser C, Windoffer R, Erséus C (1995) A comparative structural study on bacterial symbioses of Caribbean gutless Tubificidae (Annelida, Oligochaeta). Acta Zool 76:281–290

Grieshaber MK, Hardewig I, Kreutzer U, Schneider A, Völkel S (1992) Hypoxia and sulfide tolerance in some marine invertebrates. Verh Dtsch Zool Ges 85(2):55–76

Harwood CS, Canale-Parola E (1983) *Spirochaeta isovalerica* sp. nov., a marine anaerobe that forms branched-chain fatty acids as fermentation products. Int J Syst Bacteriol 33:573–579

Hespell RB, Canale-Parola E (1973) Glucose and pyruvate metabolism of *Spirochaeta litoralis*, an anaerobic marine spirochete. J Bacteriol 116:931–937

Jamieson BGM (1977) Marine meiobenthic Oligochaeta from Heron and Wistari reefs (Great Barrier Reef) of the genera *Clitellio*, *Limnodriloides* and *Phallodrilus* (Tubificidae) and *Grania* (Enchytraeidae). Zool J Linnean Soc 61:329–349

Johnson KS, Childress JJ, Hessler RR, Sakamoto-Arnold CM, Beehler CL (1988) Chemical and biological interactions in the Rose Garden hydrothermal vent field, Galapagos spreading center. Deep Sea Res 35:1723–1744

Krieger J (2000) Funktion und Übertragung endosymbiontischer Bakterien bei bakteriensymbiontischen, darmlosen marine Oligochaeten. (Function and transmission of endosymbiotic bacteria in gutless marine oligochaetes.) PhD, Dept. of Biology, University of Hamburg, Germany.

Le Bris N, Sarradin P-M, Caprais J-C (2003) Contrasted sulphide chemistries in the environment of 13°N EPR vent fauna. Deep Sea Res I 50:737–747

Leadbetter JR, Schmidt TM, Graber JR, Breznak JA (1999) Acetogenesis from H_2 plus CO_2 by spirochetes from termite guts. Science 283:686–689

Levin L, Gutièrrez D, Rathburn A, Neira C, Sellanes J, Muñoz P, Gallardo V, Salamanca M (2002) Benthic processes on the Peru margin: a transect across the oxygen minimum zone during the 1997–98 El Niño. Prog Oceanogr 53:1–27

Levin LA, Rathburn AE, Gutiérrez D, Muñoz P, Shankle A (2003) Bioturbation by symbiont-bearing annelids in near-anoxic sediments: implications for biofacies models and paleo-oxygen assessments. Palaeogeogr Palaeoclimatol Palaeoecol 199:129–140

Li L, Kato C, Horikoshi K (1999) Microbial diversity in sediments collected from the deepest cold-seep area, the Japan trench. Mar Biotechnol 1:391–400

Liebezeit G, Felbeck H, Dawson R, Giere O (1983) Transepidermal uptake of dissolved carbohydrates by the gutless marine oligochaete *Phallodrilus leukodermatus* (Annelida). Oceanis 9:205–211

Lilburn TG, Kim KS, Ostrom NE, Byzek KR, Leadbetter JR, Breznak JA (2001) Nitrogen fixation by symbiotic and free-living spirochetes. Science 292:2495–2498

Nylander J, Erséus C, Kallersjo M (1999) A test of monophyly of the gutless Phallodrilinae (Oligochaeta, Tubificidae) and the use of a 573-bp region of the mitochondrial cytochrome oxidase I gene in analysis of annelid phylogeny. Zool Scr 28:305–313

Perner M (2003) Biogeochemische und mikrobiologische Charakterisierung mariner Sedimente vor Elba - ein Beitrag zur ökosystemaren Analyse bakteriensymbiontischer Oligochaeten. (Biogeochemical and microbiological characterisation of marine sediments from Elba - a contribution to an ecosystemic analysis of oligochaetes with bacterial symbionts). Master's Thesis. Biology Dept., University of Hamburg

Pernthaler A, Pernthaler J, Amann R (2002) Fluorescence in situ hybridization and catalyzed reporter deposition for the identification of marine bacteria. Appl Environ Microbiol 68:3094–3131

Polz MF, Cavanaugh CM (1998) Bias in template-to-product ratios in multitemplate PCR. Appl Environ Microbiol 64:3724–3730

Polz MF, Distel DL, Zarda B, Amann R, Felbeck H, Ott JA, Cavanaugh CM (1994) Phylogenetic analysis of a highly specific association between ectosymbiotic, sulfur-oxidizing bacteria and a marine nematode. Appl Environ Microbiol 60:4461–4467

Richards KS, Fleming TP, Jamieson BGM (1982) An ultrastructural study of the distal epidermis and the occurrence of subcuticular bacteria in the gutless tubificid *Phallodrilus albidus* (Oligochaeta: Annelida). Aust J Zool 30:327–336

Sievert SM, Muyzer G, Küver J (1999) Novel sulfur-oxidizing bacteria from a shallow submarine hydrothermal vent, most closely related to obligate symbionts of invertebrates. PhD, University of Bremen, Germany

Southward AJ, Southward EC (1980) The significance of dissolved organic compounds in the nutrition of *Siboglinum ekmani* and other small species of Pogonophora. J Mar Biol Assoc UK 60:1005–10034

Thompson JR, Marcelino LA, Polz MF (2002) Heteroduplexes in mixed-template amplifications: formation, consequence and elimination by 'reconditioning PCR'. Nucleic Acid Res 30:2083–2088

Van den Ende FP, Meier J, van Gemerden H (1997) Syntrophic growth of sulfate-reducing bacteria and colorless sulfur bacteria during oxygen limitation. FEMS Microbiol Ecol 23:65–80

Zehr JP, Ward BB (2002) Nitrogen cycling in the ocean: new perspectives on processes and paradigms. Appl Environ Microbiol 68:1015–1024

Roles of Bacterial Regulators in the Symbiosis between *Vibrio fischeri* and *Euprymna scolopes*

Kati Geszvain, Karen L. Visick

1
Introduction

In a symbiosis, two or more evolutionarily distinct organisms communicate with one another in order to co-exist and co-adapt in their shared environment. The mutualistic symbiosis between the bioluminescent marine bacterium *Vibrio fischeri* and the Hawaiian squid *Euprymna scolopes* provides a model system that allows scientists to examine the mechanisms by which this communication occurs (McFall-Ngai and Ruby 1991). The squid, although *V. fischeri*-free (aposymbiotic) at hatching, rapidly acquires this bacterium and promotes its growth in a special symbiotic organ called the light organ (LO). In exchange for nutrients and a niche safe from competing bacteria, *V. fischeri* provides the bioluminescence used by *E. scolopes* to camouflage itself from predators.

In this chapter, we will give an overview of the early events in establishing the symbiosis and describe associated developmental changes triggered in each organism by the interaction. We will then discuss bacterial regulators and, where known, the traits they control that are necessary for a productive interaction between *V. fischeri* and *E. scolopes*. Finally, we will conclude by highlighting important directions for future investigation.

2
Early Events in the *Euprymna scolopes* – *Vibrio fischeri* Symbiosis

2.1
Vibrio fischeri strains are specifically recruited from the seawater

V. fischeri comprises less than 0.1% of the total bacterial population in the seawater inhabited by the squid (Lee and Ruby 1992), yet this organism alone

K. Geszvain, K. Visick (e-mail: kvisick@lumc.edu)
Dept. Microbiology and Immunology, Loyola University Chicago, 2160 S. First Ave. Bldg. 105, Maywood, IL 60153, USA

is found in the light organ association (Boettcher and Ruby 1990). Furthermore, inoculation in the laboratory with bacteria closely related to *V. fischeri*, including *V. harveyi* and *V. parahaemolyticus*, fails to result in colonization (McFall-Ngai and Ruby 1991; Nyholm et al. 2000). In addition to this species-specific selection, strain-specific enrichment also occurs. Both visibly luminescent and non-visibly luminescent strains of *V. fischeri* co-exist in the seawater, but only the latter strains colonize the squid LO in nature (Lee and Ruby 1994b). This strict limitation on the species and strains of bacteria capable of colonizing the LO suggests that a specific exchange of signals must occur between the squid and the bacteria early during colonization.

Within hours of hatching, *E. scolopes* recruits *V. fischeri* from the surrounding seawater. The presence of bacteria or the bacterial cell wall component peptidoglycan in the seawater causes the squid to secrete mucus (Nyholm et al. 2002), allowing *V. fischeri* cells to aggregate near pores leading into the LO (Fig. 1). Other bacteria such as *V. parahaemolyticus* also exhibit the ability to aggregate in squid mucus, suggesting that *E. scolopes* does not distinguish between *V. fischeri* and other Gram negative bacteria at this stage (Nyholm et al. 2000). However, when both *V. parahaemolyticus* and *V. fischeri* are present, the latter organism becomes the dominant species in the aggregate (Nyholm and McFall-Ngai 2003), indicating that *V. fischeri* may participate in establishing specificity at this stage.

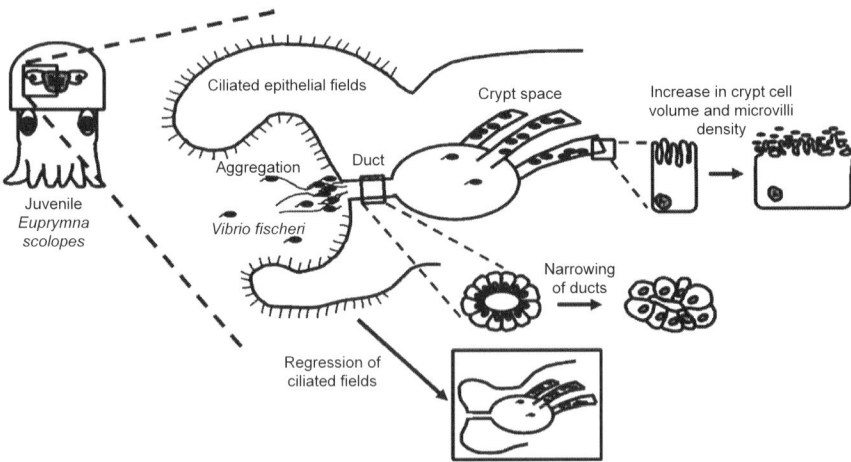

Fig. 1. Cartoon depicting the structure of and developmental changes in the juvenile squid LO during colonization. The position of the LO in a juvenile squid is shown on the *left*, while an enlarged cross section is shown on the *right*. The juvenile LO contains three pores on each side (six total), only one of which is depicted at the opening of the duct. *Arrows* indicate developmental events that occur within the first 4 days after exposure to *V. fischeri*. *Dashed lines* indicate an enlargement of the boxed area. *V. fischeri* cells are shown as *black ovals* aggregated in the mucus (depicted as *wavy lines*) outside the pore and in the crypt spaces (without flagella). This depiction of the light organ is based on Visick and McFall-Ngai (2000) and references described therein.

2.2
Vibrio fischeri cells navigate physical and chemical barriers to colonize *Euprymna scolopes*

From the aggregates, the *V. fischeri* cells migrate through LO pores, reaching ducts that ultimately lead into crypts, the sites of colonization (Fig. 1). In the ducts, the bacteria must move through mucus against an outward current generated by ciliated cells lining the passageway (McFall-Ngai and Ruby 1998). As a further barrier to colonization, the ducts contain high levels of nitric oxide, an anti-microbial agent that may function as a layer of defense against invasion by non-specific bacteria (Davidson et al. 2004). In the crypts, *V. fischeri* cells may encounter macrophage-like cells, a potential immune surveillance system (Nyholm and McFall-Ngai 1998). In addition, the bacteria may be exposed to toxic oxygen radicals such as hypohalous acid, produced by a halide peroxidase enzyme secreted by epithelial cells within the crypts (Weis et al. 1996; Small and McFall-Ngai 1999). Despite this plethora of potential host defenses, *V. fischeri* cells can enter the LO and grow to high cell density, approximately 10^{11} cells/cm^3 (Visick and McFall-Ngai 2000). Thus, *V. fischeri* must possess mechanisms by which it can evade host defenses and thrive in the LO environment.

Growth to high cell density does not represent the endpoint of the symbiosis. Rather, the symbiosis is dynamic. Each morning the squid expels between 90 and 95% of the bacterial population from its LO (Lee and Ruby 1994a). During the day, the *V. fischeri* cells retained in the squid divide to repopulate the LO. Therefore, persistent colonization actually consists of cycles of expulsion and re-growth, requiring the symbiotic bacteria to adapt to changing environments within the LO.

2.3
Both Organisms Undergo Developmental Changes in Response to the Symbiosis

The interaction between *E. scolopes* and *V. fischeri* induces a number of developmental and morphological changes in each organism (Fig. 1). Ciliated epithelial cells, present on a field that projects outward from the LO, likely function to facilitate recruitment of *V. fischeri* by drawing the bacteria-laden seawater into the mucus matrix near the LO pores. Once the symbiont has successfully migrated into the LO, apoptosis and subsequently regression of these ciliated fields results in their loss over the course of four days (Montgomery and McFall-Ngai 1994; Foster and McFall-Ngai 1998). The consequence, presumably, is a reduction in any further recruitment of additional symbiotic bacteria.

A bacterial signal that triggers some of the developmental changes in the epithelial fields is the bacterial cell wall component lipopolysaccharide (LPS). Purified LPS is sufficient to induce apoptosis in the fields (Foster et al. 2000). Most likely, the highly conserved lipid A portion of LPS is responsible, as LPS purified from many species of Gram negative bacteria can induce apoptosis. Possibly, an LPS detection pathway similar to the Toll-like receptor pathway found in many organisms (Gerard 1998) recognizes bacterial LPS and triggers apoptosis.

The LPS signal, however, is not sufficient to trigger regression of the epithelial fields; this suggests that more than one signal is required for this developmental change (Foster et al. 2000). *V. fischeri* strains that do not enter the LO fail to induce regression, suggesting signaling occurs between the bacteria and squid cells in the LO (Doino and McFall-Ngai 1995). Although regression requires a bacterial signal, the program continues regardless of the presence of bacteria: removing *V. fischeri* with antibiotic treatment after 12 h does not stop or reverse regression (Doino and McFall-Ngai 1995).

Another developmental event in the squid may also function to reduce LO accessibility. Within 12 h after symbiotic colonization, a two- to three-fold increase in actin levels occurs within the apical surface of epithelial cells lining the LO ducts (Kimbell and McFall-Ngai 2004). This increase in actin is correlated with a narrowing of the ducts, which decrease in size two-fold (Fig. 1). The narrowing of the ducts, along with the loss of the ciliated fields on the LO surface, likely limits entry into the LO. However, the LO remains at least somewhat open to the environment, as marked bacteria introduced into the seawater can subsequently be isolated from the adult LO (Lee and Ruby 1994b). Because *V. fischeri* remains the only bacterial resident, mechanisms must remain in place to prevent other species from infecting the LO.

Changes also occur in crypt epithelial cells immediately adjacent to the colonizing *V. fischeri* bacteria. Within 72 h of symbiotic colonization, these cells increase in volume as they develop from columnar to cuboidal cells (Fig. 1) (Montgomery and McFall-Ngai 1994). Concurrently, the microvilli on their surfaces increase in density and complexity (Lamarcq and McFall-Ngai 1998). These alterations likely increase the surface area available for interactions with the symbiotic bacteria. These structural changes require the persistent presence of bacteria (Doino 1998; Lamarcq and McFall-Ngai 1998), suggesting that a continual signal exchange occurs between the squid and bacteria throughout the symbiosis.

V. fischeri cells also undergo developmental changes upon colonization of the LO. Planktonic *V. fischeri* are flagellated and motile, traits that are essential for the bacteria to enter the LO (Graf et al. 1994; Millikan and Ruby 2003). Within 24 h of colonization, however, most of the bacteria lose their flagella and become non-motile. The cells re-grow flagella and regain motility shortly after expulsion from the LO (Ruby and Asato 1993). The bacteria also decrease in size in the LO and, after attaining high cell density, induce light production (Ruby and Asato 1993) to a level 100-fold higher than in culture (Boettcher and Ruby 1990; Stabb et al. 2004). This luminescence is essential for persistent infection (Visick et al. 2000). Thus, in addition to signaling *E. scolopes* to induce developmental changes during the onset of symbiosis, *V. fischeri* also recognizes signals within the LO environment and adapts accordingly.

3 Regulatory Systems Employed by *Vibrio fischeri* to Promote the Symbiosis

3.1 Two-Component Signal Transduction Systems

Many bacteria, including *V. fischeri*, recognize and respond to their environments using two-component regulatory systems (Fig. 2A, reviewed in Stock et al. 2000). These systems are composed of a sensor histidine kinase

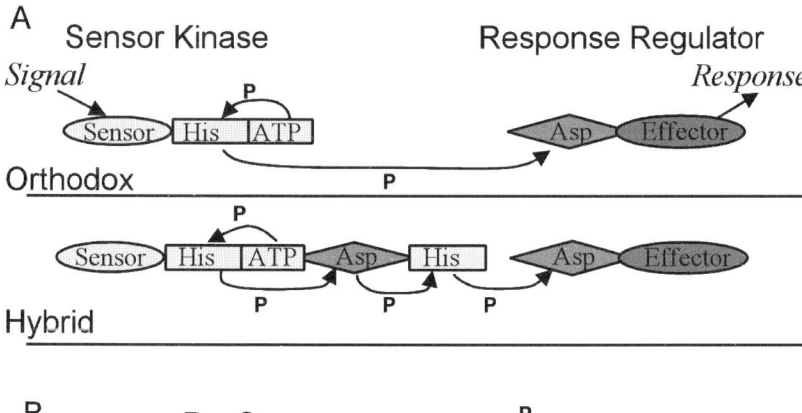

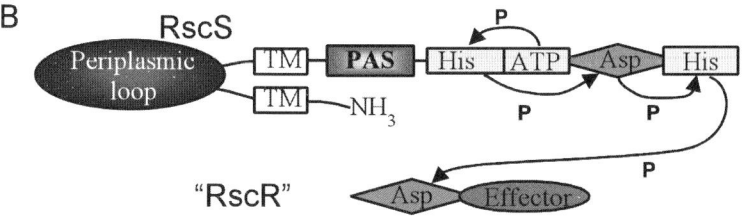

Fig. 2. Two-component regulatory systems. A. The phospho-relay in orthodox (*top*) and hybrid (*bottom*) two-component systems. Upon detection of signal, phosphate generated from a bound ATP is passed from conserved His to Asp residues until finally being transferred to an Asp in the response regulator, resulting in a response, either altered transcription or protein function. B. RscS is a hybrid sensor kinase. The sensor domain of RscS is composed of two transmembrane helices (*TM*), a large periplasmic loop and a PAS domain

protein that recognizes and transmits an environmental signal (through autophosphorylation on a His residue) to a second protein, the response regulator, which (when phosphorylated on a conserved Asp residue) carries out a response. Most frequently, the response consists of a change in gene expression; alternatively, changes in protein activity can result.

Given the changes in environmental conditions that *V. fischeri* cells experience as they travel from seawater into the LO, it is not surprising that colonization by *V. fischeri* requires two-component regulators. At least two such regulators are required for efficient initiation of symbiotic colonization: the sensor kinase RscS (Visick and Skoufos 2001) and the response regulator, GacA (Whistler and Ruby 2003). A transcriptional regulator, FlrA, which exhibits limited similarity to response regulators and is required for initiation (Millikan and Ruby 2003) will also be discussed here.

rscS. Mutations in *rscS* severely reduce the ability of *V. fischeri* to initiate symbiotic colonization: most animals remain uncolonized following exposure to *rscS* mutants, although other animals become colonized after a delay of several hours (Visick and Skoufos 2001). These results suggest that mutants are blocked at an early stage of colonization, but that they can occasionally by-pass this block and ultimately achieve what appears to be normal colonization. In culture, *rscS* mutants do not exhibit defects in growth, motility, or the timing and level of bioluminescence induction, traits known to be important for colonization (Visick and Skoufos 2001). Thus, to date no clues to *rscS* function have been garnered by phenotypes observed in culture.

The sequence of *rscS* suggests that it encodes a hybrid sensor kinase similar to ArcB and BvgS (Fig. 2B) (Visick and Skoufos 2001). These proteins contain, in addition to the conserved His residue that serves as the site of autophosphorylation, two additional domains with conserved residues (Asp and His) predicted to be sequentially phosphorylated and that may serve as sites of additional regulation (Fig. 2B). Upon receipt of a colonization signal, RscS is predicted to autophosphorylate and transfer the phosphate to an as-yet-unidentified response regulator, termed RscR, which may regulate genes or activities essential for symbiosis.

What serves as the colonization signal, and how is it detected by RscS? Clearly, many possibilities exist, and include in addition to bacterially produced molecules and seawater signals, components of the LO mucus, cell surface signals and nutrients. Determining the portion of RscS responsible for detecting the colonization signal will advance our understanding of symbiotic signal exchange. In many cases, the amino terminal periplasmic portion of sensor kinase proteins receives the environmental signal (Stock et al. 2000). For example, *Salmonella* PhoQ detects Mg^{2+} in the environment through its periplasmic domain; binding of Mg^{2+} to this domain results in a conformational change and inactivation of the response regulator PhoP (Vescovi et al. 1997). RscS is predicted to possess a periplasmic domain of ~200 residues (Visick and Skoufos 2001); the large size of this region suggests it may play a role in RscS function, possibly signal detection.

In addition to a potential periplasmic signaling domain, RscS contains a second input domain, known as PAS. In other proteins, PAS detects signals such as small ligands, or changes in light levels, oxygen concentration or redox potential (Taylor and Zhulin 1999). Whether the PAS domain contributes to signal detection by RscS during colonization remains unknown. However, the transition from seawater to the nutrient-rich LO could plausibly affect the energy status of the *V. fischeri* cells thereby altering their redox potential or oxygen concentration, which could be sensed by the PAS domain. Thus, investigations of the PAS and periplasmic domains of RscS will be fruitful for exploring bacteria-host interactions. Perhaps each domain detects a distinct condition, allowing RscS to integrate multiple signals from the squid environment to regulate the initiation of colonization.

What is the identity of the cognate response regulator, RscR, and what genes or proteins are controlled by the RscS/R regulatory system? In many cases, the genes for sensor kinases and their cognate response regulators are linked on the chromosome, and in some cases, genes controlled by the regulators are also nearby. This is not the case for *rscS* and the gene encoding its response regulator. The advent of the *V. fischeri* genome sequencing project (http://ergo. integratedgenomics.com/Genomes/VFI), has made it possible to use bioinformatics to look for RscR. Using the sequences of known regulators, we have searched and identified about 40 response regulators (Hussa and Visick, unpubl. data). At least 14 appear unlinked to sensor kinase genes, and thus represent the best candidates for RscR. Current work is aimed at mutagenizing these candidates and asking whether any mutants exhibit *rscS*-like colonization defects. If *rscR* encodes a DNA binding protein, then newly available DNA microarrays will be used to explore the regulon controlled by RscS/R. Identification of the targets of RscS/R regulation may also suggest a role for this regulon in symbiosis initiation. Once a target(s) of these regulators is identified, experiments aimed at identifying the colonization signal can be formulated.

gacA. In a number of pathogenic bacteria, the two-component system GacS/A regulates expression of virulence and host association traits, such as production of exoenzymes in *Pseudomonas* spp. (Heeb and Haas 2001) and motility in *Salmonella* (Goodier and Ahmer 2001). *V. fischeri* GacA also plays a role in host association. Mutants defective for *gacA* exhibit severe defects in initiating colonization: only about 10% of animals become colonized and those animals that become colonized exhibit a nearly 100-fold reduction in the level of colonization (i.e., the number of bacteria residing in the LO) (Whistler and Ruby 2003). The role of GacA is likely to be quite complex. In culture, it is associated with a number of phenotypes known to be important for symbiosis, including motility, nutrient acquisition, siderophore activity and luminescence (Whistler and Ruby 2003). The global control of disparate traits, all of which contribute to host-association, highlights the importance of such regulators in the evolution of symbiotic associations. As with RscS/R, neither the signal nor the gene/protein targets for GacA/S are known. Identification of targets of GacA regulation, possibly through DNA microarray experiments, will help elucidate the role of this regulator in symbiosis and potentially reveal previously unknown traits important for host-microbe interaction.

flrA. FlrA, a transcription regulator with limited sequence similarity to response regulators, functions as a master regulator of flagellar biosynthesis (Millikan and Ruby 2003). Given the absolute requirement for motility in symbiotic initiation, the requirement for FlrA seems straightforward as mutations lead to a lack of flagella. However, complemented *flrA* mutants showed restored motility but not normal colonization: initiation was delayed and the level of colonization at 48 h post-inoculation was reduced by 10-fold.

One explanation for the above result is that the timing and level of flagellar biosynthesis are critical for optimal initiation and colonization and these characteristics were not properly restored in the complemented strains. In support of this hypothesis, hyper-motile (hyper-flagellated) *V. fischeri* mutants also exhibit severe delays in initiating colonization and defects in the level of colonization 24 h post-inoculation (Millikan and Ruby 2002). Alternatively, an equally plausible explanation is that FlrA controls genes other than those involved in flagellar biosynthesis (Millikan and Ruby 2003) that are also required for colonization.

Several non-flagellar genes appear to be regulated by FlrA (Millikan and Ruby 2003). One gene that appears to be repressed by FlrA, *hvnC*, encodes a protein related to HvnA and HvnB, two secreted NAD^+ glycohydrolases found in *V. fischeri*. However, neither *hvnA* nor *hvnB* appears necessary for colonization (Stabb et al. 2001); therefore, the relevance of FlrA-mediated regulation of *hvnC* is unclear. A second putative FlrA-repressed gene is homologous to *V. cholerae kefB*. In *E.coli*, KefB is a potassium efflux protein that is important for protecting cells from toxic metabolites during growth on a poor carbon source (Ferguson et al. 2000). Possibly, the *V. fischeri* KefB homolog provides protection from a LO-specific toxin.

Are FlrA-repressed genes relevant to symbiotic colonization? FlrA-controlled flagella, which are required for initiation, become dispensable to colonized bacteria. Thus, a switch in flagella gene transcription may be coordinated with induction or repression of non-flagellar genes through FlrA. The regulation of FlrA itself may be at the level of transcription, analogous to cAMP-CRP mediated control of the master flagellar regulators *flhDC* in *E. coli* (Soutourina et al. 1999). In addition, the limited similarity of FlrA to response regulators suggests its activity could be regulated via phosphorylation by a sensor kinase. Future work will likely focus on determining whether FlrA itself is transcriptionally controlled, whether overexpression of FlrA during colonization affects the level or timing of transcription of putative FlrA-controlled genes and whether such genes themselves promote (or interfere with) colonization.

3.2
Quorum-Sensing Regulatory Systems

First described in *V. fischeri*, quorum sensing is used by many bacteria to detect the presence of other bacteria in their surroundings (reviewed in Taga and Bassler 2003). This method of monitoring the environment involves the production of a small molecule known as an autoinducer (AI) by an autoinducer synthase. Secreted into the environment, AIs can be recognized in recipient cells either by a specific two-component sensor kinase, or more frequently in Gram-negative bacteria, by a DNA-binding protein in the LuxR family. In either case, the AI signal results in transcriptional control of target genes.

V. fischeri uses both the LuxR DNA binding protein and specific sensor kinases to detect at least three AI signals (Fig. 3). Both pathways contribute to the control of bioluminescence, a trait required for symbiosis. A mutant

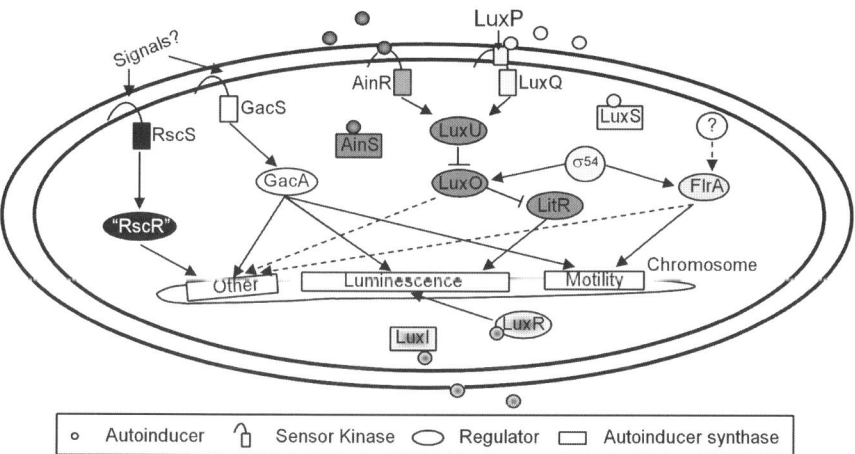

Fig. 3. Regulatory circuits required for symbiosis. *Dotted lines* represent hypothesized regulatory events. Activities required for symbiosis –"luminescence," "motility," and "other" – are represented as genes on the *V. fischeri* chromosome. Regulation of these activities may be through activation of transcription, as is the case for FlrA, through activation of transcription of a repressor, as is predicted to be the case for LuxO, or through modulating the activity of the protein product. The *V. fischeri* proteins AinS, AinR, and LitR are homologous to *V. harveyi* proteins LuxM, LuxN and LuxR, respectively

defective for *luxA*, one of two genes that encode bacterial luciferase, exhibits a three- to four-fold reduction in colonization level within 48 h post-inoculation (Visick et al. 2000). Encoded upstream of *luxA* are LuxR and LuxI, an AI synthase that produces the AI detected by LuxR. Mutations in either *luxR* or *luxI* result in a colonization defect similar to that of the *luxA* mutant, suggesting that these regulators are required for symbiosis due to their role in transcriptional control of the *lux* operon.

A second AI synthase, AinS, produces a distinct AI that also is required for symbiosis. The pathway through which the AinS-synthesized AI is detected and transmitted is predicted based on studies in the related bacterium, *V. harveyi* (reviewed in Taga and Bassler 2003). In *V. harveyi*, AIs signal through two hybrid sensor kinases, LuxN and LuxQ, using a phosphotransferase protein, LuxU, to ultimately affect the activity of a response regulator, LuxO. In the absence of AIs, LuxO negatively regulates *lux* genes by indirectly controlling transcription of a transcriptional activator (LitR in *V. fischeri* [Fidopiastis et al. 2002]). In *V. fischeri*, the AinS-produced AI likely is recognized by AinR, a sensor kinase with significant homology to *V. harveyi* LuxN (Gilson et al. 1995; Lupp et al. 2003) (Fig. 3).

A mutant defective for *ainS* exhibits a colonization level indistinguishable from that of *luxA*, *luxI* and *luxR* mutants (Lupp et al. 2003). However, whereas the *luxA*, *luxI*, and *luxR* mutants produce no symbiotic bioluminescence (at least 1000-fold decreased [Visick et al. 2000]), the *ainS* mutant produces approximately 10–20% of the wild-type bioluminescence. Thus, it seems probable that the role of *ainS* in colonization may be independent of its role in bioluminescence regulation. These phenotypes are difficult to separate, however: mutations in *luxO*, the response regulator through which the AI signals are transmitted, restore to wild-type levels both the slightly decreased symbiotic bioluminescence and the colonization defect of the *ainS* mutant (Lupp et al. 2003). Thus, an important direction will be to determine whether this pathway controls genes, other than *lux*, necessary for colonization.

V. fischeri encodes a third AI synthase, LuxS (Lupp and Ruby 2004). In *V. harveyi*, LuxS produces an AI that is detected by sensor kinase LuxQ through its interaction with the periplasmic protein LuxP (Taga and Bassler 2003). Because *V. fischeri* contains homologs for all of these genes (Lupp and Ruby 2004), it seems likely that this AI system functions similarly in the symbiotic organism (Fig. 3). A strain of *V. fischeri* in which *luxS* is mutated colonizes the LO as well as the wild-type strain; however the *luxS* mutation appreciably decreases the colonization efficiency of an *ainS* mutant, but not its per cell luminescence (Lupp and Ruby 2004). These data provide further support for a role of the AinS system in symbiosis distinct from that of luminescence. Further investigation of the three quorum sensing pathways likely will provide insight both into genes necessary for symbiotic colonization and, because the signals are known, signal transduction during symbiotic colonization.

4
Future Directions

The initiation of the symbiosis between *E. scolopes* and *V. fischeri* involves several regulatory systems, many of which affect traits known to be involved in colonization (Fig. 3). Many important questions remain. First, what is the relation of these regulatory circuits to one another? Both FlrA and the luminescence regulator LuxO modulate transcription by σ^{54}-containing RNA polymerase (Lilley and Bassler 2000; Millikan and Ruby 2003). Not surprisingly, a σ^{54} mutant is defective for colonization, motility and luminescence (Wolfe et al. 2004). Links between luminescence and motility also are found with the GacA mutant (Whistler and Ruby 2003) as well as with one class of hyper-motile mutants (Millikan and Ruby 2002), supporting coordinate regulation of motility and luminescence. Therefore, global regulation of the colonization response by *V. fischeri* may involve regulation of σ^{54} activity. Epistasis experiments could help to determine how these systems integrate to regulate colonization.

Second, what other traits, aside from luminescence and motility, are required for initiation of symbiosis? RscS is required for initiation, yet has no effect on motility or luminescence. GacA and FlrA both appear to regulate other functions as well. The major outer membrane protein OmpU is required for initiation of colonization (Aeckersberg et al. 2001) as is a recently identified gene cluster (Yip et al. 2005). These genes are possible targets of the systems described here. Identification of additional targets will be greatly aided by the *V. fischeri* genome sequence and available microarrays.

Finally, what signals are detected by the bacteria to regulate symbiosis? Aside from AIs produced by the quorum sensing systems, the signals received by the bacteria remain unknown. Predictions of the signals can be made based on our current knowledge of the environmental conditions in the LO. Furthermore, the recent sequencing of an *E. scolopes* expressed sequence tag library (http://trace.ensembl.org/) will facilitate identification of squid genes important for colonization and thus provide clues as to the conditions/signals the bacteria encounter in the LO. The answers to these questions will advance our understanding of the communication between and adaptation by *V. fischeri* and its host *E. scolopes*.

Acknowledgements. The authors would like to thank Drs. Joerg Graf, Margaret McFall-Ngai, Deborah Millikan, Eric Stabb and Cheryl Whistler and members of the Visick lab for their helpful comments in the preparation of this manuscript. We also thank Emily Yip for her drawing of juvenile light organ development shown in Fig. 1.

References

Aeckersberg F, Lupp C, Feliciano B, Ruby EG (2001). *Vibrio fischeri* outer membrane protein OmpU plays a role in normal symbiotic colonization. J Bacteriol 183:6590–6597

Boettcher KJ, Ruby EG (1990). Depressed light emission by symbiotic *Vibrio fischeri* of the sepiolid squid *Euprymna scolopes*. J Bacteriol 172:3701–3706

Davidson SK, Koropatnick TA, Kossmehl R, Sycuro L, McFall-Ngai MJ (2004) NO means 'yes' in the squid-vibrio symbiosis: nitric oxide (NO) during the initial stages of a beneficial association. Cell Microbiol 6(12):1139–1151

Doino JA (1998) The role of light organ symbionts in signaling early morphological and biochemical events in the sepiolid squid *Euprymna scolopes*, University of Southern California, Los Angeles

Doino JA, McFall-Ngai MJ (1995) A transient exposure to symbiosis-competent bacteria induces light organ morphogenesis in the host squid. Biol Bull 189:347–355

Ferguson GP, Battista JR, Lee AT, Booth IR (2000) Protection of the DNA during the exposure of *Escherichia coli* cells to a toxic metabolite: the role of the KefB and KefC potassium channels. Mol Microbiol 35:113–122

Fidopiastis PM, Miyamoto CM, Jobling MG, Meighen EA, Ruby EG (2002) LitR, a new transcriptional activator in *Vibrio fischeri*, regulates luminescence and symbiotic light organ colonization. Mol Microbiol 45:131–143

Foster JS, McFall-Ngai MJ (1998) Induction of apoptosis by cooperative bacteria in the morphogenesis of host epithelial tissues. Dev Genes Evol 208:295–303

Foster JS, Apicella MA, McFall-Ngai MJ (2000) *Vibrio fischeri* lipopolysaccharide induces developmental apoptosis, but not complete morphogenesis, of the *Euprymna scolopes* symbiotic light organ. Dev Biol 226:242–254

Gerard C (1998) Bacterial infection. For whom the bell tolls. Nature 395:217, 219

Gilson L, Kuo A, Dunlap PV (1995) AinS and a new family of autoinducer synthesis proteins. J Bacteriol 177:6946–6951

Goodier RI, Ahmer BM (2001) SirA orthologs affect both motility and virulence. J Bacteriol 183:2249–2258

Graf J, Dunlap PV, Ruby EG (1994) Effect of transposon-induced motility mutations on colonization of the host light organ by *Vibrio fischeri*. J Bacteriol 176:6986–6991

Heeb S, Haas D (2001) Regulatory roles of the GacS/GacA two-component system in plant-associated and other gram-negative bacteria. Mol Plant Microbe Interact 14:1351–1363

Kimbell JR, McFall-Ngai MJ (2004) Symbiont-induced changes in host actin during the onset of a beneficial animal-bacterial association. Appl Environ Microbiol 70:1434–1441

Lamarcq LH, McFall-Ngai MJ (1998) Induction of a gradual, reversible morphogenesis of its host's epithelial brush border by *Vibrio fischeri*. Infect Immun 66:777–785

Lee K, Ruby E (1992) Detection of the light organ symbiont, *Vibrio fischeri*, in Hawaiian seawater by using *lux* gene probes. Appl Environ Microbiol 58:942–947

Lee K, Ruby E (1994a) Effect of the squid host on the abundance and distribution of symbiotic *Vibrio fischeri* in nature. Appl Environ Microbiol 60:1565–1571

Lee KH, Ruby EG (1994b) Competition between *Vibrio fischeri* strains during initiation and maintenance of a light organ symbiosis. J Bacteriol 176:1985–1991

Lilley BN, Bassler BL (2000) Regulation of quorum sensing in *Vibrio harveyi* by LuxO and σ^{54}. Mol Microbiol 36:940–954

Lupp C, Ruby EG (2004) Vibrio fischeri LuxS and AinS: comparative study of two signal synthases. J Bacteriol 186:3873–3881

Lupp C, Urbanowski M, Greenberg EP, Ruby EG (2003) The *Vibrio fischeri* quorum-sensing systems *ain* and *lux* sequentially induce luminescence gene expression and are important for persistence in the squid host. Mol Microbiol 50:319–331

McFall-Ngai MJ, Ruby EG (1991) Symbiont recognition and subsequent morphogenesis as early events in an animal-bacterial mutualism. Science 254:1491–1494

McFall-Ngai M, Ruby EG (1998) Sepiolids and Vibrios: when first they meet. Reciprocal interactions between host and symbiont lead to the creation of a complex light-emitting organ. BioScience 48:257–265

Millikan DS, Ruby EG (2002) Alterations in *Vibrio fischeri* motility correlate with a delay in symbiosis initiation and are associated with additional symbiotic colonization defects. Appl Environ Microbiol 68:2519–2528

Millikan DS, Ruby EG (2003) FlrA, a σ^{54}-dependent transcriptional activator in *Vibrio fischeri*, is required for motility and symbiotic light-organ colonization. J Bacteriol *185*, 3547–3557

Montgomery MK, McFall-Ngai M (1994) Bacterial symbionts induce host organ morphogenesis during early postembryonic development of the squid *Euprymna scolopes*. Development 120:1719–1729

Nyholm SV, McFall-Ngai MJ (1998) Sampling the light-organ microenvironment of *Euprymna scolopes*: description of a population of host cells in association with the bacterial symbiont *Vibrio fischeri*. Biol Bull 195:89–97

Nyholm SV, McFall-Ngai MJ (2003) Dominance of *Vibrio fischeri* in secreted mucus outside the light organ of *Euprymna scolopes*: the first site of symbiont specificity. Appl Environ Microbiol 69:3932–3937

Nyholm SV, Stabb EV, Ruby EG, McFall-Ngai MJ (2000) Establishment of an animal-bacterial association: recruiting symbiotic *vibrios* from the environment. Proc Natl Acad Sci USA 97:10231–10235

Nyholm SV, Deplancke B, Gaskins HR, Apicella MA, McFall-Ngai MJ (2002) Roles of *Vibrio fischeri* and nonsymbiotic bacteria in the dynamics of mucus secretion during symbiont colonization of the *Euprymna scolopes* light organ. Appl Environ Microbiol 68:5113–5122

Ruby EG, Asato LM (1993) Growth and flagellation of *Vibrio fischeri* during initiation of the sepiolid squid light organ symbiosis. Arch Microbiol 159:160–167

Small AL, McFall-Ngai MJ (1999) Halide peroxidase in tissues that interact with bacteria in the host squid *Euprymna scolopes*. J Cell Biochem 72:445–457

Soutourina O, Kolb A, Krin E, Laurent-Winter C, Rimsky S, Danchin A, Bertin P (1999) Multiple control of flagellum biosynthesis in *Escherichia coli*: role of H-NS protein and the cyclic AMP-catabolite activator protein complex in transcription of the *flhDC* master operon. J Bacteriol 181:7500–7508

Stabb EV, Reich KA, Ruby EG (2001) *Vibrio fischeri* genes *hvnA* and *hvnB* encode secreted NAD(+)-glycohydrolases. J Bacteriol 183:309–317

Stabb EV, Butler MS, Adin DM (2004) Correlation between osmolarity and luminescence of symbiotic *Vibrio fischeri* strain ES114. J Bacteriol 186:2906–2908

Stock AM, Robinson VL, Goudreau PN (2000) Two-component signal transduction. Annu Rev Biochem 69:183–215

Taga ME, Bassler BL (2003) Chemical communication among bacteria. Proc Natl Acad Sci USA 100 [Suppl 2]:14549–14554

Taylor BL, Zhulin IB (1999) PAS domains: internal sensors of oxygen, redox potential, and light. Microbiol Mol Biol Rev 63:479–506

Vescovi EG, Ayala YM, di Cera E, Groisman EA (1997) Characterization of the bacterial sensor protein PhoQ. Evidence for distinct binding sites for Mg^{2+} and Ca^{2+}. J Biol Chem 272:1440–1443

Visick KL, McFall-Ngai MJ (2000) An exclusive contract: specificity in the *Vibrio fischeri-Euprymna scolopes* partnership. J Bacteriol 182:1779–1787

Visick KL, Skoufos LM (2001) Two-component sensor required for normal symbiotic colonization of *Euprymna scolopes* by *Vibrio fischeri*. J Bacteriol 183:835–842

Visick KL, Foster J, Doino J, McFall-Ngai M, Ruby EG (2000) *Vibrio fischeri lux* genes play an important role in colonization and development of the host light organ. J Bacteriol 182:4578–4586

Weis VM, Small AL, McFall-Ngai MJ (1996) A peroxidase related to the mammalian antimicrobial protein myeloperoxidase in the *Euprymna-Vibrio* mutualism. Proc Natl Acad Sci USA 93:13683–13688

Whistler CA, Ruby EG (2003) GacA regulates symbiotic colonization traits of *Vibrio fischeri* and facilitates a beneficial association with an animal host. J Bacteriol 185:7202–7212

Wolfe AJ, Millikan DS, Campbell JM, Visick KL (2004) *Vibrio fischeri* σ^{54} controls motility, biofilm formation, luminescence, and colonization. Appl Environ Microbiol 70:2520–2524

Yip ES, Grublesky BT, Hussa EA, Visick KL (2005) A novel, conserved cluster of genes promotes symbiotic colonization and sigma^{54}-dependent biofilm formation by *Vibrio fischeri*. Mol Microbiol. 57:1485–1498

Molecular Requirements for the Colonization of *Hirudo medicinalis* by *Aeromonas veronii*

Joerg Graf

1
Introduction

A fundamental feature of symbioses is the colonization of the host by specific microorganisms. The degree of specificity is reflected both in the number of microbial species associated with the host and in the frequency of detecting any given microbe in the host. Some associations such as the light organ symbiosis of *Vibrio fischeri* and the squid *Euprymna scolopes* are highly specific; involving one microbial species that is always detectable in the functional symbiotic association to the exclusion of all other microorganisms (McFall-Ngai and Ruby 1991). This association has been well studied and revealed multiple layers of molecular communication between the partners that allow the precise development of the association (McFall-Ngai and Ruby 1991; Visick and McFall-Ngai 2000; Chap. 13). Such two-member associations are probably an exception, especially among extracellular associations.

The microbiota in most digestive tracts consists of a large number of different species and could be considered to be less specific (Savage 1977; Moore and Moore 1995; Lilburn et al. 1999). Inside the host, these symbionts find a predictable source of nutrients and relatively constant environmental conditions that are presumably favorable for their growth. For a single host species, the composition of the microbial community inside individual animals can vary. For example, some bacterial species are regularly found in association with the host while other species are present in only some host specimen or occur only transiently, possibly depending on the age of the animal, its health, food or geographic location (Savage 1977; Moore and Moore 1995; Broderick et al. 2004). Factors such as the introduction of new microorganisms with each food consumption and changes in diet represent special challenges for the maintenance of specific microbial communities in the digestive tract (Savage 1977). Four hundred bacterial species can be cultured from the human digestive tract and between 800 and 1,200 operational taxonomic units of uncultured bacteria are present (Suau et al.

J. Graf (e-mail: joerg.graf@uconn.edu)
University of Connecticut, Department of Molecular and Cell Biology,
91 N. Eagleville Rd. U-3125, Storrs, CT 06268, USA

1999; Hayashi et al. 2002). This decreased specificity in the association does not necessarily imply that there is a lower degree of molecular communication between the symbionts and the host. Indeed, the communication may be more complex because more microbial partners need to be controlled by the host and because the microorganisms interact with each other. In this chapter, I will begin with an introduction of a relatively new model for digestive-tract associations, the digestive-tract symbiosis of *Aeromonas veronii* biovar sobria with *Hirudo medicinalis*, the medicinal leech, and present first insights into the molecular requirements for bacteria to colonize the digestive tract of the leech.

2
The Digestive Tract Symbiosis of *Hirudo medicinalis*

In contrast to the digestive tract communities residing in most animals (Savage 1977; Demaio et al. 1996; Lilburn et al. 1999), the digestive tract of *Hirudo medicinalis* appears to be colonized by a relatively simple microbial community (Lehmensick 1941; Büsing et al. 1953; Graf 2000). Only one symbiont, an *Aeromonas* sp., was consistently cultured in several investigations and culture independent approaches suggest the presence of only a few additional taxa (described below). This apparent simplicity is intriguing and suggests that several mechanisms are involved in establishing and maintaining this simplicity.

The biology of the leech is important for identifying possible factors that may contribute to the specificity of this symbiosis and I want to begin by briefly introducing this fascinating parasite that was used extensively for bloodletting in the 1800s (the biology of *H. medicinalis* is reviewed in the monograph by Sawyer (1986)). The medicinal leech feeds exclusively on blood. In a single blood meal, *H. medicinalis* can consume over five times its body weight. During the feeding, the leech releases powerful anticoagulants and vasodilators. Their activity is so potent that the blood continues to flow for about 15 min from the wound after the engorged leech falls off. This "blood-letting" ability has been utilized for centuries and the heavy collection in conjunction with habitat destruction has led to the near extinction of the medicinal leech in Western Europe (Graf 2000; Carter 2001). The medicinal use of leeches has made a recent revival to rescue tissue with venous congestion after microvascular plastic surgery (Henderson et al. 1983; Whitlock et al. 1983; de Chalain 1996). Interestingly for microbiologists, in up to 30% of these patients, wound infections with the digestive tract symbiont, *Aeromonas* sp., were reported (Whitlock et al. 1983; Dickson et al. 1984; Abrutyn 1988; de Chalain 1996;

Sartor et al. 2002). Currently, these infections are prevented with a preemptive antibiotic treatment.

The large blood meal is stored in the crop where water and salts are absorbed from the intraluminal fluid (ingested blood, Fig. 1). Within 48 h, approximately 50% of the gained weight is lost, due to the loss of water from the crop (Sawyer 1986). In addition, ions are also absorbed from the intraluminal until it is isoosmotic with the leech hemolymph (approximately 20 milliosmole). The erythrocytes are stored for months, apparently intact, in the crop despite of the presence of β-hemolytic bacteria. Whether or not a portion of the erythrocytes is lysed or punctured in the crop to release nutrients to the resident symbionts has not been determined. The physical breakdown of all of the erythrocytes occurs in the intestinum. The animals can survive for over 9 months between blood meals. This long time between feeding events is likely to be an important selection pressure on the symbionts.

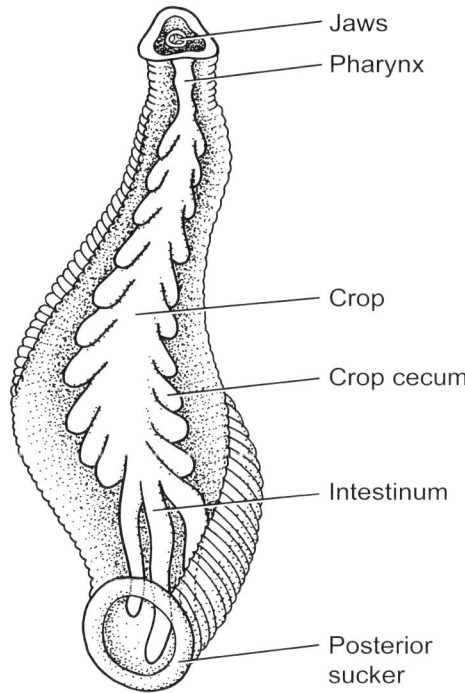

Fig. 1. Drawing of a ventrally dissected *H. medicinalis*. The crop houses the ingested blood. *A. veronii* biovar *sobria* is found in the crop and intestinum (modified from Graf 2000)

3
Characterization of the Microbiota in the Crop of the Leech

The early bacteriological investigations in the 1940s and 1950s described the isolation of a single bacterium from the crop of the medicinal leech (Lehmensick 1941; Hornbostel 1942; Büsing 1951; Büsing et al. 1953). The consistent isolation of this bacterium and the absence of other microorganisms lead the early investigators to suggest a symbiotic association. Three functions have been proposed for the symbiont (Büsing et al. 1953; Graf 2000, 2002): (1) digestion of the ingested blood meal, (2) synthesis of essential nutrients such as vitamin B for the host and (3) preventing the growth of other microorganisms in the crop of the leech. To the knowledge of this author, the only experimental evidence for an influence of the symbionts on host physiology comes from two studies (Büsing et al. 1953; Zebe et al. 1986) that used relatively high antibiotic concentrations such that a direct effect on the host physiology could not be excluded (Graf 2002). These studies indicated that release of nitrogenous metabolic end products from the animal and the consumption of oxygen by the animal were affected by antibiotic treatment, thereby suggesting a role of the microbial partners. Further studies are needed to determine the role of the symbionts, but the ability of the leech to survive for nine months without feeding makes the determination of the contribution by the symbiont more difficult.

As the taxonomy of bacteria improved, the symbiont was renamed *A. hydrophila* (Jennings and van der Lande 1967). Similarly, the bacteria isolated from leech therapy associated wound infections were identified as *A. hydrophila* (Lowen et al. 1989; Snower et al. 1989; Lineaweaver 1991). The taxonomy *Aeromonas* has become more sophisticated and with it the complexity of identifying species (Janda and Abbott 1998). In one of our studies, we identified symbionts from animals from various sources using biochemical tests and 16S rRNA gene sequences and detected only *A. veronii* biovar sobria (Graf 1999). However, some of our strains that were isolated from leeches differed in key biochemical tests from the type strain. These differences may have resulted in the misidentification of these strains using commercial test kits. Other studies also reported the presence of various other bacteria in the digestive tract of the medicinal leech in addition to *Aeromonas* (Nonomura et al. 1996; Mackay et al. 1999; Eroglu et al. 2001). In these studies, it is unclear what proportion of the isolates from one animal belongs to which species. This lack of quantitative data does not allow one to determine whether these bacteria are always part of the digestive tract community, whether they represent contaminations from other organs, were introduced with a contaminated blood meal or are passing transiently through the digestive tract.

An open question is the presence of bacteria in the crop that do not grow when standard culturing techniques are used. We are currently addressing this question by cloning and sequencing 16S rRNA genes from the crop and intestinum. The initial results suggest the presence of up to three species in the crop of which two species, *A. veronii* biovar sobria and a species belonging to the *Bacteroidetes* group, appear to be the dominant players (Worthen, P., C. Gode and J. Graf, unpubl.). These results indicate that the microbial community is more complex than we had anticipated but sufficiently simple to allow us to investigate the population dynamics and molecular requirements underlying this symbiosis. Indeed, it will be important to understand microbial consortia because monospecific associations represent the exception and simple microbial consortia are likely to reveal mechanisms that are applicable to more complex associations.

4
Specificity of the Symbiosis

A powerful tool for the molecular investigation of a symbiosis is a colonization assay that allows one to evaluate the ability of specific strains to colonize the medicinal leech (Graf 1999, 2002; Indergand and Graf 2000). In order to differentiate the introduced test strain from the native microbiota, we use spontaneous rifampin-resistant mutants as test strains. The strains are added to a blood meal that was preheated to 37°C and is fed to the leech in a parafilm covered disposable centrifuge tube. At specified time intervals animals are surface sterilized and killed by placing them in 70% ethanol before recovering the intraluminal fluid. Serial dilution of the samples and plating aliquots on agar plates containing appropriate antibiotics allows us to evaluate the ability of a specific strain to colonize the leech. The versatility of this colonization assay allows us to feed multiple strains carrying different antibiotic resistance markers in a competition assay or add antibiotics to eliminate the native microbiota.

The native symbiont colonized the crop of *H. medicinalis* rapidly, doubling approximately every 1.2 h, the symbiotic strain HM21R reached a maximum concentration of about 1×10^8 CFU/ml (Graf 1999) which was maintained for at least 7 days (Indergand and Graf 2000). The growth curve suggests the presence of a rapid proliferation phase and a persistence phase (Fig 2). During the proliferation phase, the symbiont was able to grow at a similar rate inside the animal as in a rich medium at room temperature. The cessation of growth could be due to the depletion of

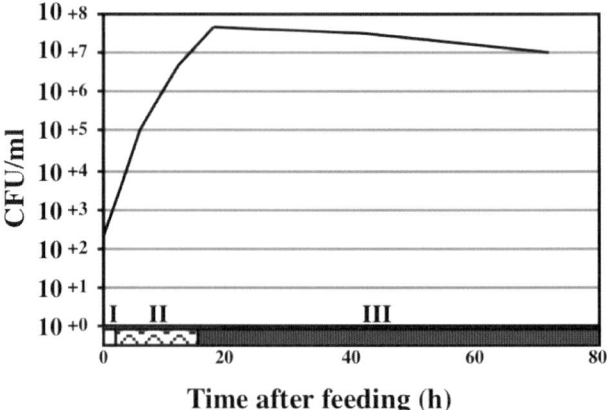

Fig. 2. A schematic representation of the growth of *A. veronii* biovar *sobria* inside the leech. The symbionts proliferate rapidly inside the leech reaching an apparent stationary phase between 12 and 24 h. The colonization process can be divided into three stages. I. Overcoming antimicrobial compounds from the ingested blood. II. Rapid growth phase. III. Persistence phase

nutrients, the accumulation of toxic byproducts, or host modification of the intraluminal fluid. Alternatively, bacteria could be removed or killed by the release of antimicrobial compounds or by phagocytosis by leech hemocytes. These results indicate that the symbiotic *A. veronii* biovar sobria is well adapted to the digestive tract environment of the leech.

5
Importance of the Vertebrate Complement System

This colonization assay allowed us to evaluate the specificity of the symbiosis experimentally. For these experiments, a portion of the inoculated blood was incubated in vitro under the same conditions as the animal. This modification allowed us to compare the growth potential of the strain in blood and perhaps detect modifications of the ingested blood inside the leech (Indergand and Graf 2000). One *Escherichia coli* strain, EcR1, had an interesting colonization phenotype (Fig. 3). The number of CFU per ml decreased both inside the leech and in the in vitro control 18 and 42 h after inoculation. At 42 h, EcR1 had decreased 1,000-fold in concentration. This concomitant decrease was consistent with an antimicrobial property of the ingested blood remaining active inside the leech and killing EcR1. Heat-treatment of the blood prior to inoculation allowed the *E. coli* strain to proliferate both inside the animal and in the in vitro control, indicating that this antimicrobial property was heat sensitive and also responsible for the reduction of viable *E. coli* inside the leech. One powerful antimicrobial property of vertebrate blood is the complement system. The complement system uses the classical or alternative pathway to activate the formation of the membrane-attack-complex, which is inserted into the membrane of sensitive cells. This leads to a permeablization of the membrane and the eventual loss of viability. One can prevent the activation of the membrane-attack-complex by pre-treating the blood with chemicals or heat. In vitro experiments, where the blood was pretreated with EDTA, EGTA with Ca^{2+} or heat treatment allowed the strain to proliferate. These treatments interfere with the activation of the complement system by the classical system and are consistent with a role of the complement system in killing *E. coli*. The genetic background of EcR1 is not known and it seemed reasonable to assume that *Aeromonas* strains that can colonize the leech would have to be resistant to the complement system.

Previously it was shown that the lipopolysaccharides (LPS) of the outer membrane is important in conferring resistance to the complement system in *Aeromonas* (Merino et al. 1992, 1994, 1996). Using well-characterized *A. veronii* biovar sobria mutants, we decided to test the hypothesis that the

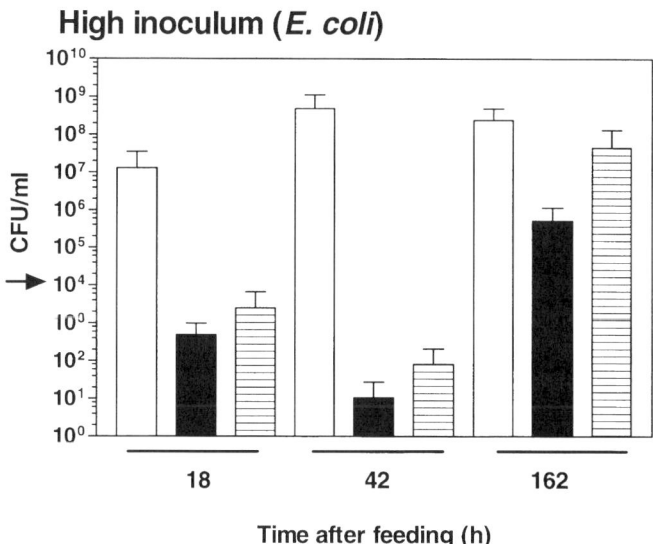

Fig. 3. Colonization of *H. medicinalis* by *E. coli* EcR1. During the first 24 h after feeding, the concentration of strain EcR1 decreased inside the animal and the in vitro control. The native microbiota and the introduced strain EcR1 was recovered on blood agar (*open bar*), strain EcR1 alone on LB containing rifampin (*solid bar*) and strain EcR1 from the in vitro control (*hatched bar*). The *arrow* depicts the cell concentration of strain EcR1 in the inoculum. Reprinted with permission from Indergand and Graf (2000)

complement system of vertebrates remains active for some time inside the leech by testing the ability of complement-sensitive *Aeromonas* mutants to colonize the leech (Braschler et al. 2003). Mutants were available that exhibited a defect in the biosynthesis of LPS and were shown to have an increased sensitivity to the complement system (Merino et al. 1992, 1994, 1996). The LPS is thought to prevent the membrane-attack-complex from reaching the lipid bilayer. These *A. veronii* biovar sobria mutants were devoid of the O34 antigen LPS and their defect could be complemented with a cosmid carrying the biosynthetic genes for rhammose which are present in the *wb* gene cluster for the O-antigen. Accordingly, we predicted that if the complement system remained active inside the leech, these mutants should have a reduced ability to colonize their host. As expected, the serum-sensitive LPS mutants had a severe colonization defect that could be reversed either by the heat-inactivation of the blood or by complementing the mutant with the biosynthetic genes for rhammose on a plasmid (Fig. 4; Braschler et al. 2003). These results provide further support to the hypothesis that the complement system of vertebrates contributes to the specificity of the association by preventing other, sensitive bacteria from colonizing the medicinal leech. The mutants with a defect in the ability to synthesize the LPS represent the first mutant class known that is defective in colonizing the digestive tract of the leech.

In a random transposon mutant screen of the symbiotic strain, HM21R, mutants that had become sensitive to antimicrobial properties in the vertebrate blood were isolated and none of the ones tested was able to colonize the leech (Rabinowitz, N., A. Silver, S. Küffer and J. Graf, unpubl. data). Initial characterization of the mutants suggests that several had transposon insertions

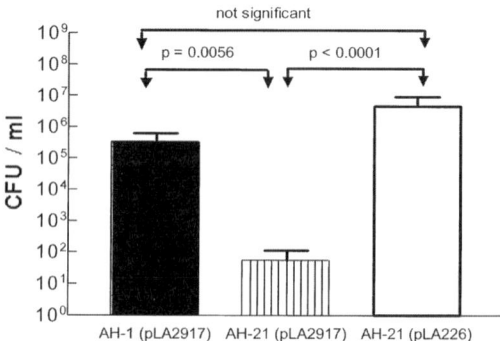

Fig. 4. The serum-sensitive mutant AH-21 had a dramatically reduced ability to colonize the leech. This defect could be restored by complementing the mutant with pLA226 which carries biosynthetic genes for the O-antigen but not by the empty vector control, pLA2917. The *P* values were calculated with an unpaired *T*-test using Welch's correction

in genes that showed significant similarity to glycosyl transferases. Preliminary comparisons of the amino acid sequence of these proteins suggest that they differ from the published ones from *A. hydrophila* and *A. veronii* biovar *sobria*.

The colonization phenotypes of *S. aureus* strain SaR1 and *P. aeruginosa* strain PaR1 differed dramatically from that of *E.coli* or the serum-sensitive mutants (Indergand and Graf 2000). Their ability to proliferate inside the leech was significantly inhibited as compared to an in vitro control; however, the mutants were able to survive at a constant level in the leech for 7 days (for example see Fig. 5). These data suggest that the ingested blood is modified in a manner that interferes with the ability of these two species to proliferate inside the leech. This is indicative of a second layer of defense that the symbionts need to overcome. The possible factors that may contribute could be simple modifications of the blood by either the host or symbionts, such as the removal of water and salts from the intraluminal fluid or that the native symbionts out compete the other bacteria for nutrients. Other possibilities are that either host or symbionts release antimicrobial compounds to which these bacteria are sensitive but that the symbionts are resistant

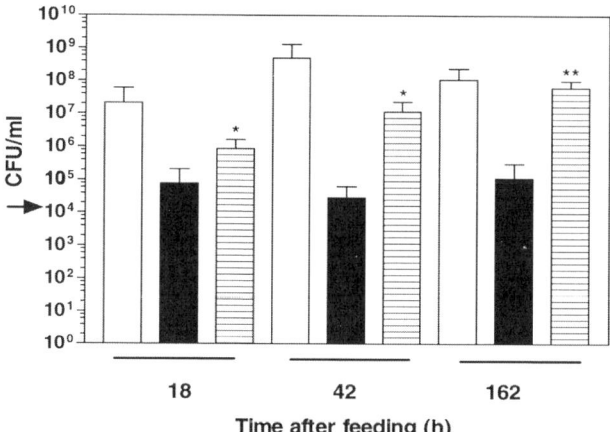

Fig. 5. Colonization of *H. medicinalis* by *S. aureus*. SaR1 was unable to increase in concentration inside the leech but was able to persist for at least seven days. The concentration reached was significantly lower than in the in vitro control. The native microbiota and the introduced strain SaR1 was recovered on blood agar (*open bar*), strain SaR1 alone on LB containing rifampin (*solid bar*) and strain SaR1 from the in vitro control (*hatched bar*). The *arrow* depicts the concentration of strain SaR1 in the inoculum. * and **, the mean concentration differed significantly from the values obtained inside the animal at $P < 0.05$ and $P < 0.005$, respectively (two-sides Mann-Whitney test). Reprinted with permission from Indergand and Graf (2000)

themselves. The results described so far begin to paint a picture of multilayered defenses with possible contributions from the leech, the symbionts as well as the vertebrate on which the leech feeds. These defenses together contribute to the unusual specificity of this symbiotic interaction.

A major goal is to identify the mechanisms that are responsible for the establishment and maintenance of the specificity. One powerful approach is to isolate mutants of the native symbiont which exhibit a reduced ability to colonize the leech. The molecular characterization of these mutants may provide us with clues that offer us with an insight into the host environment and may reveal active mechanisms that interfere with microbial growth.

6
Conclusions and Outlook

Considering the known physiology of the leech and the mutant phenotypes that we have characterized so far, possible mechanisms can be deduced that may interfere with microbial growth and thus contribute to the specificity of this symbiosis. These mechanisms are likely to play important roles at different times, immediately after feeding, during the rapid growth period and during the persistence phase (Fig. 2). The first powerful layer of defense is the antimicrobial properties of the ingested blood. It appears that these properties remain active for some time inside the animal. Our evidence suggests that the complement system is potent even inside the leech and prevents sensitive bacteria from proliferating. Other properties such as iron sequestration by transferrin are likely to be active as well.

During feeding, the leech secrets numerous compounds into the host animal and these would enter the leech during the feeding process with the blood meal. The leech has been shown to remove most of the water from the ingested blood within 48 h. It is possible that the reduced volume restricts bacterial growth. At the same time as the water is removed from the crop, salts are absorbed until the intraluminal fluid is isoosmotic with the leech hemolymph. The osmolarity of the leech hemolymph is approximately two-thirds of that of vertebrate blood. It is not clear if this level of decrease would affect bacterial growth, especially when considering the fact that the bacteria can survive in freshwater that has an even lower osmolarity. These factors or changes have been shown to be present but it is not clear at this point whether any of them contribute to the specificity. Considering that these modifications take over 48 h to occur, it would seem reasonable to assume that they will not play a dominant role during the first 12 h after feeding.

It may be speculated that other factors prevent bacteria from growing within the host. Invertebrate animals have a powerful innate immune response that has many similarities to that of vertebrates, the large exception being the components of the adaptive immunity (Mallo et al. 2002; Tzou et al. 2002). Other factors could include the release of antimicrobial peptides or enzymes into the crop by either the host or symbiont. Some compounds could be release into the ingested blood during feeding for example from the salivary gland. One would assume that these compounds would be active immediately and have an early effect on bacterial growth or survival. In addition, host hemocytes could be present and phagocytose bacteria or generate oxidative stress. Most likely, the different antimicrobial factors act at different times after feeding. The colonization process can be divided into three distinct stages. The first stage consists of overcoming the antimicrobial properties of the ingested blood. The second stage is the rapid proliferation and the third stage the long-term persistence.

The key question is how to identify the mechanisms that are active inside the leech. We are currently undertaking a random transposon mutant screen to identify bacteria that have lost the ability to colonize the leech. One can view each mutant as a sensor of the microenvironment that the bacteria encounter. The characterization of these mutants will reveal conditions that interfere with the proliferation or survival of bacteria, and hence will yield testable hypotheses to elucidate the role of the host or the native symbionts. We have already shown that the analysis of the mutants allowed us to draw conclusions about the complement system. The characterization of the mutants with a colonization defect should provide us with novel and exciting information on the mechanisms underlying the association of bacteria with their leech host.

Acknowledgements. I would like to thank the Swiss National Science Foundation (31–63775), the U.S. National Science Foundation (MCB 0334627) and the Research Foundation of the University of Connecticut for their financial support.

References

Abrutyn E (1988) Hospital-associated infections from leeches. Ann Intern Med 109:356–358
Braschler TR, Merino S, Tomas JM, Graf J (2003) Complement resistance is essential for colonization of the digestive tract of *Hirudo medicinalis* by *Aeromonas* strains. Appl Environ Microbiol 69:4268–4271
Broderick NA, Raffa KF, Goodman RM, Handelsman J (2004) Census of the bacterial community of the gypsy moth larval midgut by using culturing and culture-independent methods. Appl Environ Microbiol 70:293–300
Büsing K-H (1951) *Pseudomonas hirudinis*, ein bakterieller Darmsymbont des Blutegels (*Hirudo officinalis*). Zentralbl Bakteriol 157:478–485
Büsing K-H, Döll W, Freytag K (1953) Die Bakterienflora der medizinischen Blutegel. Arch Mikrobiol 19:52–86

Carter KC (2001) Leechcraft in nineteenth century British medicine. J R Soc Med 94:38–42
De Chalain TM (1996) Exploring the use of the medicinal leech: a clinical risk-benefit analysis. J Reconstr Microsurg 12:165–172
Demaio J, Pumpuni CB, Kent M, Beier JC (1996) The midgut bacterial flora of wild *Aedes triseriatus*, *Culex pipiens*, and *Psorophora columbiae* mosquitoes. Am J Trop Med Hyg 54:219–223
Dickson WA, Boothman P, Hare K (1984) An unusual source of hospital wound infection. Br Med J 289:1727–1728
Eroglu C, Hokelek M, Guneren E, Esen S, Pekbay A, Uysal OA (2001) Bacterial flora of *Hirudo medicinalis* and their antibiotic sensitivities in the middle black sea region, Turkey. Ann Plast Surg 47:70–73
Graf J (1999) Symbiosis of *Aeromonas veronii* biovar *sobria* and *Hirudo medicinalis*, the medicinal leech: a novel model for digestive tract associations. Infect Immun 67:1–7
Graf J (2000) The symbiosis of *Aeromonas* and *Hirudo medicinalis*, the medicinal leech. ASM News 66:147–153
Graf J (2002) The effect of the symbionts on the physiology of *Hirudo medicinalis*, the medicinal leech. Int J Reprod Biol 41:269–275
Hayashi H, Sakamoto M, Benno Y (2002) Phylogenetic analysis of the human gut microbiota using 16S rDNA clone libraries and strictly anaerobic culture-based methods. Microbiol Immunol 46:535–548
Henderson HP, Matti B, Laing AG, Morelli S, Sully L (1983) Avulsion of the scalp treated by microvascular repair: the use of leeches for post-operative decongestion. Br J Plast Surg 36:235–239
Hornbostel H (1942) Über die bakteriologischen Eigenschaften des Darmsymbionten beim medizinischen Blutegel (*Hirudo officinalis*) nebst Bemerkungen zur Symbiosefrage. Zentralbl Bakteriol 148:36–47
Indergand S, Graf J (2000) Ingested blood contributes to the specificity of the symbiosis of *Aeromonas veronii* biovar *sobria* and *Hirudo medicinalis*, the medicinal leech. Appl Environ Microbiol 66:4735–4741
Janda JM, Abbott SL (1998) Evolving concepts regarding the genus *Aeromonas*: an expanding panorama of species, disease presentations, and unanswered questions. Clin Infect Dis 27:332–344
Jennings JB, van der Lande VM (1967) Histochemical and bacteriological studies on digestion in nine species of leeches (Annelidia: Hirudinea). Biol Bull 33:166–183
Lehmensick R (1941) Ueber einen neuen bakteriellen Symbionten im Darm von *Hirudo officinalis* L. Zentralbl Bakteriol 147:317–321
Lilburn TG, Schmidt TM, Breznak JA (1999) Phylogenetic diversity of termite gut spirochaetes. Environ Microbiol 1:331–345
Lineaweaver WC (1991) *Aeromonas hydrophila* infections following clinical use of medicinal leeches: a review of published cases. Blood Coagul Fibrinolysis 2:201–203
Lowen RM, Rodgers CM, Ketch LL, Phelps DB (1989) *Aeromonas hydrophila* infection complicating digital replantation and revascularization. J Hand Surg 14:714–718
Mackay DR, Manders EK, Saggers GC, Banducci DR, Prinsloo J, Klugman K (1999) *Aeromonas* species isolated from medicinal leeches. Ann Plast Surg 42:275–279

Mallo GV, Kurz CL, Couillault C, Pujol N, Granjeaud S, Kohara Y, Ewbank JJ (2002) Inducible antibacterial defense system in *C. elegans*. Curr Biol 12:1209–1214

McFall-Ngai MJ, Ruby EG (1991) Symbiont recognition and subsequent morphogenesis as early events in an animal-bacterial mutualism. Science 254:1491–1494

Merino S, Camprubi S, Tomas JM (1992) Characterization of an O-antigen bacteriophage from *Aeromonas hydrophila*. Can J Microbiol 38:235–240

Merino S, Alvarez D, Hernandez-Alles S, Tomas JM (1994) Effect of growth temperature on complement-mediated killing of mesophilic *Aeromonas* spp. serotype O:34. FEMS Microbiol Lett 118:163–166

Merino S, Rubires X, Aguillar A, Guillot JF, Tomas JM (1996) The role of the O-antigen lipopolysaccharide on the colonization in vivo of the germfree chicken gut by *Aeromonas hydrophila* serogroup O:34. Microb Pathog 20:325–333

Moore WE, Moore LH (1995) Intestinal floras of populations that have a high risk of colon cancer. Appl Environ Microbiol 61:3202–3207

Nonomura H, Kato N, Ohno Y, Itokazu M, Matsunaga T, Watanabe (1996) Indigenous bacterial flora of medicinal leeches and their susceptibilities to 15 antimicrobial agents. J Med Microbiol 45:490–493

Sartor C, Limouzin-Perotti F, Legre R, Casanova D, Bongrand MC, Sambuc R, Drancourt M (2002) Nosocomial infections with *Aeromonas hydrophila* from leeches. Clin Infect Dis 35:E1–E5

Savage DC (1977) Microbial ecology of the gastrointestinal tract. Annu Rev Microbiol 31:107–133

Sawyer RT (1986) Leech biology and behavior. Clarendon Press, Oxford

Snower DP, Ruef C, Kuritza AP, Edberg SC (1989) *Aeromonas hydrophila* infection associated with the use of medicinal leeches. J Clin Microbiol 27:1421–1422

Suau A, Bonnet R, Sutren M, Godon JJ, Gibson GR, Collins MD, Dore J (1999) Direct analysis of genes encoding 16S rRNA from complex communities reveals many novel molecular species within the human gut. Appl Environ Microbiol 65:4799–4807

Tzou P, de Gregorio E, Lemaitre B (2002) How *Drosophila* combats microbial infection: a model to study innate immunity and host-pathogen interactions. Curr Opin Microbiol 5:102–110

Visick KL, McFall-Ngai MJ (2000) An exclusive contract: specificity in the *Vibrio fischeri–Euprymna scolopes* partnership. J Bacteriol 182:1779–1787

Whitlock MR, O'Hare PM, Sanders R, Morrow NC (1983) The medicinal leech and its use in plastic surgery: a possible cause for infection. Br J Plast Surg 36:240–244

Zebe E, Roters F-J, Kaiping B (1986) Metabolic changes in the medical leech *Hirudo medicinalis* following feeding. Comp Biochem Physiol 84A:49–55

Index

16S rRNA phylotype 216, 262, 267
16S rRNA sequencing 100, 261, 295
2-Oxoglutarate 28, 30, 188

A

Acanthamoeba 68, 69
Acetate:succinate CoA-transferase (ASCT) 125
Acetogenesis 42, 53, 55
Actin 112, 280
Adenosine 5'-phosphosulfate (APS) pathway 207
Adenylate cyclase 105, 110
ADP/ATP carriers 121, 129, 131, 138, 140
Aeromonas veronii 291–292
ainS 286
Allelopathic 65
Alpha Proteobacteria 233, 262
Alternative electron acceptor 121
Ammonium 168, 170, 172–173, 180, 182, 239–240
Amoeba 63, 66–67, 68
Annelida 230, 272–274
Anoxic 11, 31, 41, 117–118, 121, 128, 130, 133, 135, 137, 183, 203, 208, 209, 227, 256, 268
Anthoceros punctatus 167, 169
Anticoagulant 292
Apoptosis 279–280
Aseriate 180–181
ATP equivalent 6
ATPase 5, 8, 71, 148, 174, 210
Autoinducer 21, 285
Autotrophic metabolism 239

B

Bacteria 7, 21, 24, 52, 61, 78, 102, 152, 197, 233, 251
Bacteriocin 71
Bacteriocyte 204, 206, 210–211, 213–214, 235–236, 238, 240
Bacteroidales 49
Bacteroids 150
Basket tubules 103, 106
Bathymodiolus 227–229, 231–232, 235–236, 238, 241–242
Biogeography 25–27, 241, 252–253
Bioluminescence 277, 282, 285–286
Blood 200, 204, 206, 209–210, 213, 255, 292–293, 296–301
Blood meal 293–295, 300

C

Caedibacter 61–72, 104–105, 133
Calvin cycle 199, 208–210
Cardiolipin 125
Caudovirales 71
Cell cycle 107, 109, 167, 179, 186, 213–214, 217
Cellular evolution 130
Cellulases 44, 53, 85–87
Cellulose 40, 43–44, 53, 86–87
Chaperonins 123
Chemoautotroph 197, 210, 227, 230, 236, 242
Chemoautotrophy 207, 238, 267
Chemolithoautotroph 97–98
Chemosynthesis 208, 213
Chemotaxis 28, 30, 32, 54, 175
Chlorobiaceae 25, 30–31, 33
Chlorochromatium 5, 22, 29, 31, 33
Chloroplana 23–24, 32
Chytrid(iomycete) fungi 122, 124, 132
Ciliate 97–98, 107–111, 126, 129, 133, 201
Cockroaches 42, 126
Co-evolution 54, 67, 198
Cold seep 201, 227–228, 233–235, 241, 264, 268

Colonization 42, 45, 172, 176–177, 214, 216, 278–280, 282–287, 291, 296–299, 301
Colonization assay 295–296
Compartmentalisation 112
Competition 65, 67, 166, 295
Complement system 296–298, 300–301
Conjugation 65–66, 68
Consortia 21–37, 134, 295
Consortium 5, 22, 24–26, 28–29, 98, 271
Cospeciation 215, 253, 263–264
Crop 78, 293–295, 300–301
Crypt 278, 280
Crypton 120, 131
Cyclic AMP 108, 290
Cylindrogloea 24

D

Deep-sea 198, 201, 217, 227, 228, 230, 232–235, 240–242
Defective phage 71
Delta Proteobacteria 258–259, 264–266, 268–269
Didinium 68, 110
Diffusion 21, 30, 32, 80, 200, 210, 257
Digestive tract 291–292, 294, 296, 298
Diversity 25–26, 29, 41, 44, 48, 51, 54, 69, 132, 145, 167, 182, 201, 217, 228, 231, 241, 252–253, 262, 267, 270
Dome-shaped zone 102–104
Domestication 170
Duct 278

E

Ectosymbionts 77, 91, 97, 122, 263
Ectosymbiotic 82–83, 87–89, 97–98, 110, 132, 134
Ejection 108–110
Ejectisomes 105–106, 108, 110
Electron acceptor 13, 117, 121, 208, 255–256

Electron transport chain 121–122, 125–126, 135
Endocytobiont 68, 72
Endocytobiosis 193
Endoglucanases 84, 86
Endomicrobia 41, 50–51
Endonuclear bacteria 52
Endosymbiont 41, 45, 48, 50, 53, 63, 100, 106, 127–128, 130, 132–133, 199, 204, 214, 217, 227–228, 231–232, 235–240, 242, 269
Endosymbiosis 231
Entamoeba 117, 120, 121, 131
Environmental transmission 201, 215–216
Epibiont 22–34, 45–46, 48–50, 54, 100, 109, 111
Epibiotic 45, 47–48, 50, 54, 88
Epithelial fields 278, 280
Epixenosomal band 99, 102, 111–112
Epixenosome 98–112
EPS 149–151
Erythrocytes 293
Euplotidium 98–100, 103, 107, 109–112
Euprymna scolopes 277, 279, 291, 303
Evolutionary tinkering 118, 131–132, 135
Expulsion 279–280
Extracellular associations 166, 291
Extrachromosomal element 63, 69, 105
Extrachromosomal inheritance 63
Extrusive apparatus 102–104, 109
Extrusomes 99–100

F

Fermentations 2, 6
FISH 13, 238, 261–262, 264, 266–268
Flagella 31–32, 46, 83, 90–91, 216, 278, 280, 284
Flagellates 39–40, 42–45, 48–54, 77–78, 80–87, 91, 119, 132–133
Flavonoids 144, 146, 150
FlrA 282, 284–285, 287
Fluorescence in situ hybridization 25, 50, 89, 261, 266

G

Gamma proteobacteria 201–202, 227, 233, 236–237, 242, 259, 263–266
Gene transfer 63–64, 66, 71, 105, 112, 119
Giardia 117, 120–121, 131–132
Gliding motility 173, 175
Glutamine synthetase 181, 212, 240
Glycolysis 10, 44, 121, 123, 125
Glycosyl transferase 299
Gradient(s) 25, 28, 32, 41, 80, 126, 188–189, 203, 206, 208, 210–211, 213, 217
Green sulfur bacteria 24–27, 29–31, 33–34
Growth control 179
Gunnera spp. 168–169

H

Heat shock protein 120, 131
Hemocytes 296, 301
Hemoglobin 200, 209, 255–256
Hemolymph 293, 300
Heterocysts 165–168, 170, 172, 176–177, 179–189
Heterotrophic metabolism 181
hetF 171, 183–184, 186, 188
hetN 183, 186
hetR 171, 183–184, 186, 188
Hindgut 39–45, 48, 52–54, 78, 80, 82–87, 91
Hirudo medicinalis 291–292
Holospora 64, 66, 68, 133
Homoacetogens 53
Hormogonia 167–168, 172–177, 181
Hormogonium-inducing factor 171–172
Hormogonium-repressing factor 171, 177
Host specificity 67
hrmA 171, 177–179
hrmE 177–178
hrmR 177–179

Hydrogenase 9, 44, 117, 120, 122–125, 127–129, 131–132, 134
Hydrogenosome 120, 122–123, 127, 129–130, 132, 134
Hydrogen-scavenging 4, 10
Hydrothermal vent 203, 215, 217, 235, 241, 254, 264, 268
Hypermastigid 50, 51, 84
Hypermastigotes 77, 81

I

Inanidrilus 253–254, 257, 263–264
Inclusion body 102–104
Infection 65–68, 133, 144, 147–148, 150–151, 154, 171–172, 174, 176–177, 188, 261, 281
Infection thread 147, 150
Interspecies electron transfer 13–14
Interspecies hydrogen transfer 4, 9, 31
Intestinum 293, 295
Intraluminal fluid 293, 295–296, 299–300
Iron reduction 2
Iron-sulphur metabolism 118

K

K-antigen 149, 151
Kappa-particle 61, 72
Killer symbiont 63, 66, 72
Killer trait 61, 104
Klebsiella 71

L

Legionella 68
Leguminous plants 143, 168, 268
Light organ (LO) 277–278, 288, 291
Lignocellulose 40, 78
Lipo-chito-oligosaccharides 144
Lipopolysaccharide (LPS) 280, 297
luxO 285–287
luxR 285–286
Lysis 68–69, 71, 213–214, 257, 268
LysR 144

M

Marine 11, 13, 97–98, 100, 104, 166, 197, 201, 203, 222, 228, 235, 242, 251–253, 255, 257, 263, 269, 277
Mate killer 66
Medicinal leech 292, 294–295, 298
Membrane attack complex 297–298
Metabiosis 3
Methane monooxygenase 233–234, 239
Methanogenesis 2, 11, 40, 42, 52
Methanogenic degradation 1, 3, 12, 14
Methanol dehydrogenase 231–234, 239
Methanotroph 201, 227–239, 340, 342
Microbial consortia 295
Minimum energy increment 6
Mitochondrial carrier family 123, 131
Mitochondrial-remnant organelles 117–118, 120–121
Mitochondrion 118–121, 126, 130–131
Mitosome 117–118, 120–121, 131
Monospecific associations 21, 295
Motility 31–32, 50, 77, 82–83, 173, 175–176, 215, 280, 282–285, 287
Motility symbiosis 50, 82
Mucus 278–279, 282
Mutualistic 51, 63, 67, 170, 277

N

Nitrate 13–14, 40, 121, 143, 187, 200, 208, 211–212, 239–240, 255–256, 270
Nitrogen assimilation 212, 240
Nitrogen fixation 40, 42, 53, 143, 152, 162, 165–166, 170, 182–183
Nitrogen storage compounds 187
Nitrogen uptake 268
Nitrogenase 54, 165, 167–169, 177, 180–185, 188
Nod 144, 170
nod-box 144, 146
Nod-factors 143–149
Nodulation 144–147, 149, 152–153, 170
Nodules 143, 145, 149–151, 153
Nonfluorescent cells 184
Nostoc punctiforme 167–168
NtcA 171, 176, 183–184, 188
Nucleotide transporter (NTT) 64
Nutrition 40, 97, 197–199, 207, 213, 240, 251, 256–257, 268
Nyctotherus 118, 120, 122, 126–127, 130

O

Olavius 253, 263–264
Operational taxonomic units 291
Oxic-anoxic interface 209, 227
Oxygen flux 41
Oxymonad 47, 50–51, 53

P

Parabasalid 43–44, 51, 84, 129
Paracaedibacter 69
Paramecium 61–69, 71–72, 104, 110, 133
Parasite 64–65, 67–68, 72, 292
Parasomal sac 112
PAS 174, 281, 283
patA 183–184
patB 183, 186
Pathogens 33, 64, 67, 152, 216
patN 183, 186, 189
patS 183–184, 186
Pattern, disruption of 185, 189
Pattern, establishment of 185–186, 189
Pattern, maintenance of 185, 188–189
PCR bias 262
Pelochromatium 5, 23–25, 28, 30
Peptidoglycan 148, 278
Periplasmic domain 176, 282–283
Phage 62–63, 70–72
Phagocytosis 296
Photolithoautotrophy 25
Phototrophic consortia 21–34
Phycobiliproteins 181, 187
pKAP298 69–71

Index 309

Plasmid 69–71, 144, 152, 298
Pogonophora 201, 204, 224, 228, 231–232
Porifera 230
Posttranslational modification 182
Prophage 71
Protozoa 39–45, 48, 50, 52–54, 67, 72, 77–78, 81, 84, 91, 131
Pyruvate 8, 28, 44, 120, 122–125, 129
Pyruvate dehydrogenase (PDH) 120, 122, 125
Pyruvate:ferredoxin oxidoreductase (PFO) 44, 120, 122–123, 125, 129
Pyruvate:formate lyase (PFL) 120, 122, 124

Q

Quorum sensing 285–287

R

R body 63, 65, 67, 69–72, 105
R-bodies 104–106
Receptor kinase 149
Refractile bodies 61
Regression 278–280
Regulation 2, 30, 144, 167, 176, 182, 217, 257, 282–287
Resistance 67, 151, 295, 297
Resource partitioning 44
Response regulator 167, 174–175, 216, 281–283, 286
Reverse electron transport 7, 9
Rhizobia 143–144, 149, 151–152, 154–155, 168–170
Rhizobiales 71
Rhizobium 142, 144–147, 149–153
Rhizosphere 144
Rhodobacter 71
Ribulose bisphosphate carboxylase 181
Ribulose bisphosphate carboxylase/oxygenase 181
Root hair 146–147
Root nodules 159

rRNA genes 25, 41, 108, 112, 126, 215, 295
rscS 281–284, 287
RTX toxin 33
RubisCO 180–182, 200, 208, 210–211, 219, 267–268

S

Scotophobic response 29
Selection pressure 87, 293
Sensor kinase 281–286
sigH 171, 173, 176
Signal transduction 29, 175, 216, 281, 286
Spatial organization 39, 41
Spirochetes 48–49, 53, 78, 81–83, 87–89, 91, 258, 267, 269, 271
Sponge 231–232
ß-Hemolysis 293
Stable isotope 199
Succinate thiokinase (STK) 122
Succinoglycans 150
Sulfate reduction 2, 207, 255, 269
Sulfate-reducing bacteria 11, 13, 27–28, 31, 42, 207, 269
Sulfide oxidation 200, 208–210, 217
Sulfide-oxidizing bacteria 97, 258, 269
Sulfur cycle 27, 31, 266, 269
Sulfur-oxidizing bacteria 251–252, 254, 257
Surface polysaccharides 143, 149
Symbiosis 143, 148–150, 154, 170, 182, 189, 197–201, 203, 207–212, 214–217, 227, 231, 235–236, 242, 256–258, 267, 270–271, 277–287, 291–292, 295–296, 300
Symbiosis genes 33–34
Symbiotic plasmid 144, 152
Syntrophic 4, 6–11, 13–15, 21, 33, 266, 269
Syntrophic sulfur cycle 27, 31, 266, 269
Syntrophy 4

T

Terminal differentiation 185, 214
Termite 40–42, 49–51, 53–54, 77–78, 80–82, 84–87, 91, 132, 269
Termite Group 1 41, 47, 50–51
Termite gut 39–45, 48, 52–54, 77–78, 82, 133
Thioautotroph 197, 199, 207–210, 238
  Calvin cycle 199, 208–210
Toxin 33, 61, 65–67, 69, 71–72, 284
Transcription 69–70, 144, 146, 150, 153, 173–174, 176–178, 184, 188–189, 281–282, 284–287
Transmission 22, 46–47, 65–66, 198, 201, 206, 214–217, 230–232, 257–259
Transposon 63, 69–70, 167, 298, 301
Trichomonads 43, 77, 81, 119, 122, 125, 132
Trichomonas 44, 117, 120–123, 125–126, 128, 130–131, 134
Trophosome 199–200, 204–210, 212–215
Tubeworm 197–201, 203–204, 207–213, 215–217, 231, 238
Tubulin 106–107
Tubulin-like genes 112
Two-component regulatory system 281
Type III secretion system 144, 152
Type IV secretion system 152

V

Vasodilators 292
Venous congestion 292
Verrucomicrobia 100–101, 106, 108
Verrucomicrobiales 102
Vertical transmission 65, 214, 259
Vestimentifera 197–198, 201, 204, 210, 215, 238
Vibrio fischeri 277, 279, 281, 296
Virulence 64–67, 271, 283

W

Worm 199, 204, 251, 257, 259, 265–266, 269
Wound infections. 292, 294

Z

Zoochlorellae 110

DATE DUE

DUE DATE SUBJECT TO CHANGE
IF A RECALL IS REQUESTED